D0927490

METAL IONS IN BIOLOGICAL SYSTEMS

VOLUME 23
Nickel and Its Role in Biology

METAL IONS IN
BIOLOGICAL SYSTEMS

Edited by

Helmut Sigel
Institute of Inorganic Chemistry
University of Basel
CH-4056 Basel, Switzerland

with Astrid Sigel

VOLUME 23
Nickel and Its Role in Biology

MARCEL DEKKER, INC.   New York and Basel

Nickel and its role in biology / edited by Helmut Sigel with Astrid
   Sigel.
      p.   cm. -- (Metal ions in biological systems ; v. 23)
   Includes bibliographies and index.
   ISBN 0-8247-7713-1
   1. Nickel--Physiological effect. 2. Nickel--Environmental
aspects. 3. Nickel--Health aspects. I. Sigel, Helmut. II. Sigel,
Astrid. III. Series.
QP532.M47 vol. 23
[QP535.N6]
574.19'214 s--dc19
[574.19'214]   87-36498
                                                                  CIP

MARCEL DEKKER, INC.
270 Madison Avenue, New York, New York 10016

Current printing (last digit):
10 9 8 7 6 5 4 3 2 1

PRINTED IN THE UNITED STATES OF AMERICA

Preface to the Series

Recently, the importance of metal ions to the vital functions
of living organisms, hence their health and well-being, has become
increasingly apparent. As a result, the long-neglected field of
"bioinorganic chemistry" is now developing at a rapid pace. The
research centers on the synthesis, stability, formation, structure,
and reactivity of biological metal ion-containing compounds of low
and high molecular weight. The metabolism and transport of metal
ions and their complexes is being studied, and new models for complicated natural structures and processes are being devised and
tested. The focal point of our attention is the connection between
the chemistry of metal ions and their role for life.

No doubt, we are only at the brink of this process. Thus,
it is with the intention of linking coordination chemistry and
biochemistry in their widest sense that the series METAL IONS IN
BIOLOGICAL SYSTEMS reflects the growing field of "bioinorganic
chemistry." We hope, also, that this series will help to break
down the barriers between the historically separate spheres of
chemistry, biochemistry, biology, medicine, and physics, with the
expectation that a good deal of the future outstanding discoveries
will be made in the interdisciplinary areas of science.

Should this series prove a stimulus for new activities in
this fascinating "field," it would well serve its purpose and would
be a satisfactory result for the efforts spent by the authors.

Fall 1973

Helmut Sigel
Institute of Inorganic Chemistry
University of Basel
CH-4056 Basel, Switzerland

Preface to Volume 23

The abundance of nickel in the earth's crust and in seawater is
comparable to that of many other transition metal ions with a long-
established history of biological importance. However, until quite
recently no biological functions for nickel were known, nor had an
absolute requirement for this element been demonstrated. It is now
about 12 years since the discovery that nickel is a constituent of
jack bean urease; in the meantime also an absolute requirement for
this element was proven for urea metabolism and urease synthesis
in several higher plant tissues. In fact, studies on "nickel and
its role in biology" have gained momentum and blossomed into a
vibrant research area; therefore, this volume is wholly devoted
to this metal ion.

First, nickel in the environment and in aquatic systems is
considered; then its role for plants is outlined, and the metabolism
in humans and animals is summarized. Nickel ion binding to amino
acids, peptides, and proteins or enzymes, including the nickel
hydrogenases, is covered, and the interactions with nucleic acids
and their constituents are described. The toxicology of nickel
compounds and the role of nickel in carcinogenesis are discussed.
The volume closes with a chapter focusing on the analysis of nickel
in biological materials and the related difficulties.

Helmut Sigel
Astrid Sigel

Contents

Chapter 6

NICKEL IN PROTEINS AND ENZYMES   165

      Robert K. Andrews, Robert L. Blakeley, and Burt Zerner

Chapter 7

NICKEL-CONTAINING HYDROGENASES   285

      José J. G. Moura, Isabel Moura, Miguel Teixeira,
      Antonio V. Xavier, Guy D. Fauque, and Jean LeGall

Chapter 11

ANALYSIS OF NICKEL IN BIOLOGICAL MATERIALS   403

    Hans G. Seiler

Contributors

Numbers in parentheses indicate the pages on which the authors' contributions begin.

Robert K. Andrews   Department of Biochemistry, University of Queensland, St. Lucia, Queensland 4067, Australia (165)

E. L. Andronikashvili   Institute of Physics, Academy of Sciences of the Georgian SSR, Tbilisi 77, USSR (331)

Robert L. Blakeley   Department of Biochemistry, University of Queensland, St. Lucia, Queensland 4067, Australia (165)

Robert W. Boyle   Geological Survey of Canada, 601 Booth Street, Ottawa, Ontario, K1A OE8, Canada (1)

V. G. Bregadze   Institute of Physics, Academy of Sciences of the Georgian SSR, Tbilisi 77, USSR (331)

Monica M. Cole   Department of Geography, University of London, Royal Holloway and Bedford New College, Egham, Surrey, TW20 OEX, U.K. (47)

Margaret E. Farago   Department of Chemistry, University of London, Royal Holloway and Bedford New College, Egham, Surrey, TW20 OEX, U.K. (47)

Guy D. Fauque   A.R.B.S., Section Enzymologie et Biochimie Bacterienne, C.E.N., Cadarache, F-13108 Saint-Paul-les-Durance Cedex, France (285)

Jean LeGall   Department of Biochemistry, School of Chemical Sciences, University of Georgia, Athens, GA 30602, USA (285)

R. Bruce Martin   Chemistry Department, McCormick Road, University of Virginia, Charlottesville, VA 22903, USA (123, 315)

C. Rajeshwari Menon   Department of Biochemistry and Occupational Health Program, Health Sciences Centre, McMaster University, 1200 Main Street West, Hamilton, Ontario, L8N 3Z5, Canada (359)

J. R. Monaselidze Institute of Physics, Academy of Sciences of the
    Georgian SSR, Tbilisi 77, USSR (331)

Isabel Moura Centro de Quimica Estrutural, Complexo I and UNL,
    Av. Rovisco Pais, P-1096 Lisboa, Portugal (285)

José J. G. Moura Centro de Quimica Estrutural, Complexo I and UNL,
    Av. Rovisco Pais, P-1096 Lisboa, Portugal (285)

Evert Nieboer Department of Biochemistry and Occupational Health
    Program, Health Sciences Centre, McMaster University, 1200 Main
    Street West, Hamilton, Ontario, L8N 3Z5, Canada (91, 359)

Heather A. Robinson Manitoba Education, 1181 Portage Avenue,
    Winnipeg, Manitoba, R3G 0T3, Canada (1)

Franco E. Rossetto Department of Biochemistry and Occupational
    Health Program, Health Sciences Centre, McMaster University,
    1200 Main Street West, Hamilton, Ontario, L8N 3Z5, Canada
    (359)

W. (Bill) E. Sanford Department of Biochemistry and Occupational
    Health Program, Health Sciences Centre, McMaster University,
    1200 Main Street West, Hamilton, Ontario, L8N 3Z5, Canada
    (91)

Hans G. Seiler Institute of Inorganic Chemistry, University of
    Basel, Spitalstrasse 51, CH-4056 Basel, Switzerland (403)

Pamela Stokes Institute for Environmental Studies and Department
    of Botany, University of Toronto, Toronto, Ontario, M5S 1A4,
    Canada (31)

Miguel Teixeira Centro de Quimica Estrutural, Complexo I and UNL,
    Av. Rovisco Pais, P-1096 Lisboa, Portugal (285)

Rickey T. Tom Department of Biochemistry and Occupational Health
    Program, Health Sciences Centre, McMaster University, 1200 Main
    Street West, Hamilton, Ontario, L8N 3Z5, Canada (91)

Antonio V. Xavier Centro de Quimica Estrutural, Complexo I and UNL,
    Av. Rovisco Pais, P-1096 Lisboa, Portugal (285)

Burt Zerner Department of Biochemistry, University of Queensland,
    St. Lucia, Queensland 4067, Australia (165)

Contents of Other Volumes

---

*Out of print

---

*Out of print

Other volumes are in preparation.

Comments and suggestions with regard to contents, topics, and the
like for future volumes of the series would be greatly welcome.

The following Marcel Dekker, Inc. book is also of interest for any
reader dealing with metals or other inorganic compounds:

HANDBOOK ON TOXICITY OF INORGANIC COMPOUNDS edited by Hans G. Seiler
and Helmut Sigel, with Astrid Sigel

METAL IONS IN BIOLOGICAL SYSTEMS

VOLUME 23
Nickel and Its Role in Biology

1

Nickel in the Natural Environment

Robert W. Boyle
Geological Survey of Canada
601 Booth Street
Ottawa, Ontario, Canada K1A 0E8

and

Heather A. Robinson
Manitoba Education
1181 Portage Avenue
Winnipeg, Manitoba, Canada R3G 0T3

1. GENERAL CHEMISTRY AND GEOCHEMISTRY OF NICKEL

Nickel, atomic number 28, is a member of the group of the periodic
table which also includes iron, cobalt, and the platinum metals.
Nickel is a transition element and with iron and cobalt forms the
horizontal Fe triad of the periodic table. Its electron configuration is $[Ar]3d^8 4s^2$.

In seventeenth century Europe the silver miners of the Erzge-
birge of Bohemia and the cobalt miners of Helsingland, Sweden recognized a mineral which they called kupfernickel (copper devil) which
on smelting produced no copper. This mineral, now known as niccolite
(NiAs), was investigated by the Swedish chemist and mineralogist
A. F. Cronstedt in 1751. He recognized a new element in the mineral,
and after many confirmations and refutations this was proven to be
the case by another Swedish chemist, T. Bergman, in 1775. Cronstedt
named the element nickel in allusion to its devilish behavior in the
old silver and cobalt smelters.

Natural nickel comprises five stable isotopes whose average
abundance in % is as follows: ^{58}Ni (67.88), ^{60}Ni (26.23), ^{61}Ni
(1.19), ^{62}Ni (3.66), and ^{64}Ni (1.08). Only slight variations from
these abundances have been observed in nickeliferous minerals.

The properties of the Fe triad of elements are remarkably
similar. The melting and boiling points of the three elements are
uniformly high. The first ionization potentials (in kcal/g-mol) are
nearly the same for all three elements (Fe-182; Co-181, Ni-176) and
their oxidation potentials are all moderately more positive than H^+.
Their atomic radii (in Å) are similar (Fe, 1.26; Co, 1.25; Ni, 1.24),
as are also their covalent radii (in Å) (Fe, 1.17; Co, 1.16; Ni,
1.15) and their ionic radii (in Å) for the M^{2+} state (Fe, 0.76; Co,
0.74; Ni, 0.72). Their electronegativities (Pauling's) are essentially identical (Fe, 1.8; Co, 1.8; Ni, 1.8). These energetic similarities explain the marked coherency of nickel with cobalt and iron
in nature; the coherency with magnesium, manganese, and the platinum
metals is also close in certain types of rocks and mineral deposits
as explained below.

Nickel can exist in four oxidation states: Ni(0), Ni(II), Ni(III), and Ni(IV). The first two are the most common in nature; Ni(III) is perhaps possible under conditions of extreme natural oxidation, but its existence is transitory since its compounds decompose rapidly in the presence of water. However, some minerals such as nickelian heterogenite and nickeliferous wad appear to contain Ni(III); the manner of stabilization of the high oxidation state of the element in these minerals is unknown. Ni(0) is the state in native nickel, and Ni(II) forms the most common ion of the element, Ni^{2+}. In aqueous solution this state yields the green hexaquo ion, $[Ni(H_2O)_6]^{2+}$. This ion is remarkably similar in its reactions to the analogous cobalt ion, $[Co(H_2O)_6]^{2+}$, but unlike the Fe^{2+} ion in that the nickel species is stable to oxidation under most natural conditions. This feature is important in assessing some of the migration characteristics of nickel in the natural environment.

The nickel salts and other compounds of geochemical interest are essentially those of Ni(II) and its complexes. A few may be mentioned. Nickel oxide is insoluble in water but soluble in an acidic environment; the hydroxide is insoluble and has colloidal properties. The chloride, sulfate, nitrate, and acetate are all soluble to very soluble. Nickel carbonate and basic nickel carbonates are known, and a number of phosphates can be prepared; all are insoluble. The nickel sulfides, selenides, tellurides, arsenides, and antimonides are all insoluble compounds as are also the arsenates. The silicates are complex compounds that are insoluble. Nickel(II) readily forms complex ions of which those with ammonia are well characterized. Numerous Ni(II) complexes with organic ligands are known, but few of these have been characterized in the natural environment. The most important of these species are the various carboxylates, fulvates, and humates, particularly the last.

Detailed accounts of the general chemistry of nickel include those by Mellor [1], Gmelin [2], and Nicholls [3].

Nickel exhibits siderophile, chalcophile, oxyphile, and biophile characteristics in nature. Its siderophile character is manifest by its occurrence as native nickel and in a number of natural

iron-nickel alloys; these are rare. The chalcophile (sulfophile)
nature of the element is evident from the relatively large number of
natural sulfides, selenides, tellurides, arsenides, and antimonides
which the element forms. The most important of these are pentlandite,
$(Fe,Ni)_9S_8$; millerite, NiS; niccolite, NiAs; breithauptite, NiSb; and
melonite, $NiTe_2$. Nickel also substitutes for iron and cobalt in a
large number of sulfides, arsenides, etc., such as pyrite, pyrrhotite,
and cobaltite. The oxyphile character of nickel is manifest in its
natural occurrence in oxides, silicates, sulfates, arsenates, vanadates, and carbonates. The rare mineral bunsenite, NiO, is the only
common oxide. The most common silicates are those of the garnierite
group, including the serpentine minerals and phyllosilicates nepouite,
pecoraite, willemsite, schuachardtite, and nimite, and nickeliferous
montmorillonite (pimelite) and vermiculite.

The only common arsenate is annabergite, $Ni_3[AsO_4]_2 \cdot 8H_2O$; the
sulfates such as morenosite, $NiSO_4 \cdot 7H_2O$, and the vanadates and carbonates are very rare. Nickel also enters into the structure of many
iron and magnesium silicates, sulfates, and carbonates because of the
similarity in the energetic properties, particularly the ionic radii,
of the three elements (Fe^{2+}, 0.74 Å; Mg^{2+}, 0.65 Å; Ni^{2+}, 0.72 Å).
The biophile character of nickel is shown by the presence of the
element in many plants and animals and in the degradation products
of living organisms, e.g., peat, coal, and petroleum, and in certain
organic substances such as anthraxolite. An essential role for
nickel in certain living organisms has been demonstrated.

The cosmic abundance of nickel as given by Cameron [4] is
4.57×10^4 atoms Ni per 10^6 atoms of Si. The abundance of nickel in
meteorites ranges widely from more than 10% in whole-iron meteorites
to less than 50 ppm in whole-silicate meteorites. See the extensive
data in Wedepohl [5].

Various values have been given for the crustal abundance of
nickel and cobalt. Those commonly quoted are 100 ppm Ni and 40 ppm
Co, yielding a Ni/Co ratio of 2.5. In actual fact the values quoted
in the literature for the crustal abundance of nickel and cobalt

range widely, from 48 to 200 ppm for nickel and from 8 to 80 ppm for
cobalt. Using our analytical data and that from the world literature,
and adjusting for the volume percentages of the various rock types in
the crust of the earth, we estimate a value of 60 ppm Ni and 15 ppm
Co, giving a Ni/Co ratio of 4.

Detailed accounts of the geochemistry of nickel may be found in
Wedepohl [5], Fersman [6], Goldschmidt [7], Rankama and Sahama [8],
and in Boyle and others in [9].

2. NICKEL IN THE LITHOSPHERE AND
IN NICKEL DEPOSITS

The estimated abundance of nickel in the common igneous, sedimentary,
and metamorphic rocks is given in Table 1. These abundance figures
have been calculated from data in the world literature and from
analyses done in the Geological Survey of Canada over some 20 years.

Of the igneous rocks the ultrabasic suite (dunite, peridotite,
and pyroxenite) are richest in nickel followed by the basic (gabbro-
basalt), intermediate (diorite-andesite, syenite-trachyte, grano-
diorite), and acidic (granite-rhyolite) suites in that order. The
alkali-rich rocks are normally low in nickel content, ranging from
2.5 to 41 ppm. Cobalt follows the same trend in the various rock
suites, the ratio Ni/Co ranging from 1.5 in certain acidic (granite)
rocks to 10 in ultrabasic rocks. It was formerly assumed that these
trends were due to differentiation processes in ultrabasic or basic
magmas. Today these views must be modified with regard to the probability that the intrusive acidic suites may be principally of ultra-
metamorphic origin (i.e., derived from the fusion of sediments and
other rocks).

In igneous and igneous-type rocks most of the nickel is present
in two principal sites—in ferromagnesian minerals (e.g., olivine,
pyroxene, biotite, chlorite), mainly because of similar energetic
considerations between nickel and magnesium, principally the similarity in the ionic radii of the two elements (Mg^{2+}, 0.65 Å; Ni^{2+},

TABLE 1

Average Nickel Content of the Common Igneous,
Sedimentary, and Metamorphic Rocks

Rock type	Ni content (ppm)
Igneous Rocks	
Ultrabasic rocks: dunite, peridotite, pyroxenite	1500
Basic rocks:	
Intrusives: gabbro, norite, diabase	208
Extrusives: basalt	140
Intermediate rocks:	
Intrusives: syenite, diorite, granodiorite	53
Extrusives: andesite, trachyte, dacite	32
Acidic rocks:	
Intrusives: granite, aplite, pegmatite	13
Extrusives: rhyolite, obsidian	11
Alkali-rich rocks: nepheline syenite, phonolite	12
Sedimentary Rocks	
Arenites and rudites (sandstone, arkose, greywacke, conglomerate)	10
Lutites (shale, argillite)	60
Saprolites (carbonaceous pyritic shale, oil shale)	85
Precipitates (limestone, dolomite, siderite)	12
Precipitates (chert)	10
Precipitates (cherty iron formation)	80
Evaporites (anhydrite, gypsum)	2
Evaporites (halite, sylvite)	2
Tuffs (intermediate and acidic composition)	20
Phosphorites (oceanic)	30
Deep sea clays	185

TABLE 1 (continued)

Rock type	Ni content (ppm)
Metamorphic Rocks	
Quartzites, meta-greywacke, meta-conglomerate	20
Marble and crystalline limestone and dolomite	15
Phyllite, meta-argillite, slate	65
Schists (igneous parentage)	27
Schists (sedimentary parentage)	40
Amphibolites and greenstones (igneous parentage)	90
Gneisses and granulites	15
Serpentinites	1500
Hornfels (sedimentary parentage)	40
Skarn (sedimentary parentage)	30

0.72 Å); and in sulfides, principally pyrrhotite and pyrite, and in pentlandite in certain differentiated layers in ultrabasic-basic bodies. In some ultrabasic rocks nickel may occur in iron-nickel alloys such as awaruite. In all igneous rock types magnetite and other spinels may contain minor amounts of nickel.

The lutite (shale, argillite) category of the sedimentary rocks are the most enriched in nickel. Among the lutite-saprolites the black (carbonaceous) pyritic varieties are highest in nickel followed by normal shales and mudstones. The arenites (sandstones, arkose, greywacke), rudites (conglomerate, sedimentary breccia), and the precipitates (halite, sylvite, anhydrite, limestone, and dolomite) are all relatively low in nickel. Most cherts are relatively low in nickel, but some black carbonaceous varieties may contain up to 200 ppm or more. Most cherty iron formations contain an average of about 80 ppm Ni, the range being from 5 to 400 ppm Ni. In iron formations the highest amounts of nickel occur in the sulfides, mainly pyrite

and pyrrhotite, of the sulfide facies and in the ferromagnesian
minerals (chlorite, greenalite, minnesotaite) of the silicate facies.

Phosphates are normally low in nickel (average 30 ppm), but
some are enriched in amounts up to 1,000 ppm or more. Certain deep
sea clays contain higher than normal amounts of nickel (averages up
to 320 ppm). Apparently much of this nickel originated from micrometeorites and meteoric dust according to some authorities; it also
seems probable that some of this nickel is derived from submarine
hot springs and fumaroles and as a result of submarine weathering
of ultrabasic and basic rocks such as komatiites, basalts, and their
tuffaceous analogs.

Certain recent riverine, lacustrine, estuarine, and near-shore
oceanic sediments have a component of nickel contributed by manmade
agencies. Normally this component is low, less than 5 ppm Ni, but
in some areas of nickel mining and high industrial activity much
larger amounts of nickel may appear in the sediments (see Sec. 4).

In sedimentary rocks much of the nickel is present in sulfides,
mainly pyrite and pyrrhotite, precipitated under reducing conditions
where a relatively high concentration of H_2S prevailed as a result
of biogenic processes (degradation of sulfoproteins, bacterial reduction of sulfate, etc.). Hydrolysate materials and their derivatives
(limonite, hematite, wad, clay minerals, glauconite, chlorite) always
contain some nickel, the element having been coprecipitated and/or
adsorbed probably largely as the hydroxide by the precursors of
limonite (i.e., ferrous hydroxide), silica-alumina gels, and so on.
When resistates such as magnetite, amphiboles, pyroxenes, and biotite
occur in sediments, these minerals generally contain small amounts
of nickel.

Metamorphic rocks generally report about the same contents of
nickel as their parents. Thus, serpentinites derived from ultrabasic
igneous rocks have the highest nickel contents as would be expected.
Likewise among the metasediments, the highest contents of nickel are
generally found in schists derived from shales. Certain skarns
(calc-silicate bodies) are enriched in nickel mainly as a result
of metasomatic processes.

Nickel deposits can be classified under four types as follows:

(1) Massive sulfide lenses and disseminated deposits containing essentially pentlandite, pyrrhotite, pyrite, and chalcopyrite commonly with small amounts of platinoid minerals. There is great variety in these deposits; most occur in or near ultrabasic and basic sills, sheets, lenses, lopoliths, stocks, dykes, and flows; others may be somewhat removed from ultrabasic and basic igneous rocks but in a terrane where they are present. The type deposits are those at Sudbury, Ontario. The principal elements concentrated are Ni, Co, Fe, Cu, Au, Ag, Pt metals, Se, Te, As, and S. The nickel content of these deposits ranges widely, from <0.5 to 2.5% Ni with an average of about 1.5% Ni according to the world data. Some deposits in this classification contain the nickel in a very dispersed state in three principal ways: as a substitution of magnesium in various minerals, e.g., olivine and serpentine; in finely divided nickel sulfides; and in Ni-Fe alloys. The average nickel content of these bodies is usually less than 0.5% Ni; thus they are not generally economic.

(2) Veins and lenses of sulfides containing millerite, gersdorffite, chalcopyrite, pyrrhotite, and pyrite. The characteristic elements concentrated are Ni, Co, Fe, Cu, and S. Some of these deposits are intimately associated with ultrabasic and basic rocks; others exhibit only a general relationship. Many of these deposits may contain up to 5% Ni, but they are usually small and hence not economic.

(3) Veins containing niccolite, gersdorffite, chloanthite, and various other nickel-cobalt arsenides and sulfides: Associated minerals are native silver, arsenic, bismuth, calcite, galena, sphalerite, chalcopyrite, pitchblende (only in certain deposits), and numerous other minerals. The characteristic elements concentrated are Ni, Co, Ag, Fe, Cu, Pb, Zn, As, Sb, S, Bi, and U. Typical deposits of this type occur at Cobalt, Ontario, Canada and at Jachymov, Czechoslovakia. Some of the deposits may contain up to 20% Ni or more, but commercially, only small amounts of nickel are obtained from these deposits. Cobalt is obtained from some deposits. Silver is the principal economic element in most of these deposits; uranium is won from some deposits.

(4) Residual (supergene) nickel-cobalt deposits developed on
nickeliferous rocks; these are of two types not readily differentiated
from each other. One type contains the nickel essentially in garnierite, nepouite, and nickeliferous serpentine (serpophyte), and the
other (laterites) contains nickel mainly in a complex of hydrous
iron, manganese, and nickel oxides and indefinite supergene silicates.
Both types are often found together and are gradational or intermixed.
Some of these deposits occur at the present erosion surface; others
are fossil and represent ancient buried erosion surfaces. The characteristic elements concentrated are Ni, Co, Fe, Mn, and Cr. The
grade of the ores varies from about <0.5% to 2.4% Ni, with an average
of about 1.5% for many deposits. Typical deposits of this type occur
in New Caledonia, USSR, Cuba, and Australia.

3. NICKEL IN THE PEDOSPHERE

The pedosphere comprises the thin outer layer or shell of the earth
in which soil-forming processes occur and in which plants grow.
Included in the media of this sphere are the various types of soils,
certain glacial products such as till, and a variety of other products
of erosion and weathering among which silcretes, ferricretes, and
calcretes may be mentioned.

Soils are tabular bodies composed of variable amounts of
mineral, plant, animal, and decaying organic constituents, commonly
impregnated with small amounts of water and differentiated into
horizons of variable thickness that constitute the soil profile.
The horizons of soils have received various names, letters, and
numerations, some remarkably complex; for the purposes of this
article we shall designate the upper or organic horizon as A_1, the
leached horizon immediately below as A_2, the next lower horizon
characterized by maximum accumulation of Fe, Mn, and Al sesquioxides
and silicate clays as B, and the parent material as C. In unglaci-
ated regions the C horizons generally comprise weathered bedrocks or
alluvium; in glaciated regions the C horizons are till, sand, gravel,
or mixtures of these glacial products.

Soils are living bodies and hence highly variable, their properties being a function of a number of interacting variables, principally those of climate (amount of rainfall, temperature, etc.),
action of organisms, relief, nature of parent materials, and time.
Numerous classifications of soils have been proposed in recent years,
all much too detailed to be considered here. For our purposes a
simple classification will suffice. At high latitudes under cold
tundra conditions Arctic brown soils and half-bog soils predominate;
in the mid-latitude cool northern forest zones podzols and brown
earths are most common; soils of the mid-latitude warm climates with
mixed forests, steppes, and semideserts include various red, brown,
and cinnamon soils; red and yellow podzolic soils; chernozems;
chestnut soils; desert soils; and soils of the low-latitude, hot
tropical rain forest and deciduous forest savanna zones comprise
various ferrisols, vertisols, and laterites.

Contributions of nickel to soils arise from both natural and
manmade sources. Among the former are parent bedrocks, deposits
enriched in nickel, micrometeorites, and cosmic dust—the last two
of relatively little importance. Manmade sources of nickel include
smelting of nickeliferous ores; metal refining; burning of coal;
burning of petroleum products; disposal of waste, sewage, and sludge;
and fertilizer applications. The effect of manmade sources on the
nickel contents of soil is generally local, although in certain
cases industrial and other manmade plumes of pollution combined with
unusual climatic conditions may disperse nickel over large regions
of the earth.

The average nickel content of soils as given by Vinogradov [10]
is 40 ppm; the equivalent cobalt content is given as 3 ppm. Swaine
[11] gives a wide range from 5 to 500 ppm Ni. Our estimate of the
average nickel content of soils calculated from a large number of
analyses done in the Geological Survey of Canada and those in the
available world literature is 35 ppm. This average takes into
account the high contents usually found over ultrabasic rocks which
on a volume basis affect the average of soils developed on normal
rocks only slightly. Variations from this average are noticeable

in some regions. Arctic tundra and brown earth soils, podzols,
chernozems, lateritic, and other soils developed on intermediate and
acidic rocks and most tills vary relatively little from the average.
Lateritic soils and most other types of soils developed on ultrabasic
and basic rocks exhibit wide variations from the average; some of
these soils, especially those on ultrabasic rocks, may contain nickel
contents as high as 2.4%. All soils in the vicinity of nickel deposits are generally enriched in nickel, e.g., at Cobalt, Ontario we
found up to 200 ppm in soils developed on till overlying the nickel-
cobalt arsenide-native silver deposits [12].

In normal soil profiles nickel tends to follow iron. Thus the
B horizon frequently shows an enrichment in nickel compared with the
A_1 and A_2 horizons, the latter usually being poor in the element.
In certain manganiferous soils nickel tends to follow manganese in
the B horizon probably because of the formation of a Ni-Mn oxide
(asbolite). In some soil profiles the A_1 (humic) horizon may be
slightly enriched in nickel (and cobalt) compared with the A_2 and
B horizons because of the strong adsorption and organometallic
binding properties of some types of humus.

The sites of nickel concentration in soils depend essentially
on the types of parent materials (disintegrated rocks, tills, etc.),
soil types, and various chemical fractionating factors. Organic
matter (humus) binds some nickel in all soils, but the largest
amounts of the element are concentrated in the hydrolysate and oxi-
date components of soils, including clay minerals, limonite, wad,
clay-limonite, clay-humus complexes, and their various colloidal
equivalents. The layer lattice (clay) silicate minerals are marked
accumulators of nickel; such minerals (nickel chlorites, pimelite,
garnierite, nepouite, serpophyte) have especially complex structures
and chemical formulas. Small amounts of nickel occur in residual
sulfides (pyrite, pyrrhotite) and in oxides such as magnetite in
many types of soils. Soil water and soil organisms usually contain
only minor amounts of nickel.

The bioavailability of nickel in soils is a complex problem
not readily addressed in a short communication. Briefly, the factors

that control availability are (1) pH—increasing the acidity makes
nickel more available and vice versa; (2) mode and strength of
binding of nickel in cation exchange sites in organic and inorganic
materials—lightly adsorbed nickel is much more available than that
incorporated within the colloidal complex of soils and in the clay
mineral and clay mineral-humic complexes; nickel incorporated in
limonite, wad, and nickeliferous chlorites, serpentines, layer
lattice minerals, and in residual oxides and sulfides such as magnetite and pyrite is not readily available; and (3) presence of
competing cations such as Co, Cu, and Zn that may occupy available
cation exchange sites. The experimental availability of nickel in
soils depends on the soil type, total nickel content, and strength
and type of extractant. No average values can be given, but distilled water generally extracts less than 1% of the total nickel,
ammonium acetate up to 30% and 2% acetic acid up to 40% or more in
most soils developed on serpentinites. Poorly drained soils tend
to yield relatively high contents of extractable nickel, whereas
well-drained soils yield comparatively less. Marked deviations
from these general statements are normal.

Certain soils developed on ultrabasic rocks (dunites, serpentinites) are marked by a stunted, chlorotic, endemic, and otherwise
unusual flora commonly referred to as "serpentine flora." These
soils are said by some investigators to be toxic because of the high
availability of Mg, Ni, Cr, and Co and low amounts of available Ca,
K, P, N, and Mo. Their ecology is discussed at length by Whittaker
et al. [13] and Proctor and Woodell [14].

4. NICKEL IN THE HYDROSPHERE

The nickel content of natural waters as compiled from the world
literature is given in Table 2. Natural freshwaters from streams,
rivers, and lakes, uncontaminated by manmade agencies or oxidizing
nickel deposits, contain low amounts of nickel, generally less than
0.01 mg Ni/liter (0.01 ppm); the range is 0.0005-0.02 mg/liter

TABLE 2

Average and/or Range of Nickel Content of Natural Waters,
Precipitates from Natural Waters, and Drainage Sediments

Description	Ni content (ppm)
Rainwater and snow (mostly in particulate matter)	Up to 0.001
Hot springs	0.0005-0.4
Groundwaters and cold springs	0.0005-4.5
Groundwaters, cold springs, and mine waters in vicinity of nickeliferous deposits	Up to 75
Stream, river, and lake waters	0.0005-0.02
Natural stream and river waters in vicinity of nickeliferous deposits	Up to 5
Contaminated stream, river, and lake waters in vicinity of nickel mines and smelters	Up to 6.4
Ocean and seawaters	0.0015
Normal Fe-Mn precipitates (dry matter) from springs	7-100
Fe-Mn precipitates (dry matter) from springs in vicinity of nickeliferous deposits	20-2000+
Stream and river sediments (dry matter)	1-150
Natural stream and river sediments (dry matter) in vicinity of nickeliferous deposits	Up to 1000
Contaminated stream, river, and lake sediments (dry matter) in vicinity of nickel-mining areas	Up to 3000

(0.0005-0.02 ppm). Seawater has a range of 0.0005-0.004 ppm Ni. The
sources of nickel in natural waters include normal weathering of rocks
and nickeliferous deposits, various human agencies such as mining,
smelting, sewage disposal, burning of coal and petroleum, and meteoric
dust, the last of minor importance. Contributions from human agencies
are generally local considering the hydrosphere as a whole.

Groundwaters in contact with rocks tend to have higher contents
of nickel than those in contact with soils and glacial materials and
those in streams, rivers, and lakes. Oxidizing groundwaters leaching
nickeliferous deposits may report high nickel contents, up to 75 ppm

or more in our experience (see below). Most hot-spring waters are
relatively low in nickel judging from the available analyses which
are sparse.

The migration of nickel in natural waters is promoted by low
pH and the presence of chloride, nitrate, sulfate, and soluble
(colloidal) humic matter. Factors which limit the migration of
nickel in natural waters are high pH and the presence of PO_4^{3-}, CO_3^{2-},
OH^-, and H_2S, which precipitate the metal as insoluble salts; also
the presence of organic (chelating) substances, humus, hydroxides
of iron, manganese, aluminum, and silica-alumina complexes (clay
minerals), all of which may strongly adsorb and absorb the metal.

Relatively few exact data on the speciation of nickel in
natural waters are available. Our studies carried out in various
parts of Canada, based on analyses and thermodynamic considerations,
suggest that nickel in natural waters exists in the following forms:
(1) Dissolved as the hexaquo ion $[Ni(H_2O)_6]^{2+}$. In acidic waters
(pH 3.0) oxidizing nickeliferous deposits at Cobalt, Ontario we have
found up to 75 ppm Ni dissolved mainly as the sulfate [12]. The
cobalt content of these waters was up to 445 ppm Co, indicating that
cobalt in some types of waters is much more mobile than nickel. In
some waters nickel may be dissolved as chloride, nitrate, and possibly arsenate. In seawater nickel is probably largely present as
dissolved chloride and sulfate. (2) Associated with dissolved or
colloidal humic matter and other organic materials of an uncharacterized nature. Certain brown (humic) waters contain much of their
nickel in this form. We think this nickel is bound as a nickel-
humate, perhaps associated with the carboxyl, carbonyl, and other
complex organic groups, similar in most respects to the behavior of
copper [15]; some nickel may be associated with or constitute an
integral part of degraded porphyrin complexes. (3) Adsorbed to or
an integral part of various colloids, suspensoids, and fine-grained
silty material, especially Fe and Mn hydroxides, clay mineral complexes, and clay mineral-humic complexes. (4) Adsorbed to or an
integral part of aquatic microorganisms. The fractionation ratios

between all of these various forms of nickel in natural waters are
highly variable and can usually only be determined in any water
sample by detailed analyses.

Precipitates at spring orifices, including those composed of
Fe and Mn hydroxides, clay mineral and siliceous complexes, humic
complexes, etc., are frequently enriched in nickel, especially in
regions containing nickeliferous deposits or rocks containing abnormal
amounts of pyrite, pyrrhotite, and pentlandite. Normal precipitates
at spring orifices usually contain nickel contents in the range 20-100
ppm; in nickeliferous terranes, e.g., Cobalt, Ontario, nickel contents
as high as 2,000 ppm have been recorded in limonitic and manganiferous
precipitates [12].

Drainage sediments (stream, river, lake) vary widely in their
nickel content depending on the type of geological terrane and on the
presence of nickeliferous rocks and deposits; other influencing factors include the presence of nickel smelters, plating plants, and
other industrial and domestic contamination. Under pristine conditions stream sediments are commonly higher in nickel content than
river and lake sediments in that order. The normal range of the
nickel content in uncontaminated drainage sediments is 1-150 ppm;
in the vicinity of nickeliferous rocks and deposits a threefold to
10-fold increase in the nickel content, as compared with the regional
background, is not unusual.

Our investigations show that nickel in drainage sediments is
partitioned between various phases, including (1) resistate minerals
(magnetite, amphiboles, pyroxenes, chlorite, pyrite, etc.), (2)
resistate hydrolysate minerals (limonite, wad, clay minerals, etc.)
formed in the weathering profile; (3) active inorganic hydrolysate
minerals (limonite, wad, clay mineral colloids, etc.) formed in the
active sediment; (4) active organic-inorganic hydrolysate phase
(humus, humic-clay mineral complexes, humic-limonite-wad complexes,
etc.) formed in and on the active sediment; (5) microorganisms and
macroorganisms inhabiting the drainage sedimentary complex; (6) pore
water invading the sediment. The partition coefficients between

these various phases are highly variable; most of the nickel is
present in phases 2, 3, and 4 in normal drainage sediments. The
concentrations in these phases is essentially due to coprecipitation
and adsorption processes.

5. NICKEL IN THE ATMOSPHERE

Atmospheric nickel may occur in two forms, as a constituent of suspended particulate matter (aerosols) and in a gaseous form. In the
particulate matter the element may be an integral component of the
aerosols or it may be adsorbed or otherwise attached to the suspended
matter which, under natural conditions, may include meteoric dust,
rock and soil dust, sea salt, volcanic ash, exudites from growing
vegetation, and smoke particles from forest fires and volcanoes.
Anthropogenic sources of atmospheric particulates include those
arising from the burning of coal (fly ash) and petroleum, smelting
and alloying processes, waste incineration, and tobacco smoke. The
only gaseous form of nickel of environmental importance is the highly
toxic nickel carbonyl, $Ni(CO)_4$. This gas is essentially of anthropogenic origin. It may, however, be formed in very small amounts in
the throats of volcanoes. Its existence in nature would be transitory
since in the presence of moist air the compound quickly decomposes to
form nontoxic nickel carbonate and basic nickel carbonates.

Particulate nickel concentrations in remote areas of the earth
appear to range from 0.0000 to 0.0008 $\mu g/m^3$ according to data in the
world literature. In urban environments the range appears to be 0.01-
0.1 $\mu g/m^3$, and near nickel smelters the content may be as high as 3.3
$\mu g/m^3$. See also the extensive data provided by Barrie in [9].

6. NICKEL IN THE BIOSPHERE AND ITS DEGRADATION PRODUCTS

Nickel is a trace element in most plants and animals; in some organisms it is an essential constituent. The element is a trace metal

in most degradation products of the biosphere and is enriched in
some.

6.1. Nickel in Biological Systems

Biological roles have been established for a number of metals (Cr,
Mn, Fe, Co, Cu, Zn) in the first transition series. Prior to 1975,
nickel was considered to have no essential biological functions.
However, recent research now indicates that nickel is significant in
a number of plant, animal, and bacterial systems, although a well-
defined biochemical mechanism for the role of nickel is still obscure.

Nickel contents for a number of species are listed in Table 3.
Data on organisms from various trophic levels suggest that the equilibrium concentrations for nickel are highest in plants, intermediate
in invertebrates, and lowest in fish predators [16]. Animals demon-
trate selective uptake of nickel in contrast to plants in which the
nickel contents commonly reflect those in the water, soils, or sediments. Some organisms accumulate nickel without toxic effects, and
concentration factors are reported as 11.4×10^3 for macrophytes,
1.4×10^2 to 9.1×10^3 for green algae, 6.34×10^2 for zooplankton,
9.00×10^2 for crayfish, $1.34-3.99 \times 10^2$ in *Daphnia magna*, 2.60×10^2
for clams, and $2.26-3.29 \times 10^2$ for fish [16-19]. In general, accumulation for all species depends on the nickel concentration in the
environment, cell density, and presence or absence of cations (Cu^{2+},
Ca^{2+}) and anions (PO_4^{3-}).

Nickel is essential to the growth of some agricultural plants
and certain hyperaccumulators growing on nickeliferous soil ecotypes.
Nickel accumulation in plants is physicochemical in nature and the
result of combinations of ion exchange, electrolyte sorption, and
particulate entrapment [27]. Biogeochemical analyses of plant samples
from areas underlain by nickel mineralization reflect the nickel content in the overlying soils, a feature of the nickel hyperaccumulators
that is especially useful in biogeochemical prospecting surveys [30].
Because of their ability to accumulate metals such as nickel, bryophytes are also used for monitoring metal fallout in areas of known

pollution [27]. Nickel is phytotoxic at 50 mg/kg dry weight in
field plants with the exception of certain hyperaccumulators and at
<1 mg/kg for plants in solution culture [33]. Nickel is usually
toxic at concentrations of 0.5-2.0 mg/liter for aquatic macrophytes
and planktonic algae, 1.1-7.4 mg/liter for zooplankton, and LC_{50}
values have been reported at 4.6-500 mg/liter for fish [17,34].
Factors affecting toxicity include the test species, cell density,
solubility and chemical form of nickel, synergistic stresses, pH of
the environment, and the presence of heavy metals, chelating agents,
and other cations. These factors especially affect the availability
of nickel to plants and allow tolerance to higher concentrations of
the metal in soils. Low-level nickel contents cause growth decrements in plants, moderate to acute stress due to high nickel contents
induces chlorosis, and during acute nickel toxicity plant tissues
become necrotic with ensuing death [33].

Estimates of the total amount of nickel in a human body (70-kg
subject) fall in the range of 7-10 mg. The greatest concentrations
of nickel (0.1-0.2 µg/g wet tissue) occur in areas exposed to external
factors (lung, intestine, skin) [35]. Normal whole-blood, serum, and
urine nickel levels reported for humans are 3-7 µg/liter (blood) and
1-5 µg/liter (serum, urine) [36]. Nickel has been confirmed in human
fetal tissues [37], and significant amounts (mean 9-20 µg/g) of the
metal have been found in kidney stones composed largely of calcium
and/or magnesium [38]. Nickel deficiency symptoms have been studied
for vertebrates; most of the toxic responses involve problems with
iron utilization, metabolites, and lesions of organs [39,40].

Total nickel in human serum and rabbit (in parentheses) is
partitioned into three fractions: (1) 40% (16%) ultrafilterable,
(2) 34% (40%) bound to albumin, and (3) 26% (44%) in nickel metalloproteins [41,42]. The last has been identified as a 9.5S α_1-glyco-
protein (3.08 x 10^5 amµ molecular weight) in human serum [43] and as
an α_1-macroglobulin (7.0 x 10^5 amµ) in rabbit serum [41,44]. The
purified rabbit nickeloplasmin contains approximately 1 mol Ni per
mol of protein. The ultrafilterable nickel complexes may act as
carriers for renal excretion of this metal thereby playing a part in

TABLE 3

Nickel Content of Various Plants and Animals

Organism	Content [mg/kg (ppm) dry weight basis]	Source	Remarks
Brown algae	3	20	
Algae			
Periphyton	0.82	16	Contaminated river system
Freshwater	0.1-2.9	18	
Bryophytes	0.83	21	
Sphagnum angustifolium	0.3	22	
Sphagnum fuscum	0.2-0.8	22	Canada
	0.55-2.22	22	Finland
Hylocomium splendens	4.6-5.1	23	
Rhacomitrium lanuginosum	3.7	24	
Sphagnum magellanicum	2.5-4.9	25	
Pterogonium gracile	6	26	
Liverworts	0.15-0.7	21	
Lichens	1.6-3.6	27	
Fungi	1.5	20	
Ferns	1.5	20	
Gymnosperms	1.8	20	
Mountain hemlock	<0.2-0.6	28	Needles
Mountain fir	1.4	28	Needles
Western red cedar	1.5	28	Needles
Engleman's spruce	0.8	28	Needles

Angiosperms			
Alyssum sp.	2.7	20	On nickeliferous soils
Eichornia crassipes	~1000	29	On nickeliferous soils
Sebertia acuminata	500	29	On nickeliferous soils
Rinorea sp.	175-2170	29	On nickeliferous soils
Rinorea bengalensis	836-17500	30	Ultrabasic substrate
Plankton			
Zooplankton	36	20	
(1) Euphausiaceae	3-51	16	
(2) Copepoda	3.8	31	Monterey Bay, California
(3) Radiolaria	2.0	31	Monterey Bay, California
(4) Mixed	3.7	31	Monterey Bay, California
	4.0	31	Monterey Bay, California
Zooplankton and microplankton			
(1) Zooplankton	8.4	31	East of Hawaii
(2) Microplankton	11.6	31	East of Hawaii
Crayfish	12-36	16	
Clams	5-91	16	
Tubificids	4-18	32	
Fish			
Freshwater	0.1-0.28	32	
Marine (shark)	0.3	7	
Insects			
Termites	2000	30	Serpentine substrate
Beetle (*Catamerus* sp.)	2000	30	Serpentine substrate

homeostatic control, an hypothesis supported by a number of findings. These include the narrow range of nickel concentrations in human blood, serum, and urine; also five nickel-containing complexes were found in the ultrafilterable fraction of ^{63}Ni-dosed rabbit serum, of which three were subsequently found in the urine [45]. Histidine has been shown to be a dominant nickel-binding amino acid in human and rabbit serum [46]. Ni^{2+} can also compete for the Cu^{2+} site in bovine, human, rat, and rabbit serum albumins [36,47]. In addition, ^{63}Ni labeling of rabbit serum α_1-macroglobulin indicates that a controlling exchange mechanism exists between Ni^{2+} in nickeloplasmin and ultra-filterable nickel [48]. A change in this homeostatic control is evident since pathological processes lead to hypernickelemia in patients with myocardial infarction, acute stroke, extensive thermal burns, and to hyponickelemia in patients with hepatic cirrhosis or uremia resulting from hypoalbuminemia [49,50].

Some blue-green algae and bacteria exhibit an absolute requirement for nickel [51-53]. For instance, nickel stimulates nitrogen fixation in soil bacteria [52], the metal is necessary for maximal activity of rumen bacterial urease [54], and a strain of *Bacillus cereus* forms a nickel-containing intracellular pigment [55]. The metal is required for the autotrophic growth and synthesis of an active hydrogenase in knallgas bacteria [56]. Acetogenic bacteria and some clostridia require nickel for the synthesis of an active carbon monoxide dehydrogenase [57,58]. Methanogenic bacteria also contain four known nickel enzymes: methyl coenzyme m-reductase (factor F430, a nickel tetrapyrrole), hydrogenase I and II, and carbon monoxide dehydrogenase [59].

Evidence now exists for the involvement of nickel in the mechanisms of enzymes using, transferring, or producing ammonia or amide [60]. For instance, ureases from jack bean (2.0-0.3 g-atoms Ni/ 105,000 g) and other sources are known nickel proteins [54,60].

Ni^{2+} binds oxygen (carbonyl, C=O; carboxylate, R-CO-O), phosphate (R-O-PO_3^{2-} and others), nitrogen (-NH_2, RN= and =N- in the imidazole of histidine), and sulfur (-SH, -SR) centers in proteins

and various biochemicals [61]. Complexes of Ni^{2+} are prevalent in
the four-coordinate tetrahedral and square planar polyhedra, and the
six-coordinate octahedral structure, but five-coordinate square pyra-
midal and trigonal bipyramidal are also known [62]. Because Ni^{2+}
forms octahedral complexes like those of other divalent cations, it
can compete with these ions for their binding sites. Therefore,
processes of inhibition, activation, or inactivation are found for
nickel substituted in vitro in Mn, Mg, Ca, Zn, Cu, and Fe enzyme
systems [36]. For example, Ni^{2+} can act as a partial antagonist
stimulating Ca^{2+}/Mg^{2+} in some systems but inhibiting other processes
such as muscle activity, nerve transmission, and secretion [63,64].
Other citings include (1) inhibition by Ni^{2+} of brain ATPase and
myocardium phosphorylase [64,65], and (2) activation after replacement of Mn^{2+} by Ni^{2+} in allantoicase [66] and chicken liver pyruvate
carboxylase [67]. In addition, Ni^{2+} can deactivate enzymes which do
not have a metal ion requirement [36].

Significant trace concentrations of nickel have been found in
various DNA and RNA molecules [63,64]. In high concentrations the
metal is known to decrease fidelity of the DNA replication and transcription, possibly by replacing the Mg^{2+} bound to phosphate moieties
involved in the reactions, or by inducing deactivation of DNA and RNA
polymerases [68,69]. Studies of double-strand unwinding of DNA impli-
cate the weak binding of nickel to nitrogen centers of the hetero-
cyclic bases [70]. Ni^{2+} also induces helix-to-coil transition in
polyribonucleotides as well as speeding up the enzymatic in vitro
hydrolysis of RNA [70]. Therefore, since nickel interferes with DNA
synthesis and destabilizes polynucleotides, this metal is a potential
mutagen or carcinogen.

6.2. Nickel in Peat, Coal, Petroleum, etc.

Our analyses of the constituents of wet site deposits indicate that
the nickel contents of peat, gyttja, and organic mucks vary considerably depending on the proximity to nickeliferous rocks and deposits.

Normal peat contains from less than 5 to 70 ppm Ni in the material
as found (10-330 ppm Ni in the ash); near nickeliferous rocks and
deposits nickel contents as high as 2,000 ppm or more in the ash may
occur. In peat, gyttja, and other similar organic materials much of
the nickel is organically bound in a form as yet uncharacterized; in
some peats the nickel is a constituent of disseminated pyrite, marca-
site, millerite, and other minor sulfides.

Our analyses of coal, and data from the voluminous world literature, indicate that the nickel content of coals is highly variable
depending, it would seem, essentially on the rank (i.e., the geological history) and geographic location of the coal in question.
We estimate the average nickel content of anthracite and other hard
coals to be 12 ppm; bituminous coals 10 ppm; and brown coals 5 ppm.
Valković [71] gives 15 ppm Ni as the world average for all coals; he
also gives values for nickel in coal ash that range from 0.0003 to
1.02% from various world coal basins. Nickel in coals is bound in a
variety of ways including organometallic compounds, adsorbed forms,
and a constituent of sulfides, particularly pyrite, millerite,
linnaeite, and arsenopyrite.

Wedepohl [5] quotes values from the literature for the nickel
content of petroleum that range from 1.0 to 13 ppm. Contents as high
as 130 ppm Ni have been recorded from some heavy oils. Nickel in
oils is usually regarded as being present as chelated porphyrin complexes such as those deriving from chlorophyll and its degradation
products. Other details are given in [9].

Bitumen and the semihard to hard hydrocarbons such as anthraxolite, thucholite, and carburan frequently concentrate nickel which is
evidently present in most of these materials as uncharacterized
organometallic compounds and in others as a constituent of fine-
grained sulfides, mainly pyrite. Hodgson [72] recorded up to 76.6
ppm Ni in the oil of the McMurray tar sands in western Canada. Our
analyses of anthraxolite, thucholite, and other solid hydrocarbons
all indicate traces of nickel in most samples; in some samples contents as high as 150 ppm Ni or more may occur.

7. GENERAL SUMMARY

Nickel has a siderophile, chalcophile, oxyphile, and biophile character in nature and is widely disseminated in the materials of the lithosphere. The crustal abundance of nickel is about 60 ppm.

Nickel is concentrated in ultrabasic rock suites (average 1500 ppm); more acidic suites have much less nickel (gabbro-208 ppm; granite-13 ppm). Residual soils tend to reflect the nickel content of their parent rocks. In the hydrosphere the average nickel content of most natural freshwaters ranges from 0.0005 to 0.02 ppm.

Nickel deposits are of two principal types: massive and disseminated nickeliferous sulfide bodies commonly associated with ultrabasic and basic rock suites, and residual (supergene, lateritic) nickel deposits developed principally on nickel-rich rocks such as gabbros, dunites, and serpentinites. The average nickel content of both types of deposits is about 1.5%.

Nickel occurs in most plants and animals, but generally only in small amounts (normal range 0.1-91 ppm, dry weight basis). Some plants are nickel accumulators on nickel-rich soils (up to 17,500 ppm, dry weight basis). In certain organisms nickel has an essential role. The natural degradation products of the biosphere generally contain small amounts of nickel (coal average 15 ppm; petroleum average ~10 ppm).

REFERENCES

1. J. W. Mellor, "A Comprehensive Treatise on Inorganic and Theoretical Chemistry", Vol. XV, Ni, Ru, Rh, Pd, Os, Ir, Longmans, Green, London, 1936.

2. "Gmelins Handbuch der anorganischen Chemie: Nickel, Teil Al", System-Nummer 57, Verlag Chemie, GMBH, Weinheim, 1967.

3. D. Nicholls, Nickel, in "Comprehensive Inorganic Chemistry" (J. C. Bailar, H. J. Eméleus, R. Nyholm, and A. F. Trotman-Dickenson, eds.), Vol. 3, No. 42, Pergamon, Oxford, 1973, pp. 1109-1161.

4. A. G. W. Cameron, A New Table of Abundances of the Elements in the Solar System, in "Origin and Distribution of the Elements" (L. H. Ahrens, ed.), Pergamon, Oxford, 1968, pp. 125-143.

5. K. H. Wedepohl (ed.), "Handbook of Geochemistry", Vol. II-3, Nickel, R. G. Burns and V. M. Burns (28-A) and K. K. Turekian (28B-280), Springer-Verlag, Berlin, 1974-78.

6. A. E. Fersman, "Geochemistry", 4 vols., Leningrad, 1934-39.

7. V. M. Goldschmidt, "Geochemistry", Oxford Univ. Press, 1954.

8. K. Rankama and T. H. G. Sahama, "Geochemistry", Univ. Chicago Press, 1950.

9. Environmental Secretariat National Research Council of Canada, Ottawa, "Effects of Nickel in the Canadian Environment", NRCC, No. 18568, 1981.

10. A. P. Vinogradov, "The Geochemistry of Rare and Dispersed Chemical Elements in Soils", Consultants Bureau, New York, 1959.

11. D. J. Swaine, "The Trace-Element Content of Soils", Commonwealth Bureau of Soil Science, Tech. Commun., *48*, 1955.

12. R. W. Boyle, A. S. Dass, D. Church, G. Mihailov, C. Durham, J. Lynch, and W. Dyck, "Research in Geochemical Prospecting Methods for Native Silver Deposits, Cobalt Area, Ontario, 1966", Geol. Surv. Canada Paper 67-35, 1969.

13. R. H. Whittaker, R. B. Walker, and A. R. Kruckeberg, *Ecology, 35*, No. 2, 258 (1954).

14. J. Proctor and S. R. J. Woodell, *Adv. Ecol. Res., 9*, 205 (1975).

15. R. W. Boyle, *J. Geochem. Explor, 8*, 495 (1977).

16. T. C. Hutchinson, A. Fedorenko, J. Fitchko, A. Kuja, J. C. Van Loon, and J. Li, in "Environmental Biogeochemistry" (J. Nriagu, ed.), Ann Arbor Science, 1975, p. 565.

17. P. M. Stokes, in "Effects of Nickel in the Canadian Environment", National Research Council of Canada No. 18568, Ottawa, 1981, p. 77.

18. D. R. Trollope and B. Evans, *Environ. Pollut., 11*, 109 (1976).

19. T. Hall, in "Nickel Uptake, Retention and Loss in *Daphnia magna*", M. Sc. thesis, Dept. of Botany, University of Toronto, 1978.

20. H. J. M. Bowen, in "Trace Elements in Biochemistry", Academic, New York, 1966.

21. H. T. Shacklette, in "Element Content of Bryophytes", U.S. Geol. Surv. Bull. 1198-D, 1965, pp. 1-21.

22. P. Pakarinen and K. Tolonen, *Ambio, 5*, 38 (1976).

23. S. S. Groet, *Oikos, 27*, 445 (1976).

24. A. N. Rencz, in "The Relationship between Heavy Metals in the Soil and their Accumulation in Various Organs of Plants Growing in the Arctic", Ph.D. thesis, Dept. of Biology, University of New Brunswick, Fredericton, 1978.

25. A. Ruhling and G. Tyler, *Oikos, 21*, 92 (1970).

26. I. Johnsen and L. Rasmussen, *Bryologist, 80*, 625 (1977).

27. M. A. S. Burton and K. Puckett, in "Effects of Nickel in the Canadian Environment", National Research Council of Canada No. 18568, Ottawa, 1981, p. 159.

28. H. V. Warren and R. E. Delavault, *Trans. Royal Soc. Can., 48*, 71 (1954).

29. H. L. Cannon, *Science, 132*, 591 (1960).

30. R. R. Brooks, in "Biological Methods of Prospecting for Minerals", John Wiley, New York, 1983.

31. J. H. Martin and G. A. Knauer, *Geochim. Cosmochim. Acta, 37*, 1639 (1973).

32. B. J. Mathis and T. F. Cummings, *J. Water Pollut. Control Fed., 45*, 1573 (1973).

33. T. C. Hutchinson, B. Freedman, and L. Whitby, in "Effects of Nickel in the Canadian Environment", National Research Council of Canada No. 18568, Ottawa, 1981, p. 119.

34. J. Fitzsimons and J. Reinke, in "Summary of Information and References on the Toxicity of Metals to Aquatic Organisms", Ontario Ministry of the Environment, 1977.

35. K. Sumino, K. Hayakawa, T. Shibata, and S. Kitamura, *Arch. Environ. Health, 30*, 487 (1975).

36. A. Cecutti and E. Nieboer, in "Effects of Nickel in the Canadian Environment", National Research Council of Canada No. 18568, Ottawa, 1981, p. 193.

37. C. E. Casey and M. F. Robinson, *Br. J. Nutr., 39*, 639 (1978).

38. A. A. Levinson, M. Nosal, M. Davidman, E. L. Prien, E. L. Prien, Jr., and R. G. Stevenson, *Invest. Urol., 15*, 270 (1978).

39. F. W. Sunderman, Jr., *Ann. Clin. Lab. Sci., 8*, 491 (1978).

40. F. W. Sunderman, Jr., in "Disorders of Mineral Metabolism" (F. Bonner and J. Coburn, eds.), Academic, New York, 1979, Chap. 10.

41. S. Nomoto, M. D. McNeely, and F. W. Sunderman, Jr., *Biochemistry, 10*, 1647 (1971).

42. F. W. Sunderman, Jr., M. I. Decsy, and M. D. McNeely, *Ann. N.Y. Acad. Sci., 199*, 300 (1972).

43. H. Haupt, N. Heimburger, T. Kranz, and S. Baudner, *Z. Physiol. Chem., 353*, 1841 (1972).

44. S. Nomoto, M. I. Decsy, J. R. Murphy, and F. W. Sunderman, Jr., *Biochem. Med., 8*, 171 (1973).

45. M. Van Soestbergen and F. W. Sunderman, Jr., *Clin. Chem., 18*, 1478 (1972).

46. M. Lucassen and B. Sarkar, *J. Toxicol. Environ. Health, 5*, 897 (1979).

47. W. M. Callan and F. W. Sunderman, Jr., *Res. Commun. Chem. Pathol. Pharmacol., 5*, 459 (1973).

48. M. I. Decsy and F. W. Sunderman, Jr., *Bioinorg. Chem., 3*, 95 (1974).

49. M. D. McNeely, F. W. Sunderman, Jr., M. W. Nechay, and H. Levine, *Clin. Chem., 17*, 1123 (1971).

50. F. W. Sunderman, Jr., *Ann. Clin. Lab. Sci., 7*, 377 (1977).

51. C. Van Baalen and R. O'Donnel, *J. Gen. Microbiol., 105*, 351 (1978).

52. D. Bertrand, *C.R. Acad. Sci. Paris, Ser. D., 278*, 2231 (1974).

53. R. Repaske and A. C. Repaske, *Appl. Environ. Microbiol., 32*, 585 (1976).

54. J. W. Spears and E. E. Hatfield, *J. Anim. Sci., 47*, 1345 (1978).

55. R. K. Thauer, G. Diekert, and P. Schönheit, *Trends Biochem. Sci., 5*, 304 (1980).

56. B. Friedrich, E. Heine, A. Finck, and C. G. Friedrich, *J. Bacteriol., 145*, 1144 (1981).

57. G. Diekert and R. K. Thauer, *FEMS Microbiol. Lett., 7*, 187 (1980).

58. H. L. Drake, S.-I. Hu, and H. G. Wood, *J. Biol. Chem., 255*, 7174 (1980).

59. R. K. Thauer, *Biol. Chem. Hoppe-Seyler, 366*, 103 (1985).

60. N. E. Dixon, C. Gazzola, R. L. Blakeley, and B. Zerner, *Science, 191*, 1144 (1976).

61. E. Nieboer and D. H. S. Richardson, *Environ. Pollut., B1*, 3 (1980).

62. F. A. Cotton and G. F. Wilkinson, in "Advanced Inorganic Chemistry: A Comprehensive Text", 3rd ed., Interscience, New York, 1972, Chaps. 21-25.

63. F. H. Nielsen, in "Trace Element Metabolism in Animals, Vol. 2" (W. G. Hoekstra, J. W. Suttie, H. E. Gauther, and W. Mertz, eds.), Maryland Univ. Park Press, Baltimore, 1974, p. 381.

64. F. W. Sunderman, Jr., F. Coulston, G. L. Eichhorn, J. A. Fellows, E. Mastromatteo, H. T. Reno, and M. H. Samitz, in "Nickel: A Report of the Committee on Medical and Biologic Effects of Environmental Pollutants", U.S. Natl. Acad. Sci., Washington, D.C., 1975.

65. A. K. Mathur, S. V. Chandra, J. Behari, and S. K. Tandon, *Arch. Toxicol., 37,* 159 (1977).

66. C. van der Drift and G. D. Vogels, *Biochem. Biophys. Acta., 198,* 339 (1970).

67. M. C. Scrutton, P. Griminger, and J. C. Wallace, *J. Biol. Chem., 247,* 3305 (1972).

68. M. A. Sirover and L. A. Loeb, *Science, 194,* 1434 (1976).

69. F. W. Sunderman, Jr., in "Proc. 2nd Arnold O. Beckman Conference in Clinical Chemistry", San Antonio, Texas, September 6-8 (M. Fleischer, ed.), Am. Assoc. Clin. Chem., Washington, D.C., 1979, p. 265.

70. G. L. Eichhorn, in "Inorganic Biochemistry" (G. L. Eichhorn, ed.), Elsevier, Amsterdam, 1975, p. 1210.

71. V. Valković, "Trace Elements in Coal", Vol. 1, CRC, Boca Raton, Florida, 1983.

72. G. W. Hodgson, *Bull. Am. Assoc. Petrol. Geol., 38,* 2537 (1954).

2

Nickel in Aquatic Systems

Pamela Stokes
Institute for Environmental Studies
and Department of Botany
University of Toronto
Toronto, Ontario, Canada M5S 1A4

1. INTRODUCTION

This chapter addresses nickel in aquatic systems, mainly in freshwater, and emphasizes advances in our knowledge based mainly on the
literature since 1980. Several recent reviews include material on
nickel in aquatic systems [1-3] and the material in these reviews is
not repeated here in any detail.

The text is organized to address nickel as a micronutrient,
the toxicity and toxic effects of nickel, the forms and partitioning
of nickel in aquatic systems, its bioavailability and biological
uptake. Finally, recent advances and needs for research are
summarized.

2. NICKEL AS AN ESSENTIAL ELEMENT FOR AQUATIC BIOTA

Until quite recently, there were no known biological functions for
nickel, nor had an absolute requirement for this element been demonstrated. In the mid-1970s, plant physiologists discovered that
nickel was a constituent of urease (see Chap. 6) and an absolute
requirement for nickel was demonstrated for the conversion of urease
to ammonia in several higher plant tissues [4].

In 1978, Van Baalen and O'Donnel [5] isolated a nickel-
dependent blue-green alga, *Oscillatoria* sp., from marine mud, but
these workers were unable to specify a possible metabolic role for
nickel. Subsequent work [6-8] showed that for certain marine microalgae the synthesis and activity of urease was nickel-dependent.
Most recent studies suggest that nickel requirement in microalgae
is quite widespread and is consistently associated with urease
activity. Thus Oliveira and Antia [4] studied the direct response
to nickel as well as the effect of the chelator citric acid followed
by subsequent additions of excess nickel for 12 marine species.
These included diatoms, prymnesiophytes, a dinoflagellate, crypto-
phyte, crysophyte, and a red alga, most of which were already known
to use urea as a nitrogen source. They concluded that "nickel

requirement for urea utilization is of widespread occurrence among
the microalgae tested" [4]. They also considered that the endogenous
nickel level in most marine waters was adequate to support urease
activity in algae.

No reports were found for algal requirements for nickel in
freshwater algae, nor for any other aquatic organisms.

3. NICKEL TOXICITY

Although nickel is not a widespread contaminant in surface waters,
nevertheless certain types of activity especially related to mining
and smelting, as well as natural anomalies, result in elevated
nickel in water or sediments, with the potential for toxic effects
on aquatic biota. In general, the mechanisms of nickel toxicity
are scarcely known for aquatic biota, but for higher animals and
humans many of the toxic responses to nickel involve interference
with iron metabolism [2] and nickel, like most metals, binds to
proteins and nucleic acids [3]. Different groups of aquatic organisms show different ranges of sensitivity to nickel (Table 1).

In vivo and in vitro inhibition of ATPase has been demonstrated for nickel [3]. The external concentrations of nickel which
evoke a toxic response are quite variable (see Secs. 3.1-3.3 and
Tables 1-4). Of more significance to environmental toxicology are
the factors affecting nickel toxicity and the chemical forms of
nickel which are toxic. Water hardness, age and size, species, and
temperature, as well as other toxic substances, are major factors
which influence nickel toxicity and which therefore affect the level
of acute toxicity. The chemical form which is most available and
most toxic is the aqueous ion Ni^{2+} (see Sec. 4).

3.1. Toxicity of Nickel to Fish

Several laboratory studies have been reported for the effects of
nickel on fish, mostly attempting to determine acutely toxic con-

TABLE 1

Toxic Concentrations of Nickel (mg/liter) for
Various Types of Freshwater Aquatic Organisms

Organism/group	Toxic concentration
Rotifers[a]	2.74-7.4
Oligochaetes[a]	0.082-120
Gastropods[a]	11.4-138
Insects[a]	4.0-<64
Crustaceans[a]	<0.32-15.2
Fish[b]	4-20
Bacteria[b]	5.0
Filamentous fungi[b]	10.0
Blue-green algae[b]	0.6
Green algae[b]	<0.1
Aquatic macrophytes[b]	0.1

[a]LC_{50} (24-96 hr).
[b]Values are thresholds for most sensitive species.
Source: Data selected from EIFAC 1984 [3].

centrations (by LC_{50} or TL_M values); these have been summarized in
NRCC [2]. The value of such studies, where reported toxic levels
range from 0.1 to 495 mg/liter nickel (depending on species and
conditions) are quite limited in the context of determining or pre-
dicting environmental effects of nickel. Levels in water, even in
nickel-polluted regions, rarely exceed 6.0 mg/liter, and in surface
water in remote areas 0.001-0.005 mg/liter is the more usual range
[2]. EIFAC [3] reviews more recent studies on nickel toxicity to
fish and concludes that "no data are available from studies in the
field in which nickel occurs as the principal pollutant".

Physiological studies of the effects of nickel on fish have
mostly been done with rather high doses and exposures of nickel.
For example, Chaudhry and Nath [9,10] and Chaudhry [11] worked on
Colisa fasciatus in 64 mg/liter nickel (which was sublethal in their

TABLE 2

Recent Studies Showing Range of Nickel Concentrations Toxic to Fish

Fish	Conditions	Test	Concentration of nickel (mg/liter)	Ref.
Rainbow trout: *Salmo gairdneri*	Hardness 100 mg/liter	LC_{50}	0.05	16
Rainbow trout: *Salmo gairdneri*	Hardness 174 mg/liter	LC_{50}	0.09	16
American flagfish: *Jordanella floridae*	Soft water, mixture of 7 metals	Fry mortality	0.007	17
Carp: *Cyprinus carpio*	Complexing agents NTA and EDTA added to water with nickel	Toxicity and accumulation	10-30 Complexing agents decreased toxicity and accumulation	18
Pumpkinseed: *Lepomis gibbosus*	28°C, hardness 55 mg/liter	LC_{50}, 96 hr	8.0	3
Banded killifish: *Fundulus diaphanus*	28°C, hardness 55 mg/liter	LC_{50}, 96 hr	8.0	3

TABLE 3

Effect of Nickel on Some Invertebrates

Organism	Conditions	Test	Nickel concentration causing toxicity (mg/liter)	Ref.
Daphnia magna (freshwater cladoceran)	18°C 45 mg/liter	LC50	<0.32	3
Asellus aquaticus (freshwater isopod)	Soft water	LC50 48 hr	435	19
Crangonyx pseudogracilis (freshwater amphipod)	Soft water	LC50 48 hr	252	19
Allorchestes compressa (marine amphipod)	Saltwater	LC50 96 hr	35	20
Clistoronia magnificans (caddisfly)	Freshwater	Life cycle prevented from completion	0.25	21
Chironomus (freshwater midge larvae)	Freshwater	LC50 48 hr	79-169	22
Juga plicifera (freshwater snail)	Freshwater	LC50 96 hr	0.237	23
Macoma balthica (marine deposit feeder)	Saltwater	LC50	5-54	24

TABLE 4

Studies of Nickel Toxicity for Algae

Organism	Toxic con-centration of nickel (mg/liter)	Ref.	Comment
Scenedesmus	1.5	27 ⎫	
Haematococcus	0.07-0.7	27 ⎬	Review of pre-1980 literature
Bluegreens	0.6	27 ⎭	
Scenedesmus ATCC 11460	18	28	pH 7.0
Scenedesmus ATCC 11460	1.8	28	pH 4.0
Chlorococcum	2.5	29 ⎫	Growth decreased but not lethal to any of the 3 spp.
Scenedesmus obliquus	2.5	29 ⎬	
Ankistrodesmus falcatus	2.5	29 ⎭	
Anabaena inaequalis	0.125	30	Growth inhibition
Anabaena inaequalis	20	30	Nitrogenase inhibition

system), and variously observed increases in blood lactic acid, decreases in muscle glycogen, depletion of liver glycogen, and increases in blood glucose at this level. They also reported mucus secretion and behavioral distress at this external concentration but did not report tissue levels. Hughes et al. [12] showed structural changes in gills of rainbow trout, with damage at 5% of the LC_{50}. This effect was reversible, with recovery after 21 days. The reversibility of the effects of nickel on gill sialic acid was noted by Arillo et al. [13].

For nickel, like other metals, the life stage of fish has been shown to affect the response. Thus Grande and Andersen [14] demonstrated, for *Salmo salmar*, that hatching and embryo mortality was affected at 0.2% of the LC_{50}. Nebeker et al. [15] showed for early life stages of rainbow trout *Salmo gairdneri* that the no-effect level (NOEL) was <0.035 μg/liter. Early life stages varied in sensitivity to nickel with newly fertilized eggs > eyed eggs > larval fish > juvenile fish.

Table 2 summarizes some other recent fish toxicity studies with nickel, emphasizing the range of exposures which produce toxic effects [3,16-18].

3.2. Toxicity of Nickel to Invertebrates

Recent interest in aquatic organisms at trophic levels other than fish is indicated by a number of studies of the effects of nickel on benthic and planktonic invertebrates. Table 1 suggests that oligo-chaetes and crustaceans are most sensitive, and Table 3 summarizes some recent studies [3,19-24]. Since nickel accumulates in sediments (Sec. 4), the benthic organisms, depending in part on their mode of feeding, are expected to be more tolerant than planktonic organisms in the same system. Unfortunately, no studies were available to determine whether this is a correct assumption. The least nickel-sensitive invertebrates appear to be the isopod and amphipod species studied by Martin and Holdich [19]; the marine organisms appear to have intermediate sensitivity; salinity also tends to decrease the toxicity of nickel for marine macroinvertebrates [25]. No studies have revealed the mechanism of nickel toxicity to invertebrates.

3.3. Toxicity of Nickel to Algae

As shown in Table 1, microalgae appear to be among the most nickel-sensitive freshwater organisms. However, tolerance to nickel has developed in some polluted systems (e.g., [26]). Spencer [27] pro-vided a comprehensive coverage of the range of nickel which has been shown to adversely affect various algae, and some typical values are given in Table 4 [27-30].

4. PARTITIONING AND CHEMICAL FORMS OF NICKEL IN WATER

4.1. Theoretical and Modeling Studies

Most authors stress that there is a need to base toxicity on the
chemical form of nickel. Theoretically, the solution chemistry of
nickel under conditions normally encountered in freshwater is less
complex than that of many other metals. In an aqueous medium, the
divalent cation Ni^{2+} dominates over the pH and redox potential ranges
encountered in most surface waters [31]. A thermodynamic model for
nickel in solution showed that it was almost entirely in the ionic
form over pH 4.5-7.0. This included a simulation with certain dissolved organics [32]. When solid phases, suspended or bed sediments,
are included in the system, amorphous oxides of iron and manganese
are important sinks for nickel [33]. Mudroch and Arafat [34] experi-
mented with nickel-contaminated bottom sediments in the Noranda Rouyn
area of Quebec, Canada. They determined nickel speciation by the
GEOCHEM model. When they added fulvic acid, the nickel (in common
with copper and manganese) stayed mainly in the organic portion.
Mouvet and Bourg [35] looked at competition between metals and protons for surface sites, using a computer simulation, and found that
the dissolved nickel, as well as other trace metals, in the Meuse
River in Holland was controlled by adsorption rather than precipitation mechanisms.

4.2. Concentrations in Water and Sediments

Concentrations of nickel (measured as total nickel) in UK rivers
ranged from 0.0007 to 0.0037 mg/liter for clean and from 0.012 to
0.073 mg/liter for polluted waters [3]. Jenkins [36] gives estimates
of 0.003-0.017 mg/liter in North American freshwater, and Birge and
Black [16] show 0.002-0.007 mg/liter nickel in seawater. Bearing in

mind then that less than 100% of this nickel is in the Ni^{2+} form, although the proportion of the divalent ion to total is likely to be very high, the risk for biota of direct toxicity appears to be extremely low. In sediments, nickel concentrations usually exceed those of water by several orders of magnitude. For example, sediments in lakes within 100 km of Sudbury contain from 100 to 500 µg/g nickel, while surface waters in the same region contained 0.01-0.1 mg/liter nickel [2].

4.3. Analytical Methods

Atomic absorption, flame or furnace, is the routine method for determining nickel in water or other solutions. Direct chemical speciation is not carried out routinely; most data for nickel speciation rely on equilibrium models such as MINEQL, REDEQL, and GEOCHEM. Pretreatment or preconcentration methods have been used to improve the detection of nickel. Other methods for the determination of nickel in environmental samples are discussed by Jaworski [2].

5. BIOACCUMULATION OF NICKEL BY AQUATIC BIOTA

Nickel has effects beyond acute toxicity, including carcinogenic effects on animals, and it is necessary to consider the potential for accumulation of nickel in food items and through aquatic food chains. Living organisms concentrate nickel from their environment, which for aquatic organisms means water or food. A detailed study of nickel uptake by an alga and a crustacean by Watras et al. [37] illustrates the complexity of accumulation mechanisms. Using the green alga *Scenedesmus obliquus*, they determined uptake from solution, with nickel activity expressed as pNi. The concentration factor for the algae from the solution was 30-300 times. An EDTA-removable component, termed labile, consistently composed 30-50% of

the total accumulated nickel, regardless of the amount of nickel
taken up or the condition of the cells. Even when the total cell
quota of nickel increased sharply as a result of toxic effects, the
relative size of the labile pool remained the same. For *Daphnia*,
the crustacean, the concentration factor was generally much less
than for the algae, namely 2-12 times. When nickel was supplied in
food, as well as in water, there was no significant change in uptake
of nickel by the animals compared with the situation with water as
the sole source.

In other studies with algae, Wang and Wood [28] showed concentration factors of $0-3 \times 10^3$ for blue-green algae, green algae, and
a *Euglena* sp. In discussing the mechanism of nickel accumulation,
these authors suggest that most of the nickel is coordinated to
functional groups at the cell surface. Other work on nickel accumulation by algae and crustaceans is reviewed in NRCC [2]. A consensus
emerges that algae accumulate more nickel on a weight basis than do
zooplankton and that the food vector at this trophic level is not
very significant in comparison with water.

A few studies on aquatic macrophytes indicate that these plants,
like the algae, have the capacity to accumulate rather high concentrations of nickel. Clark et al. [38] using *Lemna perpusilla* in
laboratory experiments achieved levels of 500-600 µg/g in 10 days,
but the levels returned to background after depuration for 8 days.
Ipomea aquatica took up 200 µg/g nickel in 8 hr from a 5 mg/liter
solution [39]. In a field study, Mudroch and Capobianco [40] measured nickel in *Myriophyllum verticillatum* and *Elodea canadensis*,
and showed a correlation between plant nickel and total sediment
nickel, but not with "labile" extractable nickel; this was not as
predicted.

Other metals, notably copper, can enhance nickel uptake as
shown for *Lemna minor* by Hutchinson and Czyrska [41] and by Stokes
[26] for algae. For animals at higher trophic levels, concentrations seem to decrease successively. According to EIFAC [3], nickel
has little capacity to accumulate in fish tissue. After 180 days
exposure to 1 mg/liter nickel, rainbow trout tissues showed concen-

TABLE 5

Range of Bioconcentration Factors for Nickel
in Field Collections of Aquatic Organisms

Organism	Normal range	Wet/dry	Max level (smelter)	Wet/dry
Algae	0.2-15	D	160	D
Macrophytes	0.5-6.0	W	690	W
Lower invertebrates	2.0-15	D	57	W
Crustaceans	0.01-9.8	W	39	W
Molluscs	0.4-2.0	W	191	W
Freshwater fish	<0.2-2.0	W	52	W
Birds	0.01-0.8	W	5	D

Source: Collected from information given by Jenkins in Ref. 36.

tration factors of 3.1 (liver), 4.2 (kidney), and 1.0 (muscle).
Jenkins [36] gives fish nickel concentrations of 0.2-2.0 µg/g on
a wet weight basis, and Gaggino [42] showed 0.17-0.3 µg/g in chub,
and 0.15-0.27 in bleak.

The nickel which is accumulated appears to be fairly mobile
in that once the organisms are placed in "clean" medium, nickel is
lost. For example, EIFAC [3] reports losses of 59-75% from rainbow
trout after 90 days in clean water, and Hall [43] demonstrated losses
of 25-33% of the nickel in *Daphnia magna* over 20 hr of experimental
exposures. Hall provided evidence for two separate loss mechanisms,
including loss through moulting.

Table 5 summarizes the range of bioconcentration factors for
field collections of aquatic organisms [36]. Aside from plants,
molluscs attain the highest concentration factors of all the trophic
levels examined here.

6. FOOD CHAIN TRANSFER OF NICKEL

From field studies there is evidence of very little transfer of nickel through the aquatic food chain. At successive trophic levels, nickel concentrations or bioconcentration factors decrease. But field studies cannot normally provide information on the major vector for uptake or the mechanisms which control uptake.

Where laboratory studies are available, they consistently demonstrate that for the lower part of the food chain (plant—herbivore) nickel is mainly acquired from water, i.e., by direct uptake. The animals appear to regulate, either by controlled uptake or by increased efflux, the nickel content of their tissues, resulting in lower concentrations in animals than in the plants on which they feed.

For molluscs, which are deposit or suspension feeders, tissues appear to accumulate nickel from the ingested solid material as well as water, but few studies are available. Feces of such animals are conveyors of metal for sediments, i.e., they are important in metal cycling. The existence of the isotope ^{63}Ni provides a convenient means of tracing nickel cycling in simple laboratory food chains but studies of nickel cycling in field situations are lacking.

With respect to the behavior of nickel in the aquatic food chain, biomagnification does not occur. Rather, biominification is the norm.

7. CONCLUSIONS AND RECOMMENDATIONS

More studies are emerging on the toxicity and bioaccumulation of nickel by aquatic organisms other than fish, including benthic and planktonic invertebrates. These animals may be useful as environmental monitors of nickel in water and sediments, provided relationships can be established between environmental concentrations, or forms of nickel, and biological uptake.

The absence of biomagnification of nickel in aquatic food chains and the relatively low importance of food as a vector, at least for lower levels of the food chain, has been confirmed and refined. Lacking is an understanding of the mechanism of transport of nickel into cells, the basis of nickel toxicity in a fundamental sense, and a knowledge of the mechanism by which nickel takes part in the urease enzyme systems.

In comparison with other toxic metals such as Hg, Cd, Cu, and Pb, nickel is not of major concern as a contaminant in surface waters. This is due in part to its relatively limited distribution as a contaminant as well as its rather moderate toxicity.

REFERENCES

1. J. O. Nriagu (ed.), in "Nickel in the Environment", John Wiley, New York, 1980, p. 833.

2. NRCC, "Effects of Nickel in the Canadian Environment", National Research Council Canada Associate Committee on Scientific Criteria for Environmental Quality, NRCC 18568, 1981, p. 352.

3. EIFAC, European Inland Fisheries Advisory Commission, EIFAC Technical Paper 45, Food and Agriculture Organization of the United Nations, 1984, p. 20.

4. L. Oliveira and N. J. Antia, Can. J. Fish. Aquat. Sci., 43, 2427 (1986).

5. C. Van Baalen and R. O'Donnel, J. Gen. Microbiol., 105, 351 (1978).

6. P. J. Syrett, Can. Bull. Fish. Aquat. Sci., 210, 182 (1981).

7. T. A. V. Rees and I. A. Bekheet, Planta, 156, 385 (1982).

8. L. Oliveira and N. J. Antia, J. Plankton Res., 8, 235 (1986).

9. H. S. Chaudhry and K. Nath, Water Air Soil Pollut., 24, 173 (1985).

10. H. S. Chaudhry and K. Nath, Acta Hydrochim Hydrobiol., 13, 245 (1985).

11. H. S. Chaudhry, Toxicol. Lett., 20, 115 (1984).

12. G. M. Hughes, S. F. Perry, and V. M. Brown, Water Res., 13, 665 (1979).

13. A. Arillo, C. Margiocco, F. Melodia, and P. Menzi, Chemosphere, 11, 47 (1982).

14. M. Grande and S. Andersen, *Vatten, 39,* 405 (1983).

15. A. U. Nebeker, C. Savonen, and D. G. Stevens, *Environ. Toxicol. Chem., 4,* 233 (1985).

16. W. J. Birge and J. A. Black, in "Nickel in the Environment" (J. O. Nriagu, ed.), John Wiley, New York, 1986, p. 833.

17. N. J. Hutchinson and J. B. Sprague, *Can. J. Fish. Aquat. Sci., 43,* 647 (1986).

18. S. Muramoto, *J. Environ. Sci. Health, A18,* 787 (1983).

19. T. R. Martin and D. M. Holdich, *Water Res., 20,* 1137 (1986).

20. M. Ahsanullah, *Aust. J. Mar. Freshwater Res., 33,* 465 (1982).

21. A. U. Nebeker, C. Savonen, R. J. Baker, and J. K. McCrady, *Environ. Toxicol. Chem., 3,* 645 (1984).

22. C. Powlesland and J. George, *Environ. Pollut. A42,* 47 (1986).

23. A. U. Nebeker, A. Stinchfield, C. Savonen, and G. A. Chapman, *Environ. Toxicol. Chem., 5,* 807 (1986).

24. V. Bryant, D. M. Newbery, D. S. McLusky, and R. Campbell, *Mar. Ecol. Progr. Ser., 24,* 139 (1985).

25. G. R. W. Denton and C. Bardon-Jones, *Chem. Ecol., 1,* 131 (1982).

26. P. M. Stokes, *Verh. Internat. Verein. Limnol., 19,* 2128 (1975).

27. D. F. Spencer, in "Nickel in the Environment" (J. O. Nriagu, ed.), John Wiley, New York, 1980, p. 833.

28. H. Wang and J. M. Wood, *Environ. Sci. Technol., 18,* 106 (1984).

29. P. V. Devi Prasad and P. S. Devi Prasad, *Water Air and Soil Pollut., 17,* 263 (1982).

30. G. W. Stratton and C. T. Corke, *Can. J. Microbiol., 25,* 1094 (1979).

31. F. M. M. Morel, R. E. M. Duff, and J. J. Morgan, in "Trace Metals and Metal Organic Interactions in Natural Waters" (P. C. Singer, ed.), Ann Arbor Science, 1973, p. 157.

32. P. C. G. Campbell and P. M. Stokes, *Can. J. Fish. Aquat. Sci., 12,* 2034 (1985).

33. R. O. Richter and T. L. Theis, in "Nickel in the Environment" (J. O. Nriagu, ed.), John Wiley, New York, 1980, p. 833.

34. A. Mudroch and N. Arafat, *Environ. Technol. Lett., 5,* 237 (1984).

35. C. Mouvet and A. C. M. Bourg, *Water Res., 17,* 641 (1983).

36. D. W. Jenkins, in "Nickel in the Environment" (J. O. Nriagu, ed.), John Wiley, New York, 1980, p. 833.

37. C. J. Watras, J. MacFarlane, and F. M. M. Morel, *Can. J. Fish. Aquat. Sci., 42,* 724 (1985).

38. J. R. Clark, J. H. Van Hassel, R. B. Nicholson, D. S. Cherry, and J. Cavins, Jr., *Ecotoxicol. Environ. Safety, 5,* 87 (1981).

39. K. S. Low and C. K. Lee, *Pertanika, 4,*16 (1981).

40. A. Mudroch and J. A. Capobianco, *Hydrobiologia, 64,* 223 (1979).

41. T. C. Hutchinson and H. Czyrska, *Verh. Internat. Ver. Limnol., 19,* 2102 (1975).

42. G. F. Gaggino, *Inquinamento* 1984, 1982, 25-48 (Cited in EIFAC [3]).

43. T. M. Hall, *Limnol. Oceanogr., 27,* 718 (1982).

3

Nickel and Plants

Margaret E. Farago[1] and Monica M. Cole[2]
Departments of [1]Chemistry and [2]Geography
University of London, Royal Holloway and Bedford New College
Egham, Surrey TW20 OEX, U.K.

1. NICKEL—AN ESSENTIAL ELEMENT FOR PLANTS?

1.1. Inorganic Composition of Plants

The early history of plant nutrition from the fifteenth century has been discussed by Hewett and Smith [1] and the history of nickel in plants and as an essential element has been chronicled by Hutchinson [2].

Fresh plant material is 80-90% water, and of the remainder over 90% consists of C, O, and H. Organic material is removed from the dried plant samples by ashing, and the remaining 1.5% of the plant's fresh weight represents its mineral content.

Plants require at least 10 elements for healthy growth: the macronutrients N, S, P, K, Ca, Mg, C, H, O; small quantities of iron; and a number of other elements in even smaller amounts, the micro-nutrients.

1.2. Criteria of Essentiality

Not all elements detected in plant tissues are deemed to be essential.
Criteria of essentiality have been proposed by Epstein [3]: an element is essential (1) if, without it, the plant cannot complete its
life cycle, and (2) if it is part of an essential plant constituent
or metabolite. The experimental evidence for essentiality comes from
solution culture (hydroponic) methods. As the experimental techniques
and the quality of the water used in the experiments have improved,
the number of essential micronutrients has tended to increase. Other
elements are known to be beneficial, and some of these may be essential for some species (Table 1).

Hewitt and Smith [1] discussed the difficulties involved in
experiments with plants that are designed to test the essentiality
of elements. In some early experiments it was found that between 5
and 20 times as much nickel and vanadium was found in plants as was
known to have been given in residual impurities in reagents for

TABLE 1

Status of Elements in Plant Nutrition

Essential macronutrients	Essential micronutrients	Beneficial or of restricted essentiality
K	Fe	Ni?
Ca	Cu	Al
Mg	Mn	Sr
C	Zn	Sn
H	Mo	Cr
O	Co	Br
P	V	F
N	Na	
S	Rb	
	B	
	Si	
	Cl	
	I	
	Se	
	Ni?	

culture media and sand for sand culture. It is suggested that there
are many sources of contamination inherent in such experiments.
These authors also point out that seed tissue which is to propagate
a plant must contain "at least a small amount of all the elements
essential for its growth". Welch and Cary found that the nickel concentration in a number of different varieties of wheat seeds was in
the range 0.08-0.35 ppm [4]. Early work by Schroeder et al. [5], who
surveyed a large number of food items, showed that most whole grains
contained considerable concentrations of nickel and suggested that
the nickel was concentrated in the germ. Whether or not the presence
of nickel in seeds indicates essentiality is as yet unknown.

Brooks [6] suggested that since all the elements from manganese
to zinc, except nickel, are known to be essential for plant nutrition,
then it might be inferred that nickel also has an essential role.
Schroeder [7] had made a similar point some 20 years earlier and was
almost convinced that nickel had a physiological role [5,7], but he
pointed out that "definitive experiments have not been done". Brooks
[6] also suggested that "there is at present (1980) little evidence
for such an inference [of essentiality]".

The growth of a number of different species has been shown to
be promoted by the addition of nickel to growth media and/or soil
cultures. For example, clones of the metal-tolerant grass *Deschampsia
cespitosa* from smelter regions near Sudbury, Canada show growth stimulation in nutrient culture at 1.0 ppm Ni [8]. It seems unlikely that
nickel essentiality can be deduced from such results since growth
stimulation has been found with palladium [9] and with platinum and
rhodium [10,11], elements which are unlikely to be essential.

However, nickel has been shown to function as an essential
element. It is now more than 10 years since Zerner and coworkers
[12] demonstrated that urease, isolated from jack bean (*Canavalia
ensiformis*), was a nickel metalloenzyme (see Chap. 6). Later studies
by Polacco [13-15], using soybean in tissue culture, showed that urea
metabolism and urease synthesis required Ni^{2+}. Polacco demonstrated
that citrate inhibited the utilization of urea by chelation to Ni.

Urease leaf activity in tomato and soybean plants was found to increase with nickel additions by Shimada and Ando [16]. These authors further reported that low-nickel plants, grown with urea as sole nitrogen source, developed necrotic leaf tips from toxic accumulations of urea.

More recently, Eskew et al. [17] demonstrated that nickel is an essential micronutrient for legumes and suggested possible essentiality for all higher plants. In this study, soybean plants were grown in carefully purified nutrient solutions with an estimated low nickel concentration of 0.06 $\mu g/dm^3$ (Ni_0 treatment). Plants grown without added Ni showed yellow necrotic lesions on the leaf tips similar to those reported by Shimado and Ando [16]. These lesions were absent when the plants were grown with either 1 $\mu g/dm^3$ or with 10 $\mu g/dm^3$ added nickel (Ni_1 and Ni_{10} treatments, respectively). The leaf tip lesions occurred both when the plants were supplied with inorganic nitrogen and when they were dependent on nitrogen fixation. Eskew et al. [17] had suspected that nitrogen-fixing plants would have a high requirement for nickel, e.g., soybeans transport most of the fixed nitrogen as ureides, which it has been suggested [18] are broken down to urea and glyoxalate before further metabolism. It was found that seed yields from the three nickel treatments were similar but that urease activity and nickel concentrations in the seed increased when nickel was supplied: the concentration of nickel in the seeds was as follows; Ni_0, <10 ng/g; Ni_1, 53 ± 6 ng/g; Ni_{10}, 637 ± 28 ng/g. These authors suggest that their evidence meets, at least in part, Epstein's criteria for essentiality [3].

Nickel has been demonstrated to be an essential element for the blue-green alga *Oscillatoria* [19], although no specific metabolic role was suggested. Since many blue-green algae are nitrogen fixers, it seems possible that nickel may be implicated in urea metabolism.

There is some evidence that nickel is involved in microbial urea utilization. *Klebsiella aerogenes* does not produce urease when citrate is the carbon source [20]. Dietary nickel administered to ruminants stimulated nitrogen retention and protein synthesis, suggesting that rumen bacterial urease required nickel [21].

Nickel has recently been found to be an essential requirement
for several different hydrogenases [22-25]. Nickel is also most
likely to be essential for the methanogens *Methanobacterium thermo-
autrophicum* and *Methanobacterium bryantii* [26-29] (see also Chap. 7).

2. NICKEL TOXICITY

2.1. Metal Toxicity in General

Although essential and beneficial metals are important for healthy
plant life, excesses or deficiencies have profound effects on the
growth and morphology of plants. The physiology of metal toxicity
in plants has been reviewed extensively by Foy et al. [30]. The
toxicities of metals and the symptoms produced may be studied in
the laboratory by culture, pot, or plot methods [1]. Excessive concentrations of some metals in soils, which produce toxic symptoms,
may come about in a number of ways: the presence of naturally occurring ore bodies, exploitation of mineral resources, agricultural
practices, and waste disposal.

Bowen [31] has classified elements that are toxic to plants
into three groups:

1. Very toxic: Those elements such as Be, Cu, Hg, Ag, Sn,
   which are toxic at concentrations of less than 1 ppm in
   the soil.
2. Moderately toxic: Toxic symptoms become apparent at concentrations ranging from 1 to 100 ppm. These include most
   of the d-block elements and most of those from groups III,
   IV, V, and VI of the periodic table.
3. Scarcely toxic: This group includes most of the s-block
   elements and the halogens in addition to macronutrients.

The most common symptom of toxic metal poisoning in plants is
chlorosis, which is usually related to a failure of chlorophyll
development. In addition, toxicity in plants often results in
changes in morphology, such as dwarfism, abnormally shaped fruits,
necrosis of the leaves, and stunting of root growth.

In an early study [32] the effects of some metal cations were monitored and the severity of the chlorosis was observed to be in the order $Co^{2+} > Cu^{2+} > Zn^{2+} > Ni^{2+} > Cr^{3+} > Mn^{2+} > Pb^{2+}$. However, other toxic symptoms were evident, principally dwarfing and necrosis, which were distinct from the induced chlorosis. On this basis a second order was established: $Ni^{2+} > Co^{2+} \gg Zn^{2+} > Cu^{2+} \gg Cr^{3+} = Mn^{2+}, Pb^{2+}$. That the induced chlorosis was due to interference with iron metabolism was shown by painting the leaf surfaces with $FeSO_4$ solution, which restored chlorophyll production. The mechanism by which chlorosis is induced is not fully understood.

2.2. Metal Tolerance

Plants which grow on soils with metal concentrations which are normally toxic are metal-tolerant, or metallophytes. Some of these plants exclude the toxic metals from their tissues; others assimilate the metals present to such a degree that they are termed accumulators. An accumulator is defined [33] as having a metal concentration in the tissues greater than that in the soil. A number of nickel-tolerant plants and nickel accumulators will be discussed in Secs. 6 and 7.

2.3. Toxicity of Nickel

The toxicity of nickel toward plants has been reviewed by Vanselow in 1966 [34] and more recently by Hutchinson [2]. Early work had shown [1,35] that in northeast Scotland, where serpentine soils contain high concentrations of nickel, oats showed toxic symptoms. These toxicity symptoms could be reproduced [36] by growing oats in sand culture with 25 ppm added nickel, when the leaves of the plants showed irregular diagonally banded white chlorotic striations. Such symptoms with oats had been used to demonstrate that serpentine soils produce toxic levels of nickel in the early 1950s [37,38].

Nickel toxicity may also induce the typical chlorosis of iron deficiency, and that resembling manganese deficiency in potatoes [1]. Further descriptions of the symptoms of nickel toxicity have been given by Mishra and Kar [39], who described yellowing of leaves (chlorosis) followed by necrosis. Other observed effects are mottling on the leaves and distorted and stunted growth of shoots and roots.

It has been pointed out by Vanselow [34] that there is a difference in the symptoms of nickel toxicity shown by monocotyledons (plants which have one seed leaf, and to which the grasses and cereals belong) and dicotyledons (which have two seed leaves). The monocotyledons show the alternate striations, as in the oat; whereas dicotyledons show more general chlorosis and mottling. The first, single-seed leaf of monocotyledons becomes a sheath, the coleoptile, around the rudimentary shoot. Anderson et al. [40] studied the development of the alternating green and chlorotic bands of nickel toxicity in the oat (a monocotyledon) and have shown that the green bands develop under the coleoptile during the daylight, while the white, chlorotic bands develop under the coleoptile during the hours of darkness.

2.4. Critical Levels of Toxic Elements

Jarrell and Beverley [41] and Leaf [42] pointed out that the terminology relating to the quantities of a particular element in plant tissues is often ambiguous. For example, "content" sometimes means the "concentration" and sometimes the "total accumulation". These authors point out that the latter term should be used for the *total mass* of an element in a whole plant or in some specified part of a plant. It was suggested [41] that since concentration is a ratio, two other absolute responses of plants to nutrients (or toxic elements) should be reported: the total accumulation of the elements, and the biomass (either wet or dry weight). Thus decreasing concentrations could result from increased biomass, with the total elements accumulation remaining constant, or even increasing: the "dilution effect".

Beckett and Davis [43] discussed the critical concentrations of metal in plant tissues which cause toxic reactions and reduce the yield (biomass). They demonstrated the yield curves of plants grown in the presence of toxic metals. Thus:

1. There is a yield plateau Y_0 over which the element is neither toxic nor deficient, and the biomass is independent of the tissue concentration of the metal.

2. There is an upper critical tissue concentration T_c, and above this concentration in the tissue, the metal is toxic and the yield is reduced.

3. With plants grown in culture media there is a similar critical solution concentration S_c.

Beckett and Davis reported that the yield, the total concentration, and S_c vary with growing conditions, but T_c proved to be remarkably independent of growing conditions, and characteristic of each element [43,44]. The upper critical levels, T_c, on Ni in barley and ryegrass are 12 and 14 ppm (median values). Vanselow [34] reported that toxic effects occur in citrus plants at folar levels greater than 55 ppm and in oats at concentrations greater than 60 ppm in the grain and 28 ppm in the straw.

It has been pointed out [1,45] that chlorosis is a secondary characteristic and that the origin of toxic effects lies in the roots. Craig [46] used the ED_{50} method (effective dosage of metal that produces 50% of normal growth) to test the toxicity of Cu, Ni, Pb, and Zn to Zea mays. and found the order of increasing toxicity, in terms of root growth, to be Cu > Ni > Pb > Zn. The ED_{50} for nickel was 0.4 ppm. The ED_{50} for Ni and ryegrass (Lolium perenne) was found by Wong and Bradshaw [45] to be 0.18 ppm, and the toxicity orders of metals toward this species (from ED_{50} root measurements) were Cu > Ni > Mn > Pb > Cd > Zn > Al > Hg > Cr > Fe. Hogan and Rauser [47] found the toxicity orders (ED_{50}) toward tolerant clones of Agrostis gigantea to be Cu > Ni > Co > Zn.

The ED_{50} (root growth) for alfalfa and Ni is 5 ppm [48]. In this study toxic symptoms were noticed at nickel concentrations greater than 5 ppm: reduction in total biomass, chlorotic patches on the leaves, and brown coloration to the roots. At concentrations >7.5 ppm the roots became brown and showed marked clubbing. The studies of ED_{50} for root growth are not strictly comparable since some are carried out in nutrient solutions [47,48] and some are in calcium nitrate solutions [45].

2.5. Toxicity of Nickel to Lower Plants

The lower plants which will be considered here are nonvascular plants, i.e., those that do not have a vascular system of xylem and phloem; they include algae, fungi, lichens, and bryophytes. The effects of metals in general on lower plants has been reviewed by Puckett and Burton [49] and those of nickel in particular have been discussed by Richardson et al. [50].

2.5.1. Lichens

Lichens are composite plants, consisting of a fungus living symbi‑ otically with an alga. They are thallus plants, i.e., they do not have true roots, stems, and leaves. Lichens are often found growing on trees and rocks. The effects of metals on lichens and metal accumulation in lichen tissue have been reviewed by Puckett and Burton [49], Richardson et al. [50], James [51], Margot and Romain [52], and Nieboer et al. [53]. The effects of nickel have been considered by Richardson et al. [50], Nieboer et al. [54], Nash [55], Hutchinson [2], and Brooks [6].

     Photosynthesis and respiration of lichens have been shown to be affected by metals [55-57]. Puckett [56] studied the effects of metals on photosynthetic ^{14}C fixation by *Umbilicaria muhlenbergii* and *Stereocaulon paschale*. It was found that relative toxicities decrease in the orders: Ag,Hg > Co > Cu,Cd > Pb,Ni for short-term exposure, and Ag,Hg, > Cu $\geqslant$ Pb,Co > Ni for longer term exposures.

     The results are in agreement with those of Richardson et al. [57] in that Ni^{2+} had no effect on total ^{14}C fixation under the conditions used. If preincubated in 10^{-2} M Ni^{2+} then *Umbilicaria muhlenbergii* showed a 25% reduction in total ^{14}C fixation [50]. Preincubation in very dilute solutions of nickel, which resulted in a nickel uptake of less than 20 μmol/g, did not affect the reduction in ^{14}C fixation in subsequent SO_2 exposures [57]. However, preincubation in Ca^{2+} and in Sr^{2+} gave protection against SO_2 damage.

A number of stress factors lead to K^+ efflux from lichens, algae, and fungi and since potassium is largely within the cell this has been used as an assessment of membrane damage. It was also shown [56] that Ni^{2+} and Co^{2+} produced a large K^+ loss from *Umbilicaria* only at high concentrations. Puckett suggested that this demonstrates lack of toxicity of nickel with respect to membrane integrity [49,56]. In contrast large K^+ losses occurred at low concentrations of Hg, Ag, and Cu. Tl^+ is similar to Cu^{2+} with respect to potassium efflux, while Mn^{2+} is like Ni^{2+} [58]. Nieboer and coworkers discussed uptake capacity of *Umbilicaria muhlenbergii* for nickel in terms of binding sites [54,57]. Two binding sites have been demonstrated with a combined capacity of 20 μmol/g. External concentrations of nickel of up to 8 μmol/cm^3 were required to saturate the first site, whereas external concentrations of 100 μmol/cm^3 were needed to reach the total capacity. The saturation of these binding sites appears to be related to the differing K^+ effluxes and C fixation at low and high concentrations of external nickel.

Laboratory studies indicate that vanadium, either as vanadate or vanadyl, can affect surface phosphatase activity in *Cladina rangiferina* [59]. Surface phosphatase activity is not affected by Al, Ni, and Na [60].

Threshold values for metal induced damage have been discussed [49,55]. High metal concentrations in lichen tissue in field conditions may have little relevance to an assessment of metal phytotoxicity, since the total amount includes particulate entrapment and represents total accumulation over an extended time period.

2.5.2. *Algae*

Algal responses to nickel have been discussed by Hutchinson [2],
Richardson et al. [50], Birge and Black [61], Wood and coworkers
[62,63], and are also discussed in detail elsewhere in this volume
(Chap. 2).

Essentiality for nickel has been demonstrated for the blue-
green alga *Oscillatoria* [19]. Because of their morphology, color,
and ability to photosynthesize, the blue-greens (the Cyanophyta)
have been traditionally grouped with the algae. However, as prokaryotes, they have many similarities to the bacteria. The green algae
are eukaryotic, they are characterized by the presence of chlorophyll
a and b and a cellulose cell wall.

Nickel has been found to be less toxic to algae than many other
metals [64]. Nickel toxicity and nickel binding to six different
strains of nickel-tolerant algae and one *Euglena* sp. were recently
investigated by Wood and Wang [63,65]. Using radioactive ^{63}Ni it
was shown that the blue-green algae were more sensitive to nickel
toxicity than the green algae, both types concentrated Ni primarily
at the cell surface. Both toxicity and metal binding were shown to
be pH-dependent.

2.5.3. *Fungi*

It has been known for a number of years that nickel salts or formu-
lations containing nickel are effective fungicides, particularly
against rust fungi [39,66,67]. Richardson et al. [50] report that
the most widespread use of nickel as a fungicide is to control
stripe rust on blue grass in the United States. The toxicity of
metal ions in general to fungi [68] is in the order Ag > Hg > Cu >
Cd > Cr > Ni > Pb > Co > Zn > Fe > Ca.

2.5.4. *Bryophytes*

As with the lichens there is an efflux of K^+ from bryophytes on
metal uptake. The evidence suggests that the major part of metal
uptake is by ion exchange onto cell wall sites [69,70]. Lepp and

Roberts [71] demonstrated that photosynthetic and respiration rates in bryophytes are affected by Cd^{2+} and by Pb^{2+}. Richardson et al. [50] suggest that Ni^{2+} is also likely to affect these processes. These authors point out that unpublished work by Sorosiek (reported by Seaward [72], and cited in [50]) showed that Ni^{2+} can cause cell disorganization and death in more than 90% of samples of two bryophytes incubated for 6 weeks at a concentration of 5 ppb.

3. NICKEL UPTAKE AND NICKEL CONTENT OF PLANTS: BACKGROUND LEVELS

Krause [73] reviewed the older literature on levels of nickel in plants and reported that the concentrations range from 1 to 5 ppm dry weight. Values much greater than these are found in plants grown on soils derived from serpentine or other ultrabasic rocks. Nickel concentrations in uncontaminated vegetation of between 1 and 2.7 ppm dry weight were reported by Bowen [31] and concentrations of up to 5 ppm were reported by Vanselow [34]. More recent values for field crops and natural vegetation in the range 0.2-4.5 ppm were given by Connor et al. [74], whereas Hutchinson et al. [75] obtained slightly higher values of 2.8-6.3 ppm Ni for crops grown on an highly organic soil.

There is some confusion in the literature concerning the levels of nickel in plants, since some have been reported on an ash weight and some on a dry weight basis. The confusion is compounded by the conversion of literature ash weight values to dry weight values by the use of different conversion factors, e.g., 5% [6] and 3.5% [76].

Rencz and Shilts [77] suggest that although most plants contain nickel in trace amounts, the data in the literature on the distribution of nickel in various plant organs may be distorted by study conditions which are typically at low or very high concentrations of nickel; and that investigations over a broad range of nickel concentrations might be advantageous. However, Hutchinson [2] pointed out that because of a wide pH range of natural soils, highly acidic to

highly alkaline, a greater range of metal concentrations occurs in
native terrestrial species than for crop plants. In agricultural
environments the soil pH lies in a much narrower range. Nickel
uptake is enhanced in natural acid conditions. In agricultural crop
production naturally acidic soils are usually limed.

Nickel levels of between 1 and 15 ppm dry weight are reported
for a number of species of forest trees [76,78] grown in unpolluted
areas. There is some indication that there is a seasonal variation
in the elemental content of leaves. Guha and Mitchell [79] found
highest concentrations of nickel in young leaves and lowest in late
summer to fall. Background levels of nickel in lower plants from
rural, unpolluted areas are similar to those in higher plants,
usually less than 5 ppm. Tables of concentrations collected from
the literature are given in references [2,49,50,61,80,81].

Nriagu [82] discussed the global nickel cycle and estimated
that in the biosphere a nickel burden of 1.4×10^{13} g is associated
with the terrestrial biomass of plants, assuming a total plant biomass of 2.4×10^{18} g and an average nickel concentration of 6 ppm.
A further 2×10^{8} g of nickel is associated with plant litter.

4. NICKEL UPTAKE AND SOIL PARAMETERS

4.1. Nickel Content of Soils

In 1955 Swain [83] suggested that the "normal" concentration of
nickel in soils is between 5 and 500 ppm with the average about
100 ppm. The elemental constituents of soils have been extensively
reviewed by Ure and Berrow [84]. This reference has collected
together almost all of the literature reports for elemental constituents in soils. The nickel concentration of 4,625 samples from various
parts of the world are detailed; these give an overall mean of 33.7
ppm. Soils derived from serpentine and basic rocks contain concentrations of nickel in the range 1,000-5,000 ppm, and Ure and Berrow
excluded these from their table of values. Subsequently these

authors [85] gave the mean value of 34.9 ppm Ni from 14,218 samples
on a worldwide basis. Further tabulations of trace element distribution in soils on a worldwide basis have appeared in the literature
[31,86-88].

The factors affecting trace element concentrations in soils
have been reviewed by Mitchell [87] and by Swain and Mitchell [89].
Particular attention has been paid to Scottish soils [84-87,89-91].
In general it has been found that soils derived from ultrabasic rocks
have high concentrations of Ni, Co, and Cr, and problems of nickel
toxicity have arisen on these soils, which have affected crops.
Welsh soils [92] and those in England and Wales [93] have been
surveyed.

4.2. Nickel Availability to Plants

The total concentration of an element in the soil is a poor indicator of that which is available to plants. Efforts have been made
to determine the amount of soluble or extractable element by extract-
ing with a number of agents, e.g., EDTA, acetic acid, ammonium acetate, and water. Swain and Mitchell [89] pointed out that the bulk
content of most trace elements occurs in the unweathered crystal
lattices of the rock-forming minerals, where it is fixed and unavailable. One of the most important factors influencing mineral
weathering, and thus trace element availability to plants, is soil
drainage. It has been found by a number of authors that there is an
increased ease of extraction of nickel in very poorly drained soils,
particularly in soils derived from rocks with a high content of
ferromagnesian minerals [84-86]. It has been demonstrated [91] that
the increased extractability is reflected in the concentrations of
Ni, Co, Mn, and Mo in the tissues of grasses and clover, but less
in the case of Cu.

Using ammonium acetate as an extractant, Halstead et al. [94]
investigated the relationship between extractable nickel and the

available nickel as determined by the concentration in plant leaves.
They found that nickel extractability and availability was reduced
by liming and by the addition of organic matter to the soils. The
availability of nickel was also reduced by the addition of phosphate,
but this was not reflected in the extractability. Haq et al. [95]
compared a number of extractants for Cu, Ni, Zn, and Cd, with the
concentrations of the elements in *Beta vulgaris* grown in the same
soils, and came to the conclusion that the nickel extracted by
acetic acid accounted for 42% of the available nickel, followed by
water with 38%. The variability could be explained if pH and cation-
exchangeable capacity of the soil were taken into consideration.
Berrow et al. [91] also investigated a range of different extractants, including 2-ketogluconic acid.

The uptake of nickel is enhanced by low pH values, and the
available nickel has been shown to increase at pH values less than
6.5 by the breakdown of nickel complexes with iron and manganese
hydrous oxides [30,77]. Metal complexes in soil can be organic or
inorganic; organic ligands are derived from organic material in the
soil [96]. Their complexes, which may be adsorbed on clays or
hydrous oxides, are in equilibrium with chelated soluble metal ions
in the "labile pool", which are in turn in equilibrium with those
being absorbed by plants from the soil solution [97,98]. The concentrations of elements which are immediately available to plants
have been estimated [99]. The concentration of available nickel in
the soil solution has been estimated as 0.05 mg/liter [99]. This
rough estimate has been derived from the concentration of nickel in
seawater; since the reported concentrations of manganese and copper
in the soil solution are about 30 times greater than those in seawater, this factor was applied to other cations, including nickel.

Humic and fulvic acids are well-studied by as yet ill-defined
organic constituents of soils, natural waters, and sediments [100].
The overall stability constants for humic and fulvic acid complexes
with nickel (pH, 8.0; ionic strength, 0.02 M) have been given by
Mantoura et al. [101]; peat fulvic acid, log K = 4.98; soil fulvic

acid, log K = 4.3; lake humic acid, log K = 5.2; marine humic acid, log K = 5.4.

    The concentrations of nickel in plant tissues have been shown to be species-specific even when these are grown on the same soils [102,103]. Shewry and Peterson [103] found that nickel concentrations reflected those in the soil and concluded that metal concentrations in plant tissues give the best indication of availability.

5. NICKEL UPTAKE AND POLLUTION

5.1. Sources of Contamination

Farago [104] listed the ways in which excessive concentrations of metals in soils may be brought about; these are (1) natural mineralization caused by the presence of undisturbed ore bodies near the surface; (2) the exploitation of mineral resources; mining activities, ore tailings, tips, smokes, and dusts, and (3) agricultural practices, such as use of metal-containing sludge as a soil amendment; or (4) waste disposal practices. Freedman and Hutchinson [105] considered the categories of metal contamination and the resultant contamination of plants and soils. In addition to those above they included emissions from moving sources and from utilities such as generating stations.

    Toxic soil conditions and nickel contamination result from serpentine soils, from soil amendments, from mining activities and smelting, from coal-burning generating stations, and from municipal incinerators.

5.2. Sewage Sludge

It has been recognized for a number of years that the addition of metal-contaminated sewage sludge to agricultural land may cause contamination leading to increased uptake of toxic elements, low crop yields, and possibly land sterility [105-109]. Friedman and

Hutchinson [105] surveyed the reported metal contents of sewage
sludge from a variety of locations. Mean nickel contents (on a dry
weight basis) in North America are Michigan 371 ppm and Ontario 390
ppm. The mean for the UK has been given as 510 ppm [110], England
and Wales as 382 ppm [111], and in Great Britain the concentration
of sludge applied to land is reported as 188 ppm Ni [112]. Other
sludge surveys have confirmed that it is variable in nature.

     Many workers have demonstrated that the metal contamination
of land by sewage sludge is irreversible, although there are indi-
cations that in the long term there is a tendency for metals to
become less available to plants. Such studies have necessitated
experiments over a number of years. The results of a study started
in 1968 were reported in 1980 [113]. The effects of four sludges
(containing predominately Cr, Zn, Cu, and Ni) on a range of crops
were compared. Sludge with a high Cr content was not toxic, the
results for the other sludges were interpreted using multiple regres-
sion techniques, when nickel was confirmed to be the most toxic to
red beet, celery, and lettuce.

     Purves [108,109] demonstrated that contamination can persist
in the soil for a period of 6 years after a single sludge treatment.
During this time the "available" Ni (0.5 M acetic acid) in the
treated soil decreased from 2.6 to 1.9 ppm. The sludge had a relatively low Ni content of 18.5 ppm. At present the element causing
most concern in sewage sludge is cadmium, although Purves showed
that there was little fixing in the soil after sludge application.
Most national guidelines for the agricultural use of sewage sludge
assume that fixation of the toxic metals takes place in the soil in
a few years. Purves [109] pointed out that there is a dearth of
information relating to the rates at which this fixing occurs.

     Various national guidelines for metal additions to land in
sewage sludge have been listed [107,109]. For nickel in the UK, the
maximum permissable addition of Ni is 70 kg/ha. Nickel is here
totaled with copper and zinc to give the "zinc equivalent"; the
additivity implied for these three metals has been criticized [107],
since the relative toxicities vary with crop and with soil pH.

5.3. Smelter Emissions

The impact on vegetation from smelter emissions in general has been
reviewed [105] and that from nickel in smelter emissions has been
discussed by Hutchinson [2]. Contamination by nickel has been
reported in Sweden [114], in the Swansea Valley, Wales [115,116],
and a number of studies have been carried out in the vicinity of
the very large nickel-copper smelting complex at Sudbury, Ontario,
Canada.

Freedman and Hutchinson [105] point out that "the fate and
effects of the emission of pollutants from the Sudbury smelters into
the surrounding environment are better documented than any other
smelter in the world". In this region the soil and vegetation have
suffered widespread contamination principally by copper, nickel, and
cobalt [117-119]. The solubility and thus the toxicity of the three
metals is accentuated by the acidic effects of SO_2 emissions. The
aerial depositions of metals are retained in the upper soil layers
by binding to humic substances [119], where concentrations of nickel
of up to 4,900 ppm (dry weight) have been found [118,119]. The level
of pollution falls off with distance from the smelter but is still
apparent some 60 km away.

Elevated levels of nickel have been found in vascular plants,
and species differences have been found in the extent of nickel
accumulation in plant tissues. The grass *Deschampsia flexuosa* collected 1.6 km from a smelter had a folar nickel concentration of
903 ppm, compared with 37 ppm at 49.8 km. *Vaccinium angustifolium*
had concentrations of 921 and 14 ppm, respectively [2]. The folar
levels of nickel are in general substantially lower than those on
the soil. Elevated levels of Ni have been reported in fruit and
vegetables grown in the Sudbury area [120].

The toxicity of the soils in the Sudbury area have been investigated [121] by the growth of a number of species in the surface
soils. The toxic effects were due not only to nickel but also the
high concentrations of cobalt and aluminum. The Ni in the soils
was found to be toxic to test species up to a distance of 12 km from

the smelter. The soil toxicity has been found to be caused by the
high concentrations of available nickel as a result of the acid
conditions near the smelter. This in turn leads to excessive and
lethal concentrations of nickel in the roots of the seedlings.

The high nickel concentrations have led to the presence of
nickel-tolerant ecotypes. These will be considered in the next
section.

5.4. Uptake of Airborne Metal

Plants have been used to monitor airborne pollution near roads and
industrial areas. Mosses and lichens have been used for this purpose, since they accumulate metals exclusively from aerial fallout.
The use of lower plants in aerial monitoring has been reviewed [6,
50,116,122].

Nickel pollution was assessed by the use of naturally occurring
mosses in early work by Ruhling and Tyler [70,123,124]. Using *Hylo-
conium splendens*, they demonstrated that nickel pollution increased
from northeast to southwest Scandinavia, reflecting the influence of
industrialized areas. Groet [125] carried out similar work in New
England where the highest concentrations were found near industrial
conurbations. Goodman et al. [126] developed the moss bag technique
for the monitoring of airborne pollution. Samples of 2.5 g of acid-
washed moss (*Sphagnum* or *Hypnum*) in flat nylon bags are suspended so
as to intercept aerial fallout. Metal retention has been found to
be directly proportional to exposure time for up to 10 weeks [126].
Lichens can similarly be used in bags [127].

The concentrations of nickel in lichens near a smelter in
Sudbury, Ontario have been found to decrease with distance from the
smelter [50,54,128,129]. *Stereocaulon* spp. were found to contain up
to 310 ppm and *Umbilicaria* spp. up to 220 ppm nickel. A linear relationship was obtained when the concentration of nickel was plotted
vs. the reciprocal of the distance from the smelter stack. This

relationship is consistent with a simple diffusion model of aerial
particulate fallout [128,129].

Pilegaard and Johnsen investigated the use of higher plants
to monitor airborne pollution [130]. They grew plants of *Achillea
millifoleum* and *Hordeum vulgare* in pots placed at five sites of varying metallic aerial deposition in Denmark. They concluded that such
methods could be used to monitor aerial deposition of Cu and Pb, but
these plants could only be used to monitor airborne Ni and Cd if the
properties of the soils are consistent throughout the monitored area.
They found that for nickel the uptake from atmospheric fallout was
greater than uptake from the soil.

A number of studies of the metal contents of herbarium specimens of lower plants were collected at known times in the past [50,
127,131]. In this way it has been shown that there has been an increase in nickel concentrations in mosses over the last 100 years,
reflecting increased aerial fallout from industrialization.

6.   NICKEL TOLERANCE

6.1.   Tolerance in General

Plants which are diagnostic of particular environmental conditions
are known as indicator plants. The qualitative term *metal indicator
species* can be used to describe species which grow over, and thus
indicate, soils containing high concentrations of metals [132,133].
Plants which grow on soils with metal concentrations which are normally toxic are metal-tolerant, or metallophytes. Some of these
plants exclude the toxic metals from their tissues, others assimilate
the metals present to such a degree that they are termed *accumulators*
[33]. Brooks et al. [134] used the term *hyperaccumulators* of nickel
for those plants which contain a folar concentration of Ni of more
than 1000 ppm (dry weight).

Various mechanisms of tolerance have been proposed to explain
how some plants cope with toxic conditions and how some species have

developed tolerant ecotypes [3,6,77,104,135-139]. Mechanisms are
usually divided into external and internal types. The former covers
those few situations where the metal is unavailable to the root [104,
134,138,140,141]. Internal mechanisms can be grouped loosely under
four headings [104,140]:

1. *Metal is available to root but is not taken up*, e.g.,
   alteration of cell wall membrane resulting in decreased
   permeability to toxic metal ion.
2. *Metal is taken up but is rendered harmless within the
   plant*, e.g., deposition in cell walls or vacuole.
3. *Metal is taken up but excreted*, e.g., by guttation,
   leaching, or leaf fall.
4. *Metal is taken up but metabolism is altered to accommodate
   increased concentration of toxic metal*, e.g., increase of
   concentration of enzymes inhibited by metal.

The ways in which the response of plants to increasing soil
levels may be reflected in the metal concentrations in the aerial
plant parts have been discussed [77,139]. These are accumulators,
where metals are concentrated in plant tissues from high or low
background levels; indicators, where the metal in the plant parts
is proportional to that in the soil; and excluders, where the concentration of metal in the tops of the plant is low and constant,
over a wide range of soil concentrations. This last type of response
results from differential uptake and transport between root and shoot.
Both Baker [139] and Rencz and Shilts [77] agree that in the case of
excluders the metal concentration in the shoots would be maintained
at a low constant value. Rencz and Shilts [77] suggested that this
value corresponds to a toxicity threshold, which if exceeded would
kill the plant, while Baker suggested that above this value the
excluding mechanism breaks down and unrestricted transport results.
Baker [139] further suggested that when the ratio of the concentration of metal in leaf to that in root is greater than unity, then
this is characteristic of accumulators, whereas a ratio of less
than unity suggests exclusion.

Data on the nickel concentrations of *Betula glandulosa* and *Ledum groenlandicum* [77] show the leaf/root ratios to be 0.9 and 0.5, respectively, suggesting exclusion. Copper ratios for copper-tolerant *Armeria maritima* [142] growing in a highly copper-impregnated bog suggest that it is an excluder concentrating copper in the outer portions of the root. Nickel concentrations in the same bog are much lower, around 25 ppm [143]. *Armeria maritima* growing in the same bog takes up "normal" levels of Ni (about 10 ppm) in both leaves and roots. Ratios for the hyperaccumulator *Hybanthus floribundus* will be discussed in Sec. 7.

6.2. Nickel-Tolerant Plants on Contaminated Sites

It has generally been considered that tolerance is normally specific to the metals found at the site where the plant grows. However, some exceptions to this general rule have been found in the case of nickel. Populations of *Agrostis tenuis* from zinc-contaminated areas showed cotolerance to nickel without prior exposure to nickel [144]. A clone of *Agrostis gigantea* from a mine waste site in Sudbury, Ontario (Sec. 5.3) [47] was found to be nickel-tolerant although it came from a site with "normal" nickel soil content. Hogan and Rauser [47] suggest that tolerance to nickel might be more general than that to copper; they also point out that in this area, in addition to the widespread metal pollution in the Sudbury area, ore bodies lie close to the surface and tolerance may have developed prior to the aerial contamination. In the same area, Cox and Hutchinson [8,145,146] described cotolerance in the grass *Deschampsia caespitosa*, which is invading derelict sites. These clones were tolerant to the major soil contaminants, nickel, copper, and aluminum. Weaker tolerances were found to lead, cadmium, and zinc which did not contaminate the site. It now appears that for certain species at least the presence of each metal at high concentrations in the soil is not required for multiple metal tolerance.

6.3. Nickel-Tolerant Plants on Serpentine Sites

The relationships between chemistry of serpentine soils and serpentine vegetation has been reviewed [147]. Serpentine soils contain
high concentrations of Mg, Ni, Co, and Cr. They are generally low
in Ca and the other macronutrients, and such soils often have dis-
tinctive vegetations. Serpentine soils occur in many parts of the
world, notably Zimbabwe, Scotland, Australia, and Portugal. The
results of many studies show that there is a correlation between
the degree of tolerance to Ni and the "available" Ni in the soil.
Proctor and Woodell [147] point out that acid extractions probably
overestimate the plant-available Ni in such soils. The same authors
list concentrations of nickel in plants growing on serpentine soils.
The concentrations vary; some plants are hyperaccumulators.

Wiltshire [148] carried out extensive studies on the metal-
tolerant plants from Zimbabwe; plants from seeds collected from
Ni-rich sites and from populations of the same species from normal
soils were grown on soils from a nickel-rich site and on normal
soils. The toxic soils were fertilized with N as either ammonia
or as nitrate. He found that in general the species from the Ni-
rich sites were more tolerant to nickel-rich soil than populations
from normal soils. Many plants from serpentine regions are accumulators of nickel and these will be considered in the next section.

7. NICKEL-ACCUMULATING PLANTS

There have been various reviews of nickel-accumulating terrestrial
species in the literature. Brooks has extensively reviewed accumulators and hyperaccumulators [6,133,149]; those growing on serpentine
soils have been listed by Proctor and Woodell [147].

7.1. Europe

The earliest report of a plant which accumulated nickel was in 1948
by Minguzzi and Vergnano [150,151], who found high concentrations in
Alyssum bertolonii from serpentine areas in Italy. These authors
[150] attempted to find interelement relationships in *Alyssum*, they
calculated "ratios of atomic ratios", where the "atomic ratio" was
the percentage concentration of an element divided by its relative
atomic mass (RAM). In this way they found that the Ca/Ni "atomic
ratio" for one sample was between 4.2 and 4.8 for various plant
organs, and concluded that the relative uptake reflected the low Mg
in the soils. In later work Vergnano-Gambi et al. [152] suggested
that the accumulation of nickel in *A. bertolonii* was more related
to the length of the growth period than to the total or extractable
nickel in soil.

Subsequent work by Brooks and coworkers on *Alyssum* species
[153-155] included the analysis of over 400 herbarium specimens of
European *Alyssum*, when 14 species were found with Ni concentrations
>1,000 ppm (dry weight). They reported that the hyperaccumulators
are located along the ultrabasic rocks from Portugal through Corsica,
Italian Alps, Tuscany, Yugoslavia, Albania, and Greece and through
the Aegean Islands to Crete.

It has recently been reported [156] that most species of
Thlaspi from European serpentine areas have Ni concentrations of
>1,000 ppm dry weight. On serpentine soils *Thlaspi goesingense*
showed increased Ca uptake and reduced Mg uptake when compared with
growth on unmineralized soils [157].

Lychnis alpina raised from seed collected from a variety of
sites in northern Finland (in the vicinity of ultrabasic and basic
rocks) demonstrated that this species accumulates nickel and the
maximum concentration was >7,000 ppm [158]. Metal concentrations
(Ni, Cu, Pb, Zn) of over 700 herbarium specimens of this species
have been reported.

7.2. Africa

The Great Dyke in Zimbabwe is a serpentine area of some 3,000 km^2, the soils are characterized by the presence of nickel and chromium, high levels of Mg, and low macronutrient status. A number of nickel-accumulating species have been identified from this region. *Dicoma nicollofera* [159-162] and *Pearsonia metallifera* [162] were reported to accumulate nickel by Wild. *Dicoma nicollifera* is a nickel accumulator, but its status as a hyperaccumulator is in doubt since Brooks and Yang [163] reported a maximum concentration of only 552 ppm in dry leaves. It is an indicator of nickel, being confined, in Zimbabwe, to serpentine soils with high concentrations of nickel. Wiltshire [148] demonstrated that the major portion of nickel taken up by *D. nicollifera* remains in the roots and is not translocated to the shoots. 32% of the nickel in the root cortex has been shown by Ernst [164] to remain in the residue after extraction by a number of solvents; the comparative percentage for Cr is 83. He suggested that the metals in the residues were not able to be translocated to the shoots because of their low solubility and nonexchangeability.

Brooks and Yang [163] confirmed that *Pearsonia metallifera* is a hyperaccumulator of nickel (1.53% in dry leaves) and have further identified three others from this area: *Blepharis acuminata, Merremia xanthophylla*, and *Rhus wildii*; maximum concentrations in dry leaves were 1,815, 1,384, and 1,378 ppm, respectively.

Cole [165] in a biogeochemical survey of the Empress Ni/Cu Prospect area in Zimbabwe found that *Dalbergia melanoxylon* and *Combretium hereroense* accumulated relatively high concentrations of both Ni and Cu and accurately delineated the orebody. Thus these two plants act as indicators of mineralization.

7.3. Australasia

7.3.1. *New Caledonia*

Nickel-accumulating plants from New Caledonia have been reviewed by Jaffré [166]. A large number of hyperaccumulators of nickel occur

in the genus *Homalium* (Flacourtiaceae) which are specific to New
Caledonia. *H. guillainii* was reported to contain 2.9% Ni in 1974
[167]. Subsequently more accumulators were identified from analysis
of herbarium specimens [168]. Recently, extensive studies [169] have
been made of the Flacourtiaceae of New Caledonia including the genera
Homalium, Caesaria, Lasiochlamys, and *Xylosma*. A total of 19 hyperaccumulators have been identified in this family.

Hyperaccumulation of nickel has also been demonstrated in
Psychotria species from New Caledonia and other islands of the Pacific
Basin [166,167,170]. Very high concentrations (4.7%) of Ni have been
recorded in the leaves of *P. douarrei* [167]; highest concentrations
of Ni found [166] in other organs are: flowers, 2.4%; fruits, 2.8%;
trunk bark, 2.8%; trunk wood, 0.23%. *P. douarrei* appears to be unique
among the 210 *Psychotria* species investigated in being a nickel hyperaccumulator, in contrast to *Homalium*. The mean Ni concentration for
leaves from 38 individual plants of *P. douarrei* was 22,400 ± 1,110
ppm. One specimen showed 19,900 ppm growing in a soil with 3,300 ppm
Ni, thus giving a plant/soil ratio of 6.03 [171].

The genus *Hybanthus* contains three known accumulators of nickel,
two of which are from New Caledonia: *H. austrocaledonicus* (reported
to contain 1.85% Ni [170]) and *H. caledonicus* (for which 1.46% Ni has
been recorded [172]). Various polymorphs of *H. caledonicus* have been
found which include form A with small leaves, form B with large
leaves, and var. *liniarifolia* with narrow leaves. These had nickel
concentrations of 0.88, 0.17, and 1.03%, respectively [172].

7.3.2. Western Australia

The first hyperaccumulator to be discovered was *Hybanthus floribundus*
growing over outcropping serpentenite in the Eastern Goldfields area
of Western Australia (Plates 1 and 2) [173,174]. It accumulates very
high concentrations of nickel in its tissues, up to 1.35% on a dry
mass basis [173-178]. Nickel is concentrated in the leaves, and the
majority of this nickel is water-soluble (see Sec. 8) [175,176].

Plate 1 shows a closeup of the flowers and leaves of *Hybanthus
floribundus* collected from Widgiemooltha in Western Australia, clearly

PLATE 1.   Closeup of the flowers and leaves of *Hybanthus floribundus*
(Lindl). F. Muell, Violaceae, growing in skeletal soils containing
over 10,000 ppm nickel and 1,000 ppm copper, nearly 1,000 ppm chromium,
and 500 ppm cobalt over talc serpentinite after peridotite, Widgiemooltha, Western Australia.

showing it to be a member of the violaceae.   Plate 2 shows *H. flori-*
bundus growing on Widgiemooltha Hill and indicating the background
sclerophyllous woodland dominated by *Eucalyptus* spp. trees and
Eremophila shrubs.

Cole [174] posed the question as to whether the source of
nickel was nickel sulfide in the bedrock and pointed out that *H.*
floribundus contains high concentrations of sulfur; such sulfide
ores were subsequently disclosed by drilling.   Cole further commented
on the occurrence of a high concentration of *H. floribundus* in the
Eastern Goldfields area, on outcropping serpentinite, and also near
quartz-feldspar-porphyry dykes cutting through ultrabasic rocks.   She
suggested [174] that the plants depend on water held temporarily near

PLATE 2. Anomalous vegetation of *Hybanthus floribundus* (Lindl). F.
Muell. Violaceae, shrubs with small *Trymalium myrtillus* S. Moore trees
at the periphery growing in skeletal soils containing over 10,000 ppm
nickel and 1000 ppm copper, nearly 1000 ppm chromium, and 500 ppm
cobalt over talc serpentinite after peridotite forming a hill feature,
Widgiemooltha, Western Australia. Over the low ground in the background sclerophyllous woodland dominated by *Eucalyptus* spp. trees and
Eremophila spp. shrubs.

the joints of the rocks, since *Hybanthus floribundus* is ill-adapted
for the arid climate. Serverne [178] also commented on the arid
environment and suggested that metal accumulation might be a xero-
morphic adaptation, with the need for nickel accounting for the
restriction to soils with high concentrations of Ni. Table 2 shows
some data for the metal content of soil and organs of *Hybanthus
floribundus* collected from the same area.

Some tentative conclusions can be made from the data in Table
2, although they are limited and come from a number of sources:

1. The ratios of metal concentrations in the leaves to metal
    concentrations in the soil are Ni, 4.5; Co, 0.9; Cu, 0.8;

TABLE 2

Comparison of data for metal content
of *Hybanthus floribundus* [177]

	Ni	Ni	Cu	Co	Zn
Metal conc. (μg/g weight)					
Soil	770[a]	7000[b]	1000[b]	100[b]	50[b]
Root	316[a]	2925	19	53	600
Leaf	6542[a]	13250	15	46	32
Ratios of metal conc.					
Leaf/root	20.6	4.5	0.8	0.9	0.05
Leaf/soil	8.4	1.9	0.015	0.5	0.6
Root/soil	0.42	0.42	0.02	0.5	13.2

[a]Results from [178], root concentration averaged.
[b]Averaged rounded results from [174].

Zn, 0.5. Thus Co and Cu appear to move freely in the
plant. Ni accumulates in the leaves and Zn in the roots.

2. Ni:Co ratios in the root are in the same range as those
   on the soil. Thus the "selectivity" for Ni in that Ni is
   transported to and accumulates in the leaves, and does not
   reside in the roots. As far as Co is concerned *Hybanthus
   floribundus* acts as an indicator in Baker's sense [139].

3. There appears to be a maximum loading of the storage sites
   in the leaves. The ratio of nickel in roots to that in
   soil for the sample from [177] and that from [178] are
   identical (0.42). However, the ratio of Ni in leaves to
   Ni in soil is much higher (20.6) for the latter than for
   the former (4.5). The sample analyzed by Farago and
   Mahmoud [177] was collected from Widgiemooltha, where soil
   Ni is very high (7,000 ppm), and probably has reached the
   maximum Ni loading in the leaf sites.

8. PHYTOCHEMISTRY OF NICKEL-ACCUMULATING PLANTS

8.1. *Alyssum* Species

Early work on the chemistry of nickel-accumulating plants was carried out on *Alyssum bertolonii* by Vergnano-Gambi and coworkers [179,180], who showed that Ni in the leaves was associated mainly with malic and malonic acids, both of which were present with concentrations near 200 μmol/g (dry weight). They also reported that high malic acid content was observed in the Portuguese *Alyssum serpyllifolium* subsp. *lusitanicum*. The two subsp. *lusitanicum* and *malicitanum* have been further investigated by Brooks et al. [181] who found that some nickel was present in the mitochondria, but the majority was in the vacuoles associated with malic and malonic acids. They suggested that Ni in mitochondria can deactivate malic dehydrogenase which would lead to accumulation of malic acid in the vacuoles. This acid could then absorb nickel by complexation and by diffusion remove nickel from mitochondria back to the vacuoles. In this way the citric acid cycle would be unblocked until nickel again built up in the mitochondria.

8.2. *Pearsonia metallifera*

Seventy-five percent of nickel in leaves of *P. metallifera* was found to be water-soluble; this nickel complexes citric acid and possibly 2-hydroxy-1,2,4-butanetricarboxylic acid [182]. Malonic acid was also found in the crude plant extracts.

8.3. *Psychotria*

Ninety-four percent of the total nickel in plant material is water-soluble [172]. Kersten et al. [183] by using a variety of techniques, including gas-liquid chromatography combined with mass spectrometry, came to the conclusion that Ni in *P. douarrei* occurs mainly (60%) as the malato complex with the charge balanced by the nickel aquo complex. It was also shown that the binding of nickel as citrate occurred in a number of New Caledonian accumulators [184]. A strong correlation was found between the concentrations of citric acid and those of nickel. The nickel/citric acid mole ratio was 4.25 in *P. douarrei*.

8.4. *Hybanthus*

The New Caledonian *Hybanthus* spp. were shown to have a positive correlation between citrate and nickel concentrations [184]. Nickel was found as a citratonickelate(II) complex with aquo-nickel cations. The nickel citric acid mole ratios for *H. caledonicus* and *H. austrocaledonicus* were 2.04 and 2.02, respectively.

Hybanthus floribundus from Western Australia has been discussed by Farago and Mahmoud [177]. Histological examination has shown that large epidermal cells in the leaves and main stems contain remarkable concentrations of Ni and high concentrations of pectins. More than half the Ni in the green parts of the plant is water-soluble (Table 3). During rainfall, leaching of the nickel could be a mechanism by which the plant rids itself of part of the Ni burden. In the Eastern Goldfields area, the rain occurs in the cool winter period, the summers being hot and dry [174]. Thus such leaching would occur in the nongrowth period and would remove much of the Ni accumulated in the previous growing season. Chromatographic studies have identified nickel galacturonates in the extracts. The phytochemistry of this plant is not yet fully understood. The sulfur content of the plant is very high [174] and *H. floribundus* contains significant amounts of a yellow pigment which is similar to isorhamnetin 3-0-rutenoside [177]. It is not known if this compound or sulfur has any connection with nickel accumulation.

Several workers have investigated the amino acid profiles of *Hybanthus* spp. [172,177,185]. There are some differences between *H. floribundus* from Queensland, which contains only moderate concen-

TABLE 3

Nickel Extracted by Sequentially Applied Solvents
from Samples of *Hybanthus floribundus* [177][a]

Sample	Area[b] of collection	Ni in sample µg/g	Ether (A) µg/g	%	Ethanol (B) µg/g	%	Water (C) µg/g	%	Dil. HCl (D) µg/g	%
Stems	Kurrajong	3526	8	0.2	239	6.6	2031	55.9	1355	37.3
Leaves	Kurrajong	7035	7	0.1	983	13.8	3251	45.6	2890	40.5
Stems	Widgiemooltha	1869	2	0.1	75	4.0	1048	55.9	749	40.0
Leaves	Widgiemooltha	13246	27	0.2	2233	16.7	5813	43.5	5272	39.5
Turgs	Mt. Thirsty	1539	3	0.2	30	2.0	121	8.0	1367	89.9

[a]Values are in µg/g of original dry sample and % of total Ni in sample.
[b]For more information on areas of collection see [174].

trations of nickel, and the hyperaccumulator from Western Australia.
The former does not contain either isoleucine or proline [172], while
the latter contains both amino acids [177,185]. However, the roots
of the samples from the Eastern Goldfields did not contain proline.
The roots are remarkable in that 70% of the ninhydrin positive material consists of two amino acids: glutamic acid and serine. Except
for aspartic acid and histidine all others are present in small quantities. Proline has been shown to be associated with the copper
tolerance and accumulation in the roots of *Armeria maritima* [142,186].
There appears to be no such relationship in *Hybanthus*.

8.5. General Considerations

There are inherent difficulties in the search for metal complexes
from plant sources [177]. Many potential complexing agents are
released by extraction procedures. The complexes subsequently isolated may be those with the greatest thermodynamic stability in the
extraction mixtures, and may bear no relation to those which existed
in vivo. Notwithstanding these reservations it does seem possible
that organic acids are involved in some way in nickel (and zinc
[187]) tolerance and accumulation. Mathys [187] observed that zinc-
resistant herbiage plants had higher concentrations of malic acid
in their leaves than did the nontolerant ecotypes. In 1966 Torii
and Laties [188] had suggested that organic acid production is a
response to excessive cation uptake and that complexation facilitated
the transport of accumulated cations into the vacuole. It had been
known for many years that high concentrations of organic acids are
present when there are elevated levels of macronutrient cations.
        Thurman [138] speculated on how extra citrate might be produced
in zinc-tolerant plants and pointed out that for a zinc-tolerant
clone of *Deschampsia caespitosa* the stimulation of citrate production
by zinc is different from that by nitrate, which is also well known
to stimulate organic acid synthesis in plants. Nitrate produced increased concentrations of a number of acids, whereas zinc stimulated

the production of citrate and some malate. It is interesting that Wiltshire [148] reported that nitrate increased the retention of nickel in the roots of a number of plants but saw no reason at that time to suppose that the level of plant acids was related to nickel tolerance.

Still and Williams [189] discussed the selective accumulation of Ni by plants and concluded that the selectivity occurs in the uptake by the root. They have pointed out that citrate or malate complexes alone cannot constitute the selectivity mechanism. The difference in the thermodynamic stability of nickel and cobalt complexes is only a factor of 3, but for many Ni accumulators selectivity for Ni over Co is more than 2 orders of magnitude, although the normal Ni/Co ratio in serpentinite soils is about 7. They suggested that only when nickel and cobalt are coordinated to two or more nitrogen donors will the stabilities be sufficiently different to produce selectivity. They have thus come to the conclusion that selectivity for nickel resides in the uptake by the root from the soil. Table 2 shows some data for the metal content of soil and organs of *Hybanthus floribundus* [177]. The Ni/Co ratios in the root are seen to be in the same ratios as those in the soil. Thus from these limited data it does not appear that selectivity lies in root/soil uptake. Carboxylic acid complexes are possibly involved with nickel transport in the xylem. There is no need for high selectivity in the xylem [189] since high concentrations of a number of metals are transported for growth.

9. FUTURE OUTLOOK

The question as to the essentiality of nickel in plants and its exact role has not yet been elucidated. There has been progress in the phytochemistry of nickel-accumulating plants. But many problems remain: the exact nature of the specificity, the role of organic acids, and the interrelationships with other elements.

The availability of nickel to plants and the mechanisms of
uptake of nickel by the root are not well understood. The mechanisms
·by which nickel is toxic to plants also require study. This is of
agricultural importance since serpentine soils have a characteristic
lack of fertility. Brooks [6] pointed out that a study of nickel-
accumulating plants offers the possibility of solving the "serpentine
problem", i.e., making fertile the considerable area of the earth
covered by such soils. Stunted growth on serpentine soils has been
reported [148,151]. There has been some discussion in the literature
as to whether the lack of fertility is due to lack of calcium or to
high levels of magnesium and nickel and other toxic metals [103,147,
148,190-192]. Proctor and coworkers [147,192] point out that many
problems concerning the relationship of the chemistry of serpentine
soils to their vegetation remain unsolved.

REFERENCES

1. E. J. Hewitt and T. A. Smith, "Plant Mineral Nutrition", English
    Univ. Press, London, 1975.

2. T. C. Hutchinson, in "Effect of Heavy Metal Pollution on Plants.
    Volume 1: Effects of Trace Metals on Plant Function" (N. W.
    Lepp, ed.), Applied Science, London, 1981, p. 171 ff.

3. E. Epstein, "Mineral Nutrition in Plants", John Wiley, London,
    1972.

4. R. M. Welch and E. E. Cary, *J. Agric. Food Chem.*, *23,* 479
    (1975).

5. H. A. Schroeder, J. J. Balassa, and I. H. Tipton, *J. Chron.
    Dis.*, *15,* 51 (1962).

6. R. R. Brooks, in "Nickel in the Environment" (J. O. Nriagu,
    ed.), John Wiley, New York, 1980, p. 407 ff.

7. H. A. Schroeder, in "Metal Binding in Medicine" (M. J. Seven,
    ed.), Lippincott, Philadelphia, 1960, p. 59 ff.

8. R. M. Cox and T. C. Hutchinson, *New Phytol.*, *84,* 631 (1980).

9. M. Sarwar, R. J. Thibert, and W. G. Benedict, *Can. J. Plant
    Sci.*, *50,* 91 (1970).

10. M. E. Farago and P. J. Parsons, in "Trace Substances in Environ-
    mental Health", Vol. XIX (D. D. Hemphill, ed.), Univ. Missouri,
    Columbia, 1985, p. 397 ff.

11. M. E. Farago and P. J. Parsons, *Environ. Technol. Lett.*, *3*, 147 (1986).

12. N. E. Dixon, C. Gazzola, R. L. Blakeley, and B. Zerner, *J. Am. Chem. Soc.*, *97*, 4137 (1975).

13. J. C. Polacco, *Plant Physiol.*, *58*, 350 (1976).

14. J. C. Polacco, *Plant Sci. Lett.*, *10*, 249 (1977).

15. J. C. Polacco, *Plant Physiol.*, *59*, 827 (1977).

16. N. Shimada and T. Ando, *Nippon Dojo Hiryogaku Zasshi*, *51*, 493 (1980).

17. D. L. Eskew, R. M. Welch, and E. E. Cary, *Science*, *222*, 621 (1983).

18. R. M. Welch, *J. Plant Nutr.*, *3*, 345 (1981).

19. C. Van Baalen and R. O'Donnell, *J. Gen. Microbiol.*, *105*, 351 (1978).

20. B. Friedrich and B. Magasanic, *J. Bacteriol.*, *131*, 446 (1977).

21. J. W. Spears, C. J. Smith, and E. E. Hatfield, *J. Dairy Sci.*, *60*, 1073 (1977).

22. E. G. Graf and R. K. Thauer, *FEBS Lett.*, *136*, 165 (1981).

23. J. R. Lancaster, *Science*, *216*, 1324 (1982).

24. J. LeGall, P. O. Ljungdahl, I. Moura, H. D. Peck, A. V. Xavier, J. J. G. Moura, B. H. Huynh, and D. D. DerVartanian, *Biochem. Biophys. Res. Commun.*, *106*, 610 (1982).

25. S. P. J. Albracht, M. L. Kalkman and E. C. Slater, *Biochim. Biophys. Acta*, *724*, 309 (1983).

26. W. B. Whitman and R. S. Wolfe, *Biochem. Biophys. Res. Commun.*, *92*, 1196 (1980).

27. G. Diekert, B. Weber, and R. K. Thauer, *Arch. Microbiol.*, *127*, 273 (1980).

28. R. E. Speece, G. F. Parkin, and D. Gallagher, *Water Res.*, *17*, 677 (1983).

29. M. Canovas-Diaz and J. A. Howell, *Biotechnol. Lett.*, *8*, 287 (1986).

30. C. D. Foy, R. L. Chaney, and M. C. White, *Ann. Rev. Plant Physiol.*, *29*, 511 (1978).

31. H. J. M. Bowen, "Trace Elements in Biochemistry", Academic, London, 1966.

32. E. J. Hewitt, *Nature*, *161*, 498 (1948).

33. P. J. Peterson, "International Symp. Uptake. Util. Metals by Plants", Phytochem. Soc., Hull, 1981.

34. A. P. Vanselow, in "Diagnostic Criteria for Plants and Soils" (E. D. Chapman, ed.), Univ. of California, Riverside, 1966, p. 302 ff.

35. W. M. Crooke and R. H. E. Inkson, *Plant Soil, 6,* 1 (1955).

36. E. J. Hewitt, *J. Exp. Bot., 4,* 59 (1953).

37. J. G. Hunter and O. Vergnano, *Ann. Appl. Biol., 39,* 279 (1952).

38. O. Vergnano and J. G. Hunter, *Ann. Bot., 17,* 317 (1953).

39. D. Mishra and M. Kar, *Bot. Rev., 49,* 395 (1974).

40. A. J. Anderson, D. R. Meyer, and F. K. Mayer, *Ann. Bot., 43,* 271 (1979).

41. W. M. Jarrell and R. B. Beverley, *Adv. Agron., 34,* 197 (1981).

42. A. L. Leaf, in "Soil Testing and Plant Analysis" (L. M. Walsh and J. D. Beaton, eds.), Soil. Sci. Soc. Am., Madison, Wisconsin, 1973, p. 427 ff.

43. P. H. T. Beckett and R. D. Davis, *New Phytol., 79,* 95 (1977).

44. R. D. Davis and P. H. T. Beckett, *New Phytol., 80,* 23 (1978).

45. H. W. Wong and A. D. Bradshaw, *New Phytol., 91,* 255 (1982).

46. G. C. Craig, *Trans. Rhodesia Sci. Assoc., 58,* 9 (1978).

47. G. D. Hogan and W. E. Rauser, *New Phytol., 83,* 665 (1979).

48. M. E. Farago, P. W. C. Barnard, and Z. Khan, 1985, unpublished results.

49. K. J. Puckett and M. A. S. Burton, in "Effect of Heavy Metal Pollution on Plants", Vol. 2, "Metals in the Environment", Applied Science, London, 1981, p. 213 ff.

50. D. H. S. Richardson, P. J. Becket, and E. Nieboer, in "Nickel in the Environment" (J. O. Nriagu, ed.), John Wiley, New York, 1980, p. 367 ff.

51. P. W. James, in "Air Pollution and Lichens" (B. W. Ferry, M. S. Baddeley, and D. L. Hawksworth, eds.), Athlone, London, 1973, p. 143 ff.

52. J. Margot and M. T. Romain, *Mem. Soc. Royal Bot. Belg., 7,* 25 (1976).

53. E. Nieboer, D. H. S. Richardson, and F. D. Tomassini, *Bryologist, 81,* 226 (1978).

54. E. Nieboer, K. J. Puckett, and B. Grace, *Can. J. Bot., 54,* 724 (1976).

55. T. H. Nash, *Ecol. Monog., 45,* 183 (1975).

56. K. J. Puckett, *Can. J. Bot., 54,* 2965 (1976).

57. D. H. S. Richardson, E. Nieboer, P. Lavoie, and D. F. Padovan, *New Phytol., 82,* 633 (1979).

58. M. A. S. Burton, P. Leseuer, and K. J. Puckett, *Can. J. Bot., 59,* 91 (1981).

59. P. Leseuer and K. J. Puckett, *Can. J. Bot.*, *58*, 502 (1980).

60. I. Lane and K. J. Puckett, *Can. J. Bot.*, *57*, 1534 (1979).

61. W. J. Birge and J. A. Black, in "Nickel in the Environment" (J. O. Nriagu, ed.), John Wiley, New York, 1980, p. 349 ff.

62. B. R. Folsom, A. Popescu, P. B. Kingsley-Hickman, and J. M. Wood, in "Frontiers in Bioinorganic Chemistry" (A. V. Xavier, ed.), VCH, Weinheim, 1986, p. 391 ff.

63. J. M. Wood and H.-K. Wang, in "Environmental Inorganic Chemistry" (K. J. Irgolic and A. E. Martell, eds.), VCH, Deerfield Beach, Florida, 1985, p. 487 ff; *Environ. Sci. Technol.*, *17*, 582A (1983).

64. T. C. Hutchinson and P. M. Stokes, *Spec. Tech. Pub. 573*, American Society for Testing and Materials, Philadelphia, 1975, p. 320 ff.

65. H.-K. Wang, *Environ. Sci. Technol.*, *18*, 106 (1984).

66. C. S. Venkata Ram, *Phytopathology*, *53*, 276 (1963).

67. J. R. Hardison, *Phytopathology*, *53*, 209 (1963).

68. R. J. Lukens, "Chemistry of Fungicidal Action", Springer-Verlag, New York, 1971.

69. D. H. Brown and J. W. Bates, *J. Bryol.*, *7*, 187 (1972).

70. A. Ruhling and G. Tyler, *Oikos*, *21*, 92 (1970).

71. N. W. Lepp and M. J. Roberts, *Bryologist*, *80*, 533 (1977).

72. M. R. D. Seaward, Proc. 2nd Int. Conf. Bioindication, 1978, Czechoslovakia.

73. W. Krause, in "Handbuch der Pflanzenphysiologie" (W. Ruhland, ed.), Springer-Verlag, Berlin, 1958, p. 755 ff.

74. J. J. Connor, H. T. Shacklette, R. J. Ebens, J. A. Erdma, A. T. Miesch, R. R. Tidball, and H. A. Tourtelot, "Background Geochemistry of Some Soils, Plants and Vegetables in the Conterminous United States", U.S. Geol. Surv. Prof. Pap., 1975, p. 574 ff.

75. T. C. M. Hutchinson, M. Czuba, and L. M. Cunningham, in "Trace Elements in Environmental Health, Vol. VIII" (D. D. Hemphill, ed.), Univ. of Missouri, Columbia, 1974, p. 81 ff.

76. H. Heinrichs and R. Mayer, in "Nickel in the Environment" (J. O. Ngriagu, ed.), John Wiley, New York, 1980.

77. A. N. Rencz and W. W. Shilts, in "Nickel in the Environment" (J. O. Ngriagu, ed.), John Wiley, New York, 1980, p. 151.

78. G. Tyler, *Ambio*, *1*, 52 (1972).

79. M. M. Guha and R. L. Mitchell, *Plant Soil*, *24*, 90 (1966).

80. D. W. Jenkins, in "Nickel in the Environment" (J. O. Ngriagu, ed.), John Wiley, New York, 1980, p. 283 ff.

81. D. F. Spencer, in "Nickel in the Environment" (J. O. Ngriagu, ed.), John Wiley, New York, 1980, p. 339 ff.

82. J. O. Ngriagu, in "Nickel in the Environment" (J. O. Ngriagu, ed.), John Wiley, New York, 1966, p. 1 ff.

83. D. J. Swain, in "Trace-Element Content of Soils", Commonwealth Bureau of Soil Science, Tech. Commun. No. 48, HMSO, London, 1955.

84. A. M. Ure and M. L. Berrow, in "Environmental Chemistry", Vol. 2 (H. J. M. Bowen, Snr. Reporter), Royal Society of Chemistry, London, 1982, p. 94 ff.

85. M. L. Berrow and A. M. Ure, Environ. Geochem. Health, 8, 19 (1986).

86. M. L. Berrow and T. A. Reeves, in "Environmental Contamination", Proceed. Int. Conf., London, CEP Consultants Ltd., Edinburgh, 1984, p. 333 ff.

87. R. L. Mitchell, in "Chemistry of the Soils", 2nd ed. (F. E. Bear, ed.), Reinhold, New York, 1964, p. 320 ff.

88. H. Aubert and M. Pinta, in "Trace Elements in Soils", Elsevier, New York, 1977, p. 395 ff.

89. D. J. Swain and R. L. Mitchell, J. Soil Sci., 11, 347 (1960).

90. M. L. Berrow and R. L. Davidson, Trans. Roy. Soc. Edin.; Earth Sci., 71, 103 (1980).

91. M. L. Berrow, M. S. Davidson, and J. C. Burridge, Plant Soil, 66, 161 (1982).

92. B. E. Davies and C. F. Paveley, in "Trace Substances in Environmental Health", Vol. XIX (D. D. Hemphill, ed.), Univ. of Missouri, Columbia, 1985, p. 87 ff.

93. S. P. McGrath, C. H. Cuncliffe, and A. J, Pope, in "First International Conference on Geochemistry and Health" (I. Thornton, ed.), Science Reviews, Northwood, 1987, in press.

94. R. L. Halstead, B. J. Finn, and A. J. MacLean, Can. J. Soil Sci., 49, 335 (1969).

95. A. V. T. Haq, T. E. Bates, and Y. Y. Soon, Soil Sci. Soc. Am. J., 44, 772 (1980).

96. W. L. Lindsay, in "The Plant Root and its Environment" (E. W. Carson, ed.), Charlottesville Univ. Press, Virginia, 1974, p. 509 ff.

97. J. E. Miller, J. E. Hassett, and D. E. Koeppe, J. Environ. Qual., 5, 157 (1976).

98. R. E. Graham, Soil Sci. Soc. Am. Proc., 37, 70 (1973).

99. H. L. Bohn, B. L. McNeal, and G. A. O'Connor, in "Soil Chemistry", Wiley-Interscience, New York, 1979.

100. P. H. Given, in "Environmental Chemistry", Vol. 1 (G. Eglington, Snr. Reporter), Royal Society of Chemistry, London, 1975.

101. R. F. C. Mantoura, A. Dickson, and A. P. Riley, *Estuar. Coastal Mar. Sci.*, *6*, 387 (1978).

102. T. C. Hutchinson and L. M. Whitby, *Environ. Cons.*, *1*, 123 (1974).

103. P. R. Shewry and P. J. Peterson, *J. Ecol.*, *64*, 195 (1976).

104. M. E. Farago, *Coord. Chem. Rev.*, *36*, 155 (1981).

105. B. Freedman and T. C. Hutchinson, in "Effect of Heavy Metal Pollution on Plants, Vol. 2, Metals in the Environment" (N. W. Lepp, ed.), Applied Science, Essex, 1981, p. 35 ff.

106. J. B. E. Patterson, in "Trace Elements in Soils and Crops", Ministry of Agriculture, Tech. Bull. No. 21, London, 1971, p. 193 ff.

107. J. Webber, in "Effect of Heavy Metal Pollution on Plants, Vol. 2, Metals in the Environment" (N. W. Lepp, ed.), Applied Science, Essex, 1981, p. 159 ff.

108. D. Purves, in "Trace Element Contamination of the Environment", 2nd ed., Elsevier, Amsterdam, 1985.

109. D. Purves, *Environ. Geochem. Health*, *11*, 8 (1986).

110. M. L. Berrow and J. Webber, *J. Sci. Food Agric.*, *23*, 93 (1972).

111. R. O. Williams, *Water Pollut. Control*, *74*, 607 (1975).

112. Department of Environment, UK, "Disposal of Sewage Sludge to Land", HMSO, 1981.

113. M. J. Marks, J. H. Williams, L. V. Vaidyanathan, and C. J. Chumbly, in "Inorganic Pollution and Agriculture", MAFF Ref. Book 326, HMSO, 1980, p. 235 ff.

114. A. Ruhling and G. Tyler, *Oikos*, *24*, 402 (1973).

115. T. M. Roberts and G. T. Goodman, in "Trace Substances in Environmental Health", Vol. VIII (D. D. Hemphill, ed.), Univ. of Missouri, Columbia, 1974, p. 117 ff.

116. G. T. Goodman and T. M. Roberts, *Nature*, *231*, 287 (1971).

117. B. Freedman and T. C. Hutchinson, *Can. J. Bot.*, *58*, 108 (1980).

118. B. Freedman and T. C. Hutchinson, *Can. J. Bot.*, *58*, 1722 (1980).

119. G. K. Rutherford and C. R. Bray, *J. Environ. Qual.*, *8*, 219 (1979).

120. W. D. McIlveen and D. Balsillie, in "Air Quality Assessment Studies in the Sudbury Area", Vol. 2, Ontario Ministry of the Environment, Sudbury, Ontario, 1978.

121. L. M. Whitby and T. C. Hutchinson, *Environ. Conserv.*, *1*, 191 (1974).

122. H. Gydeson, K. Pilegaard, L. Rassmussen, and A. Ruhling, National Swedish Environmental Protection Board, Bull. SNV PM 1670, 1983.

123. A. Ruhling and G. Tyler, *J. Appl. Ecol.*, *8*, 497 (1971).

124. A. Ruhling and G. Tyler, *Water Soil Air Pollut.*, *2*, 445 (1973).

125. S. S. Groet, *Oikos*, *27*, 445 (1976).

126. G. T. Goodman, S. Smith, G. D. R. Parry, and M. J. Inskip, "Proc. 41st Conf. Natl. Soc. Clean Air", Brighton, 1974, p. 1 ff.

127. I. Johnsen and L. Rassmussen, *Bryologist*, *80*, 625 (1977).

128. E. Nieboer, H. M. Ahmad, K. J. Puckett, and D. H. S. Richardson, *Lichenologist*, *5*, 292 (1972).

129. J. A. Campbell, *Lichenologist*, *8*, 83 (1976).

130. K. Pilegaard and I. Johnsen, *Ecol. Bull.* (Stockholm), *36*, 97 (1984).

131. A. Ruhling and G. Tyler, *Bot. Not.*, *122*, 248 (1969).

132. H. L. Cannon, *Science (NY)*, *132*, 591 (1960).

133. R. R. Brooks, "Geobotany and Biogeochemistry in Mineral Exploration", Harper and Row, New York, 1972.

134. R. R. Brooks, J. Lee, R. D. Reeves, and T. Jaffre, *J. Geochem. Explor.*, *7*, 49 (1977).

135. J. Antonovics, J. D. Bradshaw, and R. G. Turner, *Adv. Ecol. Res.*, *7*, 1 (1971).

136. S. Wainwright and H. W. Woolhouse, in "Ecology and Resource Degredation and Renewal" (M. J. Chadwick and G. T. Goodman, eds.), Blackwell, Oxford, 1975, p. 231 ff.

137. H. W. Woolhouse, *Chem. Br.*, *16*, 72 (1980).

138. D. A. Thurman, in "Effect of Heavy Metal Pollution on Plants, Vol. 2, Metals in the Environment" (N. W. Lepp, ed.), Applied Science, London, 1981, p. 231 ff.

139. A. J. M. Baker, *J. Plant Nutrit.*, *3*, 643 (1981).

140. M. E. Farago, in "Frontiers in Bioinorganic Chemistry" (A. V. Xavier, ed.), VCH, Weinheim, 1986, p. 106 ff.

141. W. H. O. Ernst, in "Proc. Int. Conf. on Metals in the Environment", Vol. 2 (T. C. Hutchinson, ed.), Toronto, 1975, p. 121 ff.

142. M. E. Farago, W. A. Mullen, M. M. Cole, and R. F. Smith, *Environ. Pollut.* (Ser. A), *21*, 225 (1980).

143. M. E. Farago and A. Mehra, 1986, unpublished results.

144. R. P. D. Gregory and A. D. Bradshaw, *New Phytol.*, *64*, 131 (1965).

145. R. M. Cox and T. C. Hutchinson, *Nature, 279*, 231 (1979).

146. R. M. Cox and T. C. Hutchinson, *Water Air Soil Pollut., 16*, 83 (1981).

147. J. Proctor and S. R. J. Woodell, *Adv. Ecol. Res., 9*, 256 (1975).

148. G. H. Wiltshire, *J. Ecol., 62*, 501 (1974).

149. R. R. Brooks and F. Malaisse, "The Heavy Metal Tolerant Flora of South Central Africa", Balkema, Rotterdam, 1985.

150. G. Minguzzi and O. Vergnano, *Atti Soc. Tosc. Sci. Nat., 55*, 49 (1948).

151. O. Vergnano, *Nuovo G. Bot. Ital., 65*, 133 (1958).

152. O. Vergnano Gambi, L. Pancaro, and C. Formica, *Webbia, 32*, 175 (1977).

153. R. R. Brooks and C. C. Radford, *Proc. Royal Soc. Lond. B, 191*, 217 (1978).

154. R. S. Morrison, R. R. Brooks, and R. D. Reeves, *Plant Sci. Lett., 17*, 451 (1980).

155. R. S. Morrison, R. R. Brooks, R. D. Reeves, T. R. Dudley, and Y. Akman, *Proc. Royal Soc. London. B, 203*, 387 (1979).

156. R. D. Reeves and R. R. Brooks, *J. Geochem. Explor., 18*, 275 (1983).

157. R. D. Reeves and J. M. Baker, *New Phytol., 98*, 191 (1984).

158. R. R. Brooks and H. M. Crooks, *Plant Soil, 54*, 491 (1980).

159. H. Wild, *Kirkia, 7*, 1 (1970).

160. H. Wild, *Mitt. Bot. Staatssamml., München, 10*, 266 (1971).

161. H. Wild in "Biogeography and Ecology of Southern Africa" (M. J. A. Werger, ed.), Junk, The Hague, 1978, pp. 1301 ff.

162. H. Wild, *Kirkia, 9*, 233 (1974).

163. R. R. Brooks and X. H. Yang, *Taxon, 33*, 392 (1984).

164. W. Ernst, *Kirkia, 8*, 125 (1972).

165. M. M. Cole, *J. S. Afr. Inst. Min. Metall., 71*, 199 (1971).

166. T. Jaffré, in "Etude ecologique du peuplement végétal des sols derivés de roches ultrabasiques en Nouvelle Caledonie", Traveaux et Documents de l'O.R.S.T.R.O.M., No. 124, Paris, 1980.

167. T. Jaffré and M. Schmid, *C.R. Acad. Sci., Ser. D, Paris, 278*, 1727 (1974).

168. R. R. Brooks, J. Lee, R. D. Reeves, and T. Jaffré, *J. Geochem. Explor., 7*, 49 (1977).

169. X. Y. Yang, R. R. Brooks, T. Jaffré, and J. Lee, *Plant Soil, 87*, 281 (1985).

170. T. Jaffré, R. R. Brooks, J. Lee, and R. D. Reeves, *Science*, *193*, 579 (1976).

171. A. J. M. Baker, R. R. Brooks, and W. J. Kersten, *Taxon, 34*, 89 (1985).

172. P. C. Kelly, R. R. Brooks, S. Dilli, and T. Jaffré, *Proc. Royal Soc. Lond. B, 189*, 69 (1979).

173. B. C. Serverne and R. R. Brooks, *Planta (Berlin)*, *103*, 91 (1972).

174. M. M. Cole, *J. Appl. Ecol., 10*, 269 (1972).

175. M. E. Farago, A. J. Clark, and M. J. Pitt, *Inorg. Chim. Acta, 24*, 53 (1977).

176. M. E. Farago, A. J. Clark, and M. J. Pitt, *Coord. Chem. Revs., 16*, 1 (1975).

177. M. E. Farago and I. E. D. A. W. Mahmoud, *Min. Environ., 5*, 113 (1983).

178. B. C. Serverne, *Nature, 248*, 807 (1974).

179. P. Pelosi, C. Galoppine, and O. Vergnano-Gambi, *L'Agrigolt. Ital., 29*, 1 (1974).

180. P. Pelosi, R. Fiorentine, and C. Galoppine, *Agric. Biol. Chem., 40*, 1641 (1976).

181. R. R. Brooks, S. Shaw, and A. A. Marfil, *Physiol. Plant, 51*, 167 (1981).

182. E. Stockley, (1980), reported in Ref. 149, pp. 144-5.

183. W. J. Kersten, R. R. Brooks, R. D. Reeves, and T. Jaffré, *Phytochemistry, 19*, 1963 (1980).

184. J. Lee, R. D. Reeves, R. R. Brooks, and T. Jaffré, *Phytochemistry, 17*, 1033 (1978).

185. M. E. Farago, I. Mahmoud, and A. J. Clark, *Inorg. Nucl. Chem. Lett., 16*, 481 (1980).

186. M. E. Farago and W. A. Mullen, *Inorg. Nucl. Chem. Lett., 17*, 9 (1981).

187. W. Mathys, *Physiol. Plant., 40*, 130 (1977).

188. K. Torii and G. G. Laties, *Plant Cell Physiol., 7*, 395 (1966).

189. E. R. Still and R. J. P. Williams, *J. Inorg. Biochem., 13*, 35 (1980).

190. J. Proctor, *J. Ecol., 59*, 297, 375 (1871).

191. R. H. Marrs and J. Proctor, *J. Ecol., 64*, 953 (1976).

192. W. R. Johnson and J. Proctor, *J. Ecol., 69*, 855 (1981).

4

Nickel Metabolism in Man and Animals

Evert Nieboer, Rickey T. Tom, and W. (Bill) E. Sanford
Department of Biochemistry and Occupational Health Program
Health Sciences Centre, McMaster University
1200 Main Street West
Hamilton, Ontario L8N 3Z5 Canada

1. INTRODUCTION

The material discussed in this chapter is crucial to a treatment of
the toxic effects of nickel compounds described later in this volume.
Specifically, the objectives are to provide a concise, up-to-date
summary of the current knowledge of nickel biochemistry and metabolism and a guide to the extensive nickel literature on this topic.
Although human data will be emphasized, the results of animal experiments are integrated where appropriate.

2. ROUTES OF NICKEL ABSORPTION

Inhalation and ingestion constitute the major routes of nickel intake
in humans. Although dermal uptake is insignificant by comparison,
penetration of the nickel(II) ion into the skin is of considerable
importance when considering nickel contact dermatitis. Other parenteral routes are of importance to humans only in isolated situations
(e.g., hemodialysis, contaminated intravenous fluids, and prosthesis),
although they are common in animal experimental work (e.g., intravenous, intraperitoneal, intramuscular, and intrarenal injections;
subcutaneous implants etc.). Consequently, only ingestion and inhalation will be discussed in detail here.

2.1. Uptake and Retention by the Respiratory Tract

Deposition of particles in the respiratory tract during breathing
is a complex process [1,2]. Particle size determines the site of
deposition and thus the toxicologic effect, be it in the nasal

passages, the tracheobronchial (conductive) airways, or the alveolar
zone. Generally speaking [2], particles smaller than 10 μm in diameter penetrate the alveolar region of the lung and constitute the
respirable fraction (proportion deposited is estimated as 50%);
particles larger than 0.1 μm deposit in the nasal passages (deposition is estimated to approach 100% for diameters > 10 μm); and particles larger and smaller than 0.5 μm deposit in the tracheobronchial
airways (deposition estimates range from 5 to 25%).

Nickel-containing particulates in ambient air and workplace
environments have diameters below 50 μm, with a substantial fraction
being smaller than 10-15 μm [3,4]. The degree of absorption and
clearance may be expected to depend on the physicochemical properties
of the particles, such as crystallinity, water solubility, and surface characteristics including charge and area. Clearance of solid
particulates from the respiratory tract (e.g., deposits in the nasal
passages and tracheobronchial tree) is aided by mucociliary action
which propels the material toward the pharynx resulting in ingestion.
The mechanisms of clearance of solids from the alveolar zone is not
completely understood, but does involve phagocytosis by interstitial
cells such as monocytes, macrophages, and polymorphonuclear leukocytes [2,5].

Direct evidence for absorption and clearance of nickel compounds by the human lung is not available. Indirect evidence
based on renal excretion patterns of nickel and nickel content of
lung tissue at necropsy are considered after the animal data are
reviewed.

Experimental inhalation and intratracheal instillation studies
confirm that the physicochemical properties of nickel compounds
determine the absorption and clearance from the lung. Removal of
$NiCl_2$ from the rat and mouse lung occurs relatively rapidly, with a
half-life ($t_{\frac{1}{2}}$) of 1-2 days [6-8]. In inhalation experiments, animals
were typically exposed for a single or repeated 2-hr period at ambient concentrations between 100 and 1000 $\mu g/m^3$ with a particle size
<3 μm (mass median aerodynamic diameter, MMAD). The bulk of the
clearance was by absorption since the nickel was primarily excreted

by the renal route (e.g., 90%) [9,10]. Menzel and colleagues [7,8] demonstrated that Michaelis-Menten kinetics describe the lung removal processes. Nevertheless, scrutiny of their rat data [8] and that of Graham et al. [6] for mice indicates that for low and moderate exposures (e.g., single 2-hr exposure to 600 µg Ni/m^3 as $NiCl_2$) first order decay approximates the dependence of clearance on time.

Results similar to $NiCl_2$ have been reported for nickel(II) carbonate administration by intratracheal instillation in mice, although only 55% of the dose appeared in the urine in 6 days [11]. Intratracheal instillation of ^{63}Ni-labeled nickel subsulfide (Ni_3S_2) in mice resulted in a biphasic removal with half-lives of 1.2 days (rapid phase) and 12.4 days (slow phase) [12]. Even though absorption dominated since 60% of the administered nickel appeared in the urine, mucociliary clearance was indicated by the 40% fecal excretion. (Biliary excretion of exogenous nickel(II) in rats has been shown to be quantitatively unimportant [13].)

Green nickel(II) oxide appears to be poorly absorbed. Estimated half-lives range from 3 months for Syrian golden hamsters (1.0 ± 1.6 µm MMAD, 99.99% < 7 µm; 14-hr exposure to 10-150 mg Ni/m^3 as NiO) to 12 months (1.2 µm MMAD) and 21 months (4 µm MMAD) in rats exposed for 140 hr to 0.5-55 mg Ni/m^3 as NiO [14-16]. Very little of the administered dose appeared in other tissues and in the body fluids.

Autopsy data for industrially nonexposed individuals illustrate clearly that nickel accumulates in the human lung with age. Recent and earlier data have been summarized by Rezuke et al. [17] and are described by the relationship: Lung nickel (µg/kg dry wt) = 3.25 age (years) - 38.4, with r = 0.77 and p < 0.001. It is known that a significant fraction of ambient air contains nickel(II) oxide, perhaps as much as 50%; water-soluble nickel(II) sulfate is the other major component [3]. By analogy to the animal experiments, it is likely that the relatively inert nickel(II) oxide accumulates in the lung tissue. For workers who have been exposed to particulates in nickel-smelting operations, the half-life of nickel released from the nasal mucosa has been estimated at 3.5 years [18]. Similar estimates (2-3 years) have

been derived from the decline with time since retirement in urinary
and plasma nickel levels [19,20]. Workers exposed to aerosols of dissolved nickel salts in plating shops have been shown to clear nickel,
as measured by its appearance in plasma and urine, with a half-life
of 1-1½ days [19,21]. Since respiratory uptake may be assumed to be
the major route for such workers, the clearance from the respiratory
tract of water-soluble salts appears to be regulated by pharmacodynamic parameters similar to those reviewed for rodents.

It is obvious that pulmonary absorption provides a major route
of nickel uptake in humans.

2.2. Gastrointestinal Absorption of Nickel

The most recent reports on average dietary nickel content assign a
value of 160 µg/day [22-24]. However, since certain foods such as
soya products, cocoa, and dried legumes are relatively rich in nickel
(2-10 µg/g), earlier reports of average dietary intakes of 300-600
µg/day remain realistic [25-27]. Calculations show that nickel
intake may reach 900 µg/day after the replacement and supplementation of certain items in the average diet [23,24].

The matter of gastrointestinal absorption is complicated,
although the recent measurement of nickel in human bile helps to
clarify the matter. Nickel concentrations in bile specimens obtained
postmortum from the gallbladders of five nonoccupationally exposed
subjects averaged 2.3 ± 0.8 µg/liter [17], which is comparable to
normal urinary nickel levels of 2.2 ± 1.2 µg/liter (or 2.6 ± 1.4
µg/day) [28]. By contrast, experiments with rats have shown that
biliary excretion is quantitatively unimportant as an elimination
route [13]. Assuming the absence of significant intestinal reabsorption of bile-excreted nickel, Rezuke et al. [17] estimated that net
biliary excretion of nickel in nonexposed persons may average 2-5
µg/day. This route of elimination when combined with the urinary
pathway suggests a net average excretion of nickel of about 6 µg/day.
Depending on the dietary intake, gastrointestinal absorption of

nickel may thus be estimated to fall between a maximum of 4% (165 µg daily intake) and a minimum of 1% (600 µg/day intake). (Other routes of excretion may be considered minor unless profuse sweating or lactation occurs.) These absorption estimates concur with the cumulative urinary excretion of nickel after a single oral dose of nickel sulfate ($\equiv$ 5.6 mg Ni) administered to nonfasting human volunteers. In the absence of fecal data, a minimum GI absorption of 3% could thus be assigned [29]. In similar human supplementation studies, Solomons et al. [30] showed that mixing of the nickel sulfate with food prior to consumption apparently suppressed its absorption. In accord with this finding, Cronin et al. [31] calculated from urinary excretion during the first 24 hr that 4-20% of administered nickel sulfate (equivalent to 0.6-2.5 mg of Ni) had been absorbed by fasting human subjects.

Absorption of water-soluble nickel salts administered to rats and mice is also in the range of 3-6% [32]. Foulkes and McMullen [33] recently demonstrated by in situ perfusion of rat jejunum that Ni(II) is taken up biphasically. The first step involved crossing of the brush border membrane and was saturable, while the subsequent transfer of the metal from the mucosa into the body was not. The first step was nonspecific and was depressed by Zn(II) or by the constituents of dried skimmed milk, e.g., Ca(II), while the second was passive in nature with ion flow occurring in both directions.

3. DISTRIBUTION OF NICKEL

Extensive studies with guinea pigs, mice, rabbits, and rats indicate clearly that intravenous or intraperitoneal injection of $^{63}NiCl_2$ dissolved in sterile medium results in significant accumulation in kidney, endocrine glands, lung, and liver. Relatively speaking, there is little compartmentalization of nickel in the bone and neural tissue. As already implied by clearance from the lung, the nickel(II) ion is also readily cleared from the whole organism, with little retention in soft or mineral tissue. Onkelinx et al. [34] found that a two-compartment pharmacodynamic model with first-order elimination kinetics ($t_{\frac{1}{2}}$ = 7 hr) described the distribution and excretion of ^{63}Ni(II) after intravenous injection in both rats and rabbits. Compartment I was the largest in both animals (90% of the total volume in rat and 70% in the rabbit) and included the blood. Exchange of nickel between compartments I and II was consistent with the model. Excretion was primarily (80-90%) by the renal route [34,35].

By contrast to water-soluble nickel salts, nickel carbonyl is lipid-soluble. It is a volatile organometallic compound (b.p. 43°C), is extremely toxic [36], and is used in nickel refining to produce ultrapure nickel [37]. The most remarkable feature following inhalation or intraperitoneal administration of $Ni(CO)_4$ is the accumulation of nickel in the lung [38,39]. Because of its volatility and lipophilicity, nickel carbonyl can pass through the alveolar walls and a significant proportion is exhaled [35,40]. It is decomposed in the animal body to nickel and carbon monoxide. The nickel becomes localized eventually as Ni(II) in various tissues and the carbon monoxide is bound to hemoglobin [38,40]. Compartmentalization of nickel after nickel carbonyl exposure in mice and rats by inhalation or intraperitoneal administration occurs primarily in the lung, and to a lesser extent in the heart, diaphragm, central nervous system, kidney, liver, adipose tissue, and blood [35,38,39].

Although available data are limited, the distribution pattern of nickel in humans parallels, to some extent, that observed in animal experiments. In Table 1, recently assessed nickel concentrations for tissues of nonexposed (occupationally) individuals demonstrate that the highest levels occur in lung, thyroid, and the adrenal glands. Interestingly, these same tissues are subject to nickel compartmentalization in animal parenteral studies. Autopsy specimens from victims of nickel carbonyl gassing show the highest concentrations of nickel in the lung, with lower levels in kidneys, liver, and brain [41]. Not only is the nickel distribution in this case comparable to the animal experience, but so is the pathogenesis. The lung is the major target, and death has been attributed primarily to respiratory failure [19,36,41].

TABLE 1

Nickel Concentrations in Human Tissues by Analysis
of Postmortem Specimens from Occupationally
Unexposed Individuals[a]

Tissue[b]	No. of subjects	Wet weight (µg/kg) Mean ± SD	Dry weight (µg/kg) Mean ± SD
Lung	9[c]	18 ± 12	173 ± 94
Thyroid	8	20 ± 10	141 ± 83
Adrenal	10	26 ± 15	132 ± 84
Kidney	10	9 ± 6	62 ± 43
Heart	9[d]	8 ± 5	54 ± 40
Liver	10	10 ± 7	50 ± 31
Brain	7	8 ± 2	44 ± 16
Spleen	10	7 ± 5	37 ± 31
Pancreas	10	8 ± 6	34 ± 25

[a]Adapted from [17]; only one of the subjects had any occupational
exposure to nickel compounds or alloys as a machinist.
[b]A small number of additional measurements (2-4) of other tissues
showed possible accumulation in hilar lymph nodes, testis, and ovary.
[c]Excluding the machinist whose levels were 241 µg/kg wet weight and
2,060 µg/kg dry weight.
[d]Excluding a patient who died of acute myocardial infarction with
levels of 78 µg/kg wet weight and 758 µg/kg dry weight.

Reliable and rapid analytical methods are now available for
the detection of nickel in body fluids [42,43] and tissues [44].
Until recently the accepted normal values in healthy, industrially
unexposed persons were: whole blood (3-7 µg/liter), serum (1-5 µg/
liter), urine (0.7-5.2 µg/liter), saliva (2.2 ± 1.2 µg/liter), sweat
(52 ± 36 µg/liter), and feces (14.2 ± 2.7 µg/g or 258 ± 126 µg/day)
[28,36,45]. Sunderman et al. [42] recently reported much lower
values for serum (0.46 ± 0.26 µg/liter) and whole blood (1.26 ± 0.33
µg/liter). Reference values for whole blood have subsequently been
lowered to 0.34 ± 0.28 µg/liter [28]. Although lower serum and
plasma nickel values are now commonly reported, urine concentrations
appear to hold near 2 µg/liter. For example, Sunderman et al. [43],

by direct quantitation of urine diluted with dilute nitric acid, found urinary nickel levels to be 2.0 ± 1.5 µg/liter (range 0.5-6.1 µg/liter) for 34 nonexposed healthy adults. The consensus is that part of this reduction in serum levels is due to better contamination prevention and the availability of atomic absorption spectrometers with modern background correction accessories (mostly by Zeeman technique). For workers occupationally exposed to nickel and its compounds, values of plasma or serum concentrations up to 35 µg/liter and urine levels up to 400 µg/liter have been reported [19].

4. NICKEL TRANSPORT AND CELLULAR UPTAKE

4.1. Nickel Transport

The recent average whole-blood and serum nickel concentrations suggest that nickel is not preferentially concentrated in blood cells. Furthermore, the two-compartment model mentioned earlier suggests that cellular and plasma nickel are in equilibrium [34]. Nickel(II) is transported in the body by way of plasma where it is bound to carriers of both high and low relative molecular mass. The high-molecular-mass ligand is human albumin, and possibly a nickel metalloprotein which has been characterized as an α_2-macroglobulin, also called nickeloplasmin [36,46,47]. The low-molecular-mass component is likely a nickel(II)-amino acid complex as shown by in vitro ^{63}Ni(II) addition experiments. Of the 22 amino acids tested, the predominant Ni(II)-binding amino acid was L-histidine [48,49]. The exact distribution between these fractions has not been firmly established, and a recent kinetic study suggests that facile equilibrium between these species, as suggested by Sarkar and colleagues [50,51], is complicated by relatively slow exchange kinetics [49].

The metalloprotein nickeloplasmin has not been well characterized. Work by Nomoto et al. [52] and Sunderman et al. [53], employing atomic absorption spectrometry measurements on serum fractions that had been separated successively by ultrafiltration and column chromatography, has demonstrated that approximately 40% of

nickel(II) present in rabbit serum was associated with albumin, 44%
with the nickeloplasmin fraction, and 16% with the ultrafilterable
fraction; whereas approximately 34% of the nickel in human serum was
associated with albumin, 26% with nickeloplasmin, and 40% was ultrafilterable. By contrast, Asato et al. [54] found the amount of
^{63}Ni(II) (introduced by i.v. injection) associated with the ultrafilterable fraction of rabbit serum to be 15%. Nomoto [46] separated
α_2-macroglobulin by affinity column chromatography from the serum of
healthy humans who had no industrial exposure to nickel. By atomic
absorption spectrometry measurement of nickel in this fraction, he
found that 43% of the total nickel content was bound to an α_2-macro-
globulin. By comparison, Lucassen and Sarkar [48] found that only
about 0.1% of the total Ni(II) in human serum was bound to nickeloplasmin, with 95.7% bound to albumin and 4.2% to L-histidine. The
values reported by them are for ^{63}NiCl added to serum and do not
necessarily represent the endogenous nickel distribution. Possible
discrepancy between in vivo and in vitro results is emphasized by
the work of Asato et al. [54]. They found that the ultrafiltration
fraction in in vitro addition experiments was 36% compared to 15%
for in vivo results. Caution in interpretation is thus warranted
since the added nickel in the Lucassen and Sarkar work resulted in
concentrations about 35,000-fold higher than those observed naturally
in serum. The in vitro distribution studies mentioned above [48,49]
involving the 22 amino acids were also carried out under similar
artificial conditions. It is obvious that there is no agreement on
the exact magnitude of the ultrafilterable or high molecular mass
fraction of Ni(II) in human serum. Recent work in the authors'
laboratory assessed the endogenous low- and high-molecular nickel
mass fractions by ultracentrifugation and atomic absorption spectrometry in sera of electrolytic nickel refinery workers. The ultrafilterable fraction contained 24 ± 6% (n = 6) of the plasma nickel
and constitutes an estimate of the amount available for renal excretion. Combination of these data with the recent Nomoto α_2-macro-
globulin results [46] suggests that human and rabbit sera may have
similar nickel distribution patterns.

4.2. Cellular Uptake of Nickel

The data in Figs. 1 and 2 illustrate that the cellular uptake of
Ni(II) is regulated by the extracellular ligand concentration and
are in accord with the "Equilibrium" model suggested by Williams [55]
for metal ion uptake under steady-state conditions (i.e., at fixed
pH, redox potential, intracellular and extracellular ligand concentrations). The distribution of metal ions among the various compartments of a cell is then determined by thermodynamic parameters such
as the pK_a values of ligands, the binding constants of the metal-
ligand complexes, their solubilities in aqueous and lipid phases

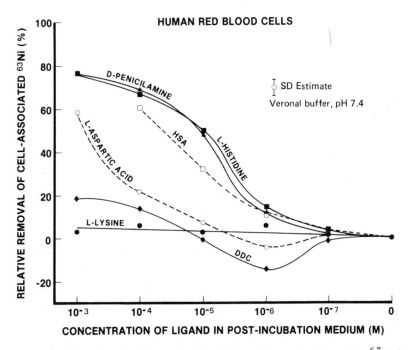

FIG. 1. Relative abilities of chelating agents in removing [63]Ni(II)
in 30 min from human erythrocytes (5 x 10[6] cells in 1 ml) preloaded
with it. Removal data for EDTA were not plotted as they overlapped
with those shown for L-histidine and D-penicillamine. The total
nickel concentration employed in the 2-hr preincubation step was
4.1 µg/liter (7.0 x 10[-8] M), which is comparable to serum levels
observed in moderately exposed individuals. The experiment was
carried out at 37°C. (Reproduced from [57] with permission.)

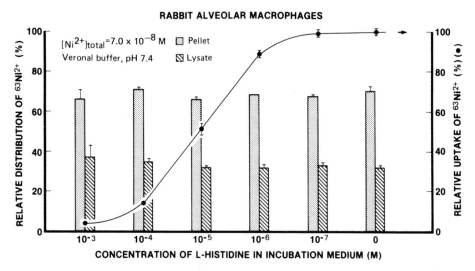

FIG. 2. Influence of L-histidine on the uptake and cellular distribution of ^{63}Ni(II) by rabbit alveolar macrophages. Cells were incubated concurrently with the ligand and ^{63}Ni(II) for 2 hr at 37°C, were washed repeatedly with buffer, and subsequently lysed by freeze-thawing; the residual pellet was isolated by centrifugation after appropriate washing with distilled water. (Reproduced from [57] with permission.)

(i.e., distribution coefficients), and the effective ligand concentrations in the various compartments [56]. It may be concluded from the data in Figs. 1 and 2 that physiologic concentrations in serum of L-histidine (7.4 x 10^{-5} M) and human serum albumin (HSA; 6 x 10^{-4} M) are effective in regulating the amount of ^{63}Ni(II) accumulated by cells. By contrast, L-aspartic acid (average serum level of 7.5 x 10^{-6} M) and L-lysine (1.5 x 10^{-4} M) are ineffective [57]. Note that at low concentrations, diethyl dithiocarbamate (DDC), which forms lipophilic complexes with Ni(II), promotes the cellular association of this metal ion. This DDC-enhanced uptake has been confirmed in other cell systems [56].

Like the uptake of Ni(II) by rat jejunum, cellular uptake appears to be a passive process. Although Abbracchio et al. [58] observed a temperature dependence of Ni(II) accumulation by Chinese hamster ovary cells (CHO), its magnitude is indicative of a passive process such as diffusion (the temperature coefficient Q_{10} is estimated to be near 1.0; $Q_{10} = (k_2/k_1)^{10/(T_2-T_1)}$, where $T_2 > T_1$). These authors also reported that 70-80% of the cell-associated nickel was removed on trypsinization (i.e., removal of cell surface protein), suggesting that 20-30% was internalized. This compares well with the 70% appearing in the cell pellet and 30% in the lysate for the rabbit alveolar macrophage data in Fig. 2 (no L-histidine). Intuitively, such cellular distribution of Ni(II) might be expected to depend on cell type and this is indeed observed [56].

Other uptake mechanisms have been identified. Cells with phagocytic capacity such as CHO cells [59,60], macrophages [61], and rat rhabdomyosarcoma cells [62] have been shown to take up particulates ($\leq$5 μm in diameter) of nickel compounds (e.g., Ni_3S_2, NiO, and NiS). Crystalline compounds are taken up most readily and dependence on particle size [64], surface charge [59,60], and medium Ca(II) concentration [63] has also been established. Nickel(II) chloride and Ni(II)-albumin complex solutions encapsulated in liposomes have been demonstrated to facilitate Ni(II) uptake by CHO cells [65]. These pathways illustrate the importance of endocytosis as an important uptake mechanism.

5. NICKEL BIOCHEMISTRY

5.1. Interactions of Ni(II) with Biomolecules

Based on the system of classification of metal ions proposed by Nieboer and Richardson [66], Ni(II) is designated as a borderline metal ion and thus it exhibits both class A (oxygen-seeking) and class B (nitrogen/sulfur-seeking) ligand preferences. As expected, Ni(II) effectively competes with the endogenous class A ions Ca(II) and Mg(II) and the borderline ions Mn(II) and Zn(II). The consequence of such substitution is either stimulatory or antagonistic as reviewed previously [45,67,68]. For example, Ni(II) replaces Zn(II) in carboxypeptidase A (peptidase function) without loss of enzymatic function, but in carbonic anhydrase and liver alcohol dehydrogenase (LAD) this

substitution results in inhibition [69,70]. It is well recognized
that Ni(II) often interferes with the role of Ca(II) in excitable
tissue experiments [67], including, among others, blocking of calcium
channels [71], inhibition of pacemaker activity [72], increasing
resting tension in uterine contraction [73], and increasing coronary
artery resistance in isolated rat hearts and anesthetized dogs in
vivo [74]. Replacement of Mg(II) and Mn(II) has also been documented
[67,68], e.g., retention of enzyme activity occurs in calcineurin on
exchange with either ion [75,76], and inhibition on substituting for
Mg(II) in DNA synthesis [77].

Nonisomorphous replacement of endogenous metal ions is a well-
recognized mechanism of metal toxicity [78] and has been postulated
to have a role in metal carcinogenesis [68]. Differential stereo-
chemical preferences, mismatches in ion size, donor atom predilection
and complex stability, as well as noncompatible kinetic factors are
employed to rationalize the altered biochemical and physiologic
responses. In carboxypeptidase A, the Zn(II), Co(II), and Ni(II)
complexes according to x-ray diffraction are all five-coordinate [79]
and enzymatic activity is known for all three ions [69]. The inhibition observed for the Ni(II) forms of carbonic anhydrase and LAD is
explained in terms of distortion of the optimum geometry and an
inability to undergo a required rapid ligand exchange [69]. Based
on water exchange rates [80], the formation of simple complexes of
Ni(II) are predicted to be slower by about two to four log units compared to Ca(II), Mn(II), Fe(II), Cu(II), and Zn(II); and one log unit
for Mg(II). This lower kinetic lability may well have significant
biological consequences. The simplest explanation of the Ni(II)
ion's antagonistic role in Ca(II) systems is inappropriate size [72]
and enhanced binding [66,71].

Cu(II) and Ni(II) exhibit a unique ability to form very stable
complexes with deprotonated amide linkages. Their interactions with
peptides involving these bonds are discussed in the next section and
comments here are limited to a description of the primary Cu(II)/
Ni(II) binding site of human serum albumin (HSA). It consists of a
square-planar chelate ring formed by the N-terminus α-amino nitrogen,

the first two peptide nitrogens (deprotonated), and the 3-nitrogen
of the imidazole ring of residue 3. NMR evidence suggests that the
side-chain carboxylate group of the N-terminus aspartic acid interacts axially with the metal center [51,81]. This agrees with spec-
trometric data (UV/VIS and CD) which suggest the existence of an
octahedral form in rapid equilibrium with the square-planar complex
[49]. At physiologic pH, it is estimated that these two forms,
respectively, account for 70 and 30% of the total Ni(II)-albumin
concentration [49]. The unique specificity of this primary albumin-
binding site is highlighted by the relative affinities of other
divalent ions: Ni(II) and Cu(II) bind most tightly, Co(II) does so
considerably less effectively, while Zn(II) and Mn(II) have an alto-
gether different mode of attachment to this protein [82,83].

True to its borderline character, Ni(II) occurs in a range of
donor atom environments. Its agonistic/antagonistic relationship to
Ca(II) attests to its affinity for oxygen donor centers (e.g., carboxylate, peptide carbonyl, and phosphate). Its affinity for mixed
oxygen/nitrogen coordination polyhedra is typified by its substitution for Zn(II) in proteases and carbonic anhydrase. There is considerable evidence that in polynucleotides such as DNA and RNA Ni(II)
attachment involves both phosphate and base binding (e.g., N-7 of
guanine; see Chap. 8). The albumin- and peptide-binding sites illustrate a pure nitrogen surrounding. Interestingly, the nickel-containing F_{430} chromophore of bacterial methylreductases appears to have
Ni(II) in a similar pure nitrogen environment [84,85]. The exact biological function of this nickel tetrapyrrole is still unclear. Like
Zn(II), Ni(II) exhibits considerable partiality for the imidazole ring
nitrogen(s) of histidine. Histidine residues have also been suggested
as contributing to a distorted octahedral environment in Ni(II)-urease
[86; Chap. 6]. Again, by analogy to Zn(II), Ni(II) is known to include
sulfur donor atoms in its coordination polyhedron in LAD [70] and in
nickel-requiring bacterial hydrogenases [87; see Chap. 7]. Interestingly, recent computer simulation models for low-molecular-weight
complexes in human plasma reflect this affinity for sulfur. The
predominating plasma complexes are (% in parentheses): $Ni(His)_2^0$

(50.6%), Ni(Cys)(His)$^-$ (23.9%), Ni(Cys)$_2^{2-}$ (11.3%), and Ni(His)$^+$
(4.4%) [88]. The illustrated catholic affinity of Ni(II) for biological metal donor groups may be considered an important determinant
in nickel toxicology.

5.2. Observed Oxidation States and Redox Properties

The occurrence of nickel as an essential cofactor in bacterial hydrogenases (see Chap. 7) highlights the biological importance of oxidation
states other than 2+. Model coordination compounds have illustrated
that nickel can exist in oxidation states ranging from Ni(0) to Ni(IV)
[80,89,90]. It appears that anionic centers provided by deprotonated
peptides (see below) or sulfur in hydrogenases (e.g., S$^-$, R-S$^-$?) sta-
bilize the trivalent state in biomolecules [91,92], while monovalent
nickel appears to be sustained by the presence of reducing ligands
(H$^-$, RS$^-$?) [87,89]. Neither the spectroscopic and electrochemical
characterization nor the postulated function of nickel in dehydrogenases is discussed here. Details are provided in Chap. 7. To illustrate the biological importance of the redox properties of nickel, some
recent work by the authors involving novel reactions of nickel(II)-
oligopeptide complexes with dioxygen species are now summarized.

A decade ago, Margerum and his colleagues characterized Ni(III)-
peptide complexes by electron paramagnetic resonance (EPR) and standard
electrode potential (E°) measurements [93,94]. E° values of 0.8–1.0 V
vs. SHE were observed for the Ni(III)/Ni(II) redox couple in complexes
with tripeptides, tetrapeptides, and peptide amides. The E° values
decreased with the number of deprotonated peptide groups. More recently [95], Ni(II) complexes of macrocyclic polyamines have been
assigned E° (Ni(III)/Ni(II)) values as low as 0.5 V vs. SHE. These
potentials are within the range to elicit biological consequences.
In 1977, Bossu and Margerum [93] predicted: "The accessibility and
relative stability in aqueous solution of nickel(III)-peptide complexes indicate that the trivalent state of nickel should be considered an attainable oxidation state for biological redox reactions and

other nickel-catalyzed reactions". The bacterial hydrogenase developments certainly fulfil this anticipation. The reports by Kimura
et al. [96,97] that Ni(II)-macrocyclic polyamine complexes potentiate
molecular oxygen and catalyze the hydroxylation of benzene in vitro
open up the possibility of unexplored toxicologic consequences.

Because glycylglycyl-L-histidine (GGH) provides a peptide
sequence similar in binding capability to the Ni(II)-binding sequence
of HSA, we selected this tripeptide for detailed solution studies
involving molecular oxygen, superoxide anion (O_2^-), and hydrogen peroxide [98]. The data in Fig. 3 illustrate that Ni(II)GGH at pH 7.4
is effective in decreasing the O_2^- flux generated by the hypoxanthine/
xanthine oxidase enzyme system. A concomitant increase in hydrogen
peroxide levels was identified, suggesting that Ni(II)GGH catalyzes
a superoxide dismutase (SOD) type of activity. Interestingly, the
Ni(II)HSA complex also showed a weak O_2^--scavenging effect. Ni(GGH)
was also capable of disproportionating H_2O_2 in phosphate buffer at
pH 7.4. Identified products included O_2, O_2^-, and an intermediate
capable of mediating hydroxylation of p-nitrophenol. Under these
conditions, Ni(II)HSA was only able to generate O_2^- from H_2O_2. In the
presence of H_2O_2 and in the same buffer, Ni(II)GGH degraded a two- to
threefold excess of uric acid to allantoin (detected as glyoxylate).
Uric acid has been proposed to function as an in vivo scavenger of
radicals [99]. Under the physiologic conditions of these experiments,
cyclic voltammetry confirmed the reversible oxidation of Ni(II)GGH
and EPR measurements indicated the presence of Ni(III) when Ni(II)GGH
was oxidized by controlled-potential electrolysis. Consequently,
redox cycling involving the Ni(III)/Ni(II) couple may be assumed to
govern the observed catalytic effects. The toxicologic implications
of these observations are discussed in Chap. 10.

6. ROUTES OF ELIMINATION

It is obvious from the discussion in Secs. 2.1 and 2.2 that the
manner of nickel excretion depends on the route of exposure and the

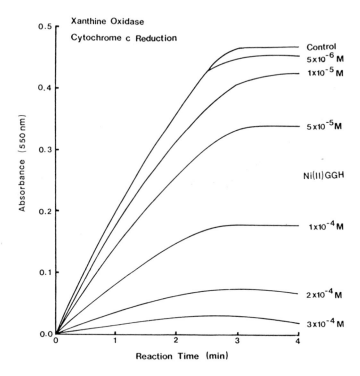

FIG. 3. Reduction in the flux of superoxide anions generated by the
enzyme system hypoxanthine/xanthine oxidase and detected by cytochrome
c reduction in the presence of increasing concentrations of Ni(II)-
glycylglycyl-L-histidine [Ni(II)GGH]. The reaction medium contained
33 µM hypoxanthine, 0.1 µM xanthine oxidase, 58 µM ferricytochrome c,
and 43 µg/ml catalase (to remove H_2O_2) in 0.1 M KH_2PO_4 (pH = 7.4);
Ni(II)GGH was added at the concentrations indicated. The control
sample contained 500 µM GGH, and 25 µg/ml superoxide dismutase was
present in all reference samples to assure that the measured reduction of cytochrome c was due to the superoxide anion.

physicochemical properties of the nickel-containing material.

Removal from the respiratory tract occurs by absorption [as dissolved

Ni(II)] or by mucociliary clearance followed by ingestion (of particu-

lates). Mobilization of particulates by phagocytes with subsequent

transfer to the lymphatic system can also occur (see note b of Table

1). As indicated earlier, the major excretion pathway for absorbed

nickel in rodents is renal; in humans there is indirect evidence of a significant biliary contribution. The enhanced levels of nickel found in feces of nickel workers exposed to nickel hydroxide in an alkaline battery factory may be interpreted to indicate mucociliary clearance from the respiratory tract [100]. Nickel levels in body fluids correlate with occupational exposure to both water-soluble and partially soluble nickel compounds, and thus whole-blood, plasma, serum, and urinary nickel constitute useful exposure indices [19,28]. In addition, statistically significant correlations between nickel in serum and in urine have been observed [19,36,101].

There is good evidence that in humans Ni(II) is reabsorbed by the renal tubular system (likely in the proximal tubule). Unlike creatinine, but like urea and other solutes, nickel excretion increases with urine flow rate [101,102]. We have recently shown that the average volume of human plasma from which ultrafilterable nickel is removed is 42 ml/min, which indicates that about 65% of the nickel in the glomerular filtrate is reabsorbed [101].

Nickel deposition in hair is not well understood, since it is complicated by external adsorption from water and air [19]. Background levels of 0.22 ± 0.08 µg/g dry weight have been reported [28]. Significant concentrations of nickel also appear in saliva (1.9 ± 1.0 µg/liter), sweat (51 ± 38 µg/liter), and mother's milk (17 ± 2 µg/kg) [28]. However, the latter two excretory pathways will only be significant during lactation or profuse sweating.

7. NICKEL AS AN ESSENTIAL ELEMENT

7.1. Criteria of Essentiality

An element is commonly considered essential [103] "when a deficient intake consistently results in an impairment of a function from optimal to suboptimal and when supplementation with physiological levels of this element, but not of others, prevents or cures this impairment" [104,105]. "Essentiality is generally acknowledged when it has been demonstrated by more than one independent investigator and in more

than one animal species" [103]. As for many other ultratrace elements, deficiency of nickel in humans has not been demonstrated. To
facilitate an assessment of the available information for a particular
element, Mertz [106] formulated five criteria that are implied by the
given definition of essentiality. These are summarized in brief below.

1. The element should be present in living matter before acci-
    dental outside contamination. For example, presence in the
    fetus is compatible with but not indicative of essentiality.

2. There should be some homeostatic regulation of the micro-
    nutrient.

3. Pools of the element in the body may be influenced by
    stress, hormones, or other substances.

4. The constant occurrence in stoichiometric amounts of the
    element in a purified protein or enzyme is considered
    positive proof.

5. Induced deficiency should produce reproducible symptoms in
    the organism which can be alleviated by administering trace
    amounts of the micronutrient.

To assist in extrapolating from animals and other living organisms to
humans, the data supporting these five criteria are examined. Only
selected references are provided because this topic has been extensively reviewed [3,45,104,105,107].

7.2. Presence of Nickel in Human Tissues

As illustrated in Sec. 3 and Table 1, nickel occurs in low concentrations in all human tissues and fluids that have been assessed by sensitive and reliable analytical techniques (strict measures to avoid
and control extraneous or adventitious contamination were included
in the protocols of specimen collection and handling) [28,108]. The
presence of this metal has also been reported in human fetal tissues
[45], although the magnitude of the concentrations reported implies
that the available data are not reliable. For example, Casey et al.
[109] reported a range of 120-1,000 μg/kg dry matter for fetal lung

tissue, compared to 71-371 µg/kg in adult lung tissue by Sunderman
[17]. As discussed in Chap. 10, there is considerable experimental
evidence for transplacental transfer of nickel to the fetus in
rodents.

7.3. Homeostatic Regulation

The narrow ranges of nickel concentrations observed in human body
fluids, including bile, of healthy individuals (Sec. 3) imply some
form of regulation. The significant renal reabsorption in humans is
a further manifestation of such control (Sec. 6). A similar deduc-
tion follows from the relatively rapid excretion and return to normal
nickel levels when individuals are exposed to water-soluble nickel
salts by ingestion or inhalation ($t_{\frac{1}{2}}$ = 1-1½ days). The consistent
association of Ni(II) in serum with albumin and α_2-macroglobulin and
the identification of a reproducible ultrafilterable fraction (see
Sec. 4) may be viewed as another indication of homeostasis. By
analogy to humans, Schroeder et al. in 1974 [110] were unable to
demonstrate additional accumulation in rat tissues of nickel despite
lifelong supplementation of 5 ppm Ni(II) in drinking water; neither
were there extensive alterations in the tissue levels of Cu, Zn, Cr,
and Mn. However, caution is warranted about these results, since
the seriousness of contamination problems in ultratrace analysis was
not fully realized in the early 1970s.

7.4. Mobilization of Tissue Nickel Pools

A number of diseased states and other physiologic stresses are
reported to alter equilibrium and tissue distribution of nickel in
humans. Elevated concentrations of nickel in serum have been documented in cases of myocardial infarction, unstable angina pectoris,
stroke, burns, hepatic cirrhosis, and uremia [111,112]. For example,
hypernickelemia ($\geq$ 1.2 µg/liter) was observed 72 hr after admission

in 76% of patients with acute myocardial infarction and in 48% of patients with unstable angina pectoris without infarction [112]. On the average, the increase in serum nickel concentration reached a maximum in about 1 day after the myocardial infarction but remained moderately elevated in angina pectoris patients for at least 3 days. Mechanisms and sources of release of nickel into the serum of patients under pathologic distress are still conjectural.

7.5. Occurrence of Nickel in Enzymes

The criterion requiring the presence of stoichiometric quantities of nickel on protein purification as proof of essentiality is fulfilled only for enzymes isolated from plants and bacteria (see Chaps. 6 and 7). In purified jack bean urease, 2.00 ± 0.12 nickel ions occur per subunit [86], while a 1:1 stoichiometry is observed for nickel in the methylreductase factor F_{430} of methane bacteria [e.g., 84]. In bacterial hydrogenases, close to a 1:1 nickel-to-enzyme mole ratio is observed [e.g., 87,91]. There is also evidence that nickel is an integral component of some carbon monoxide dehydrogenases of aerobic bacteria [113,114]. The requirement of a nickel ion or a nickel-containing factor appears to be a prerequisite for enzyme activity, as expected for an essential cofactor.

7.6. Induced Deficiencies

Nickel deficiency research prior to 1975 was plagued by analytical limitations and lack of rigid control of nickel contamination and other environmental factors [104]. However, signs of nickel deprivation have now been unequivocally established in six animal models, including chicks, cows, goats, minipigs, rats, and sheep [3,45,104, 105,107]. The most prominent and consistent symptoms are reduced hematopoiesis, depressed growth, and metabolic alterations.

"The interaction between nickel and iron can be synergistic or antagonistic depending on the form of iron and level of dietary

iron and nickel" [115]. Nickel-deficient rats exhibit poor iron
absorption and develop anemia [107]. Apparently, Ni(II) enhances
the intestinal absorption of Fe(III) and not of Fe(II) [104]. In
studies with rats and pigs receiving diets marginal or inadequate in
iron, nickel supplements of 5-25 mg/kg generally increased or had no
effect on hepatic iron concentrations; in calves and young lambs a
5 mg/kg supplementation reduced lung and liver iron [115]. Nickel
deficiency in minipigs, goats, and rats has produced a concomitant
deprivation of zinc in organs and was manifested as skin alterations
[107].

Decrements in weight gain have occurred in nickel-deficient
goats, minipigs, and rats [107]. For a number of animals minor
effects on reproductive performance have been noted as well as histo-
logic changes such as skin pigmentation and scaling [105,107]. Ultra-
structural changes in liver cells, reduced levels of tissue and serum
enzymes, and decreased liver concentrations of fat, glucose, and glycogen have primarily been substantiated in nickel-deprived rats [107].

7.7. Overall Assessment

Of the five criteria of essentiality listed in Sec. 7.1, the first
three are satisfied by human evidence: (1) nickel is present in all
human tissues tested; (2) there is evidence for its homeostatic control; and (3) pathologic disturbances of serum nickel levels have
been documented. As yet, no nickel-requiring enzymes or proteins
are known in vertebrates, although biological roles of nickel enzymes
and cofactors have been identified in plants and bacteria. The final
criterion of induced deficiency symptoms has been satisfied in a
number of animal species and, as required, have been observed by
different researchers. Although the role of nickel in human physiology has not been confirmed directly, collectively the evidence
reviewed strongly suggests that nickel is required by humans. Based
on nickel requirements by monogastric animals, Nielsen [104] proposed
that a dietary intake of nickel by humans of 35 µg/day ought to

suffice. This appears to be an underestimate since the considerations in Sec. 2.2 of net excretion and the percentage of gastrointestinal absorption predict that the average dietary nickel content of 160 μg/day approximates the optimal uptake required for
mass balance.

8.   CONCLUDING REMARKS AND SUMMARY

It is concluded that pulmonary absorption provides a major route of
nickel uptake in humans. Deposition, clearance, and absorption in
the respiratory tract depends on physicochemical properties of nickel
compounds. Gastrointestinal absorption is estimated to be in the
1-4% range, and the average daily dietary intake of nickel is 160 μg.
Although in rodents the renal route is the major excretory pathway,
in humans urinary and biliary excretion appear to be of comparable
magnitude. There is little long-term compartmentalization nor accumulation of water-soluble nickel(II) compounds in animals and humans.
More inert and less reactive nickel-containing substances are known
to accumulate with time, such as in human lungs. Nickel concentrations in whole blood, plasma, serum, and urine provide good indices
to exposure. Reference nickel levels for these body fluids and for
tissues have been drastically lowered in the past decade because of
improved analytical techniques and contamination control during collection, storage, handling, and analysis of specimens. In human
plasma much of the nickel is bound to high-molecular-mass protein
components, while 24% is associated with the ultrafiltrate and thus
is available for urinary excretion. Partial reabsorption of Ni(II)
occurs in the tubular assembly of the kidney. Cellular uptake of
the nickel(II) ion appears to be a passive process and is explained
by an "Equilibrium" model. Uptake by phagocytosis has also been documented in cell culture and animal studies. In biomolecules, Ni(II)
is found in pure oxygen, nitrogen, sulfur, and mixed donor atom
environments. This is consistent with its borderline reactivity
recognized in coordination chemistry. Agonistic and antagonistic

replacements of Ca(II), Mg(II), Mn(II), and Zn(II) in enzymes and physiologic processes have been documented. Oxidation states other than 2+ are now recognized in biological systems, especially 3+. This development is important from both the biochemical and toxico-logic perspectives. Finally, it is concluded that nickel is essen-tial for humans. Human and/or animal data satisfy four of the five criteria of essentiality: presence in biological tissues; homeo-static regulation; mobilization of tissue pools in response to dis-ease and physiologic stress; and well-defined nickel deficiency symptoms. Although no specific biological or physiologic role has been characterized in animals, the constant occurrence of nickel in stoichiometric amounts in purified enzymes and its concomitant requirement for enzyme activity is considered positive proof of essentiality in plants and bacteria.

ABBREVIATIONS

CHO - Chinese hamster ovary cells
DDC - diethyl dithiocarbamate
EPR - electron paramagnetic resonance
GGH - glycylglycyl-L-histidine
GI   - gastrointestinal tract
HSA - human serum albumin
LAD - liver alcohol dehydrogenase
MMAD - mass median aerodynamic diameter
SHE - standard hydrogen electrode
SOD - superoxide dismutase

ACKNOWLEDGMENTS

Financial support from the Natural Sciences and Engineering Research Council of Canada and from the Occupational Health and Safety Division of the Ontario Ministry of Labour is gratefully acknowledged.

REFERENCES

1. D. C. F. Muir, and D. K. Verma, in "Aerosols in Medicine: Prin-
    ciples, Diagnosis and Therapy" (F. Morén, M. T. Newhouse, and
    M. B. Dolovich, eds.), Elsevier, Amsterdam, 1985, pp. 313-332.

2. I. T. T. Higgins (ed.-in-chief), "Airborne Particles", Univer-
    sity Park Press, Baltimore, 1979, pp. 107-145.

3. EPA, "Health Assessment Document for Nickel and Nickel Com-
    pounds", EPA/600/8-83/ 012 FF, United States Environmental
    Protection Agency, Research Triangle Park, NC, 1986.

4. International Nickel Company, "Nickel and its Inorganic Com-
    pounds (Including Nickel Carbonyl)", a supplementary submission
    to the National Institute for Occupational Safety and Health by
    International Nickel (U.S.) Inc., October 1976.

5. J. D. Crapo, B. E. Barry, P. Gehr, M. Bachofen, and E. R. Weibel,
    Am. Rev. Resp. Dis., *125*, 332 (1982).

6. J. A. Graham, F. J. Miller, M. J. Daniels, E. A. Payne, and D. E.
    Gardner, *Environ. Res.*, *16*, 77 (1978).

7. D. B. Menzel, R. L. Wolpert, C. R. Shoaf, and D. L. Deal, in
    "Aerosols" (S. D. Lee, T. Schneider, L. D. Grant, and P. J.
    Verkerk, eds.), Lewis, Chelsea, MI, 1986, pp. 637-648.

8. D. B. Menzel, D. L. Deal, M. I. Tayyeb, R. L. Wolpert, J. R.
    Boger, III, C. R. Shoaf, J. Sandy, K. Wilkinson, and R. Franco-
    vitch, *Toxicol. Lett.*, *38*, 33 (1987).

9. J. C. English, R. D. R. Parker, R. P. Sharma, and S. G. Oberg,
    Am. Ind. Hyg. Assoc. J., *42*, 486 (1981).

10. S. M. M. Carvalho, and P. L. Ziemer, *Arch. Environ. Contam.
    Toxicol.*, *11*, 245 (1982).

11. A. Furst and H. Al-Mahrouq, *Proc. West. Pharmacol. Soc.*, *24*,
    119 (1981).

12. R. Valentine and G. L. Fisher, *Environ. Res.*, *34*, 328 (1984).

13. A. Marzouk and F. W. Sunderman, Jr., *Toxicol. Lett.*, *27*, 65
    (1985).

14. A. P. Wehner and D. K. Craig, *Am. Ind. Hyg. Assoc. J.*, *33*, 146
    (1972).

15. Y. Kodama, S. Ishimatsu, K. Matsuno, I. Tanaka, and K. Tsuchiya,
    Biol. Trace Element Res., *7*, 1 (1985).

16. I. Tanaka, S. Ishimatsu, K. Matsuno, Y. Kodama, and K. Tsuchiya,
    Biol. Trace Element Res., *8*, 203 (1985).

17. W. N. Rezuke, J. A. Knight, and F. W. Sunderman, Jr., *Am. J.
    Ind. Med.*, *11*, 419 (1987).

18. W. Torjussen and I. Andersen, *Ann. Clin. Lab. Sci.*, *9*, 289
    (1979).

19. E. Nieboer, A. Yassi, A. A. Jusys, and D. C. F. Muir, "The Tech-
     nical Feasibility and Usefulness of Biological Monitoring in the
     Nickel Producing Industry", special document, McMaster Univer-
     sity, Hamilton, Ontario, Canada, 1984, 285 pp. (Available from
     the Nickel Producers Environmental Research Association, Tribune
     Tower, 435 North Michigan Avenue, Chicago, IL 60611.)

20. M. Boysen, L. A. Solberg, W. Torjussen, S. Poppe, and A. C.
     Høgetveit, *Acta Otolaryngol.*, *97*, 105 (1984).

21. A. Tossavainen, M. Nurminen, P. Mutanen, and S. Tola, *Br. J.
     Ind. Med.*, *37*, 285 (1980).

22. D. R. Myron, T. J. Zimmerman, T. R. Shuler, L. M. Klevay, D. E.
     Lee, and F. H. Nielsen, *Am. J. Clin. Nutr.*, *31*, 527 (1978).

23. G. D. Nielsen and M. Flyvholm, in "Nickel in the Human Environ-
     ment", IARC Sci. Publ. 53 (F. W. Sunderman, Jr., ed.-in-chief),
     Oxford Univ. Press, Oxford, U.K., 1984, pp. 333-338.

24. M. A. Flyvholm, G. D. Nielsen, and A. Andersen, *Z. Lebensm.
     Unters. Forsch.*, *179*, 427 (1984).

25. H. A. Schroeder, J. J. Balassa, and I. H. Tipton, *J. Chron.
     Dis.*, *15*, 51 (1962).

26. G. K. Murthy, U. S. Rhea, and J. T. Peeler, *Environ. Sci. Tech-
     nol.*, *7*, 1042 (1973).

27. D. C. Kirkpatrick and D. E. Coffin, *Can. J. Publ. Health*, *68*,
     162 (1977).

28. F. W. Sunderman, Jr., A. Aitio, L. G. Morgan, and T. Norseth,
     Toxicol. Indust. Health, *2*, 17 (1986).

29. O. B. Christensen and V. Lagesson, *Ann. Clin. Lab. Sci.*, *11*,
     119 (1981).

30. N. W. Solomons, F. Viteri, T. R. Schuler, and F. H. Nielsen,
     J. Nutr., *112*, 39 (1982).

31. E. Cronin, A. D. DiMichiel, and S. S. Brown, in "Nickel Toxicol-
     ogy" (S. S. Brown and F. W. Sunderman, Jr., eds.), Academic,
     London, 1980, pp. 149-152.

32. W. Ho and A. Furst, *Proc. West. Pharmacol. Soc.*, *16*, 245 (1973).

33. E. C. Foulkes and D. M. McMullen, *Toxicology*, *38*, 35 (1986).

34. C. Onkelinx, J. Becker, and F. W. Sunderman, Jr., *Res. Commun.
     Chem. Path. Pharmacol.*, *6*, 663 (1973).

35. F. W. Sunderman, Jr., and C. E. Selin, *Toxicol. Appl. Pharmacol.*,
     12, 207 (1968).

36. F. W. Sunderman, Jr., *Ann. Clin. Lab. Sci.*, *7*, 377 (1977).

37. L. G. Morgan, *J. Soc. Occup. Med.*, *29*, 33 (1979).

38. A. Oskarsson and H. Tjälve, *Br. J. Ind. Med.*, *36*, 326 (1979).

39. H. Tjälve, S. Jasim, and A. Oskarsson, in "Nickel in the Human Environment", IARC Sci. Publ. 53 (F. W. Sunderman, Jr., ed.-in-chief), Oxford Univ. Press, Oxford, UK, 1984, pp. 311-320.

40. K. S. Kasprzak and F. W. Sunderman, Jr., *Toxicol. Appl. Pharmacol.*, *15*, 295 (1969).

41. F. W. Sunderman, Jr. (ed.-in-chief), "Nickel," National Academy of Sciences, Washington, D.C., 1975, 277 pages.

42. F. W. Sunderman, Jr., M. C. Crisostomo, M. C. Reid, S. M. Hopfer, and S. Nomoto, *Ann. Clin. Lab. Sci.*, *14*, 232 (1984).

43. F. W. Sunderman, Jr., S. M. Hopfer, M. C. Crisostomo, and M. Stoeppler, *Ann. Clin. Lab. Sci.*, *16*, 219 (1986).

44. F. W. Sunderman, Jr., A. Marzouk, M. C. Crisostomo, and D. R. Weatherby, *Ann. Clin. Lab. Sci.*, *15*, 299 (1985).

45. A. Cecutti, and E. Nieboer, in "Effects of Nickel in the Canadian Environment", NRCC Document 18568, National Research Council of Canada, Ottawa, Ontario, 1981, pp. 193-216.

46. S. Nomoto in "Nickel Toxicology" (S. S. Brown and F. W. Sunderman, Jr., eds.), Academic, London, UK, 1980, pp. 89-90.

47. B. J. Scott and A. R. Bradwell, *Clin. Chem.*, *29*, 629 (1983).

48. M. Lucassen and B. Sarkar, *J. Toxicol. Environ. Health*, *5*, 897 (1979).

49. S. H. Laurie and D. E. Pratt, *J. Inorg. Biochem.*, *28*, 431 (1986).

50. J. D. Glennon and B. Sarkar, *Biochem. J.*, *203*, 15 (1982).

51. B. Sarkar, *Chemica Scripta*, *21*, 101 (1983).

52. S. Nomoto, M. D. McNeeley, and F. W. Sunderman, Jr., *Biochemistry*, *10*, 1647 (1971).

53. F. W. Sunderman, Jr., M. I. Decsy, and M. D. McNeeley, *Ann. N.Y. Acad. Sci.*, *199*, 300 (1972).

54. N. Asato, M. van Soestbergen, and F. W. Sunderman, Jr., *Clin. Chem.*, *21*, 521 (1975).

55. R. J. Williams, *Phil. Trans. Royal Soc. London, Ser. B.*, *294*, 57 (1981).

56. C. R. Menon and E. Nieboer, *J. Inorg. Biochem.*, *28*, 217 (1986).

57. E. Nieboer, A. R. Stafford, S. L. Evans, and J. Dolovich, in "Nickel in the Human Environment", IARC Sci. Publ. 53 (F. W. Sunderman, Jr., ed.-in-chief), Oxford Univ. Press, Oxford, UK, 1984, pp. 321-331.

58. M. P. Abbracchio, R. M. Evans, J. D. Heck, O. Cantoni, and M. Costa, *Biol. Trace Element Res.*, *4*, 289 (1982).

59. M. P. Abbracchio, J. D. Heck, and M. Costa, *Carcinogenesis*, *3*, 175 (1982).

60. J. D. Heck and M. Costa, *Cancer Res.*, *43*, 5652 (1983).

61. K. Kuehn, C. B. Fraser, and F. W. Sunderman, Jr., *Carcinogenesis*, *3*, 321 (1982).

62. J. P. Berry, M. F. Poupon, J. C. Judde, and P. Galle, *Ann. Clin. Lab. Sci.*, *15*, 109 (1985).

63. J. D. Heck and M. Costa, *Toxicol. Lett.*, *12*, 243 (1982).

64. M. Costa, M. P. Abbracchio, and J. Simmons-Hansen, *Toxicol. Appl. Pharmacol.*, *60*, 313 (1981).

65. P. Sen and M. Costa, *Toxicol. Appl. Pharmacol.*, *84*, 278 (1986).

66. E. Nieboer and D. H. S. Richardson, *Environ. Pollut.* (Ser. B), *1*, 3 (1980).

67. E. Nieboer, R. I. Maxwell, and A. R. Stafford, in "Nickel in the Human Environment", IARC Sci. Publ. 53 (F. W. Sunderman, Jr., ed.-in-chief), Oxford Univ. Press, Oxford, UK, 1984, pp. 439-458.

68. K. Wetterhahn-Jennette, *Environ. Health Persp.*, *40*, 233 (1981).

69. R. J. P. Williams, *J. Mol. Catal.*, *30*, 1 (1985).

70. H. Dietrich, W. Maret, H. Kozlowski, and M. Zeppezauer, *J. Inorg. Biochem.*, *14*, 297 (1981).

71. S. Hagiwara and L. Byerly, *Fed. Proc.*, *40*, 2220 (1981).

72. A. L. F. Gorman, A. Hermann, and M. V. Thomas, *Fed. Proc.*, *40*, 2233 (1981).

73. G. Rubányi and I. Balogh, *Am. J. Obstet. Gynecol.*, *142*, 1016 (1982).

74. G. Rubányi, L. Ligeti, and A. Koller, *J. Mol. Cell. Cardiol.*, *13*, 1023 (1981).

75. C. J. Pallen and J. H. Wang, *J. Biol. Chem.*, *261*, 16115 (1986).

76. M. M. King, K. K. Lynn, and C. Y. Huang, in "Progress in Nickel Toxicology" (S. S. Brown and F. W. Sunderman, Jr., eds.), Blackwell, Oxford, UK, 1985, pp. 117-120.

77. M. A. Sirover, D. K. Dube, and L. A. Loeb, *J. Biol. Chem.*, *254*, 107 (1979).

78. E. Nieboer and W. E. Sanford, in "Reviews in Biochemical Toxicology", Vol. 7 (E. Hodgson, J. R. Bend, and R. M. Philpot, eds.), Elsevier, New York, 1985, pp. 205-245.

79. K. D. Hardman and W. N. Lipscomb, *J. Am. Chem. Soc.*, *106*, 463 (1984).

80. F. A. Cotton and G. Wilkinson, "Advanced Inorganic Chemistry: A Comprehensive Text", 4th ed., John Wiley, New York, 1980, pp. 785, 1051, and 1188.

81. J. P. Laussac and B. Sarkar, *Biochemistry*, *23*, 2832 (1984).

82. J. Dolovich, S. L. Evans, and E. Nieboer, *Br. J. Ind. Med.*, *41*, 51 (1984).

83. E. Nieboer, S. L. Evans, and J. Dolovich, *Br. J. Ind. Med.*, *41*, 56 (1984).

84. W. L. Ellefson, W. B. Whitman, and R. S. Wolfe, *Proc. Natl. Acad. Sci. USA*, *79*, 3707 (1982).

85. R. S. Wolfe, *Trends Biol. Sci.*, *10*, 396 (1985).

86. R. L. Blakeley and B. Zerner, *J. Mol. Cat.*, *23*, 263 (1984).

87. S. P. J. Albracht, A. Kröger, J. W. van der Zwaan, G. Unden, R. Böcher, H. Mell, and R. D. Fontijn, *Biochim. Biophys. Acta.*, *874*, 116 (1986).

88. A. Cole, C. Furnival, Z. X. Huang, D. C. Jones, P. M. May, G. L. Smith, J. Whittaker, and D. R. Williams, *Inorg. Chim. Acta*, *108*, 165 (1985).

89. K. Nag and A. Chakravorty, *Coord. Chem. Rev.*, *33*, 87 (1980).

90. R. I. Haines and A. McAuley, *Coord. Chem. Rev.*, *39*, 77 (1981).

91. M. Teixeira, I. Moura, A. V. Xavier, B. H. Huynh, D. V. DerVartanian, H. D. Peck, Jr., J. LeGall, and J. J. G. Moura, *J. Biol. Chem.*, *260*, 8942 (1985).

92. Y. Sugiura, J. Kuwahara, and T. Suzuki, *Biochem. Biophys. Res. Commun.*, *115*, 878 (1983).

93. F. P. Bossu and D. W. Margerum, *Inorg. Chem.*, *16*, 1210 (1977).

94. A. G. Lappin, C. K. Murray, and D. W. Margerum, *Inorg. Chem.*, *17*, 1630 (1978).

95. E. Kimura, A. Sakonaka, and M. Nakamoto, *Biochim. Biophys. Acta*, *678*, 172 (1981).

96. E. Kimura, A. Sakonaka, and R. Machida, *J. Am. Chem. Soc.*, *104*, 4255 (1982).

97. E. Kimura and R. Machida, *J. Chem. Soc. Chem. Commun.*, 499 (1984).

98. R. T. Tom, "Novel Reactions of Nickel(II)-Oligopeptide Complexes with Dioxygen Species", M.Sc. thesis, McMaster University, May 1987.

99. B. N. Ames, R. Cathcart, E. Schwiers, and P. Hochstein, *Proc. Natl. Acad. Sci. USA*, *78*, 6858 (1981).

100. E. Hassler, B. Lind, B. Nilsson, and M. Piscator, *Ann. Clin. Lab. Sci.*, *13*, 217 (1983).

101. W. E. Sanford, "The Renal Clearance of Nickel in Man: Implications for Biological Monitoring", Ph.D. thesis, University of Surrey, UK, November 1987.

102. D. P. Mertz, R. Koschnick, and G. Wilk, *Z. Klin. Chem. Klin. Biochem.*, *8*, 387 (1970).

103. W. Mertz, *Science, 213,* 1332 (1981).

104. F. H. Nielsen, *Ann. Rev. Nutr., 4,* 21 (1984).

105. E. J. Underwood, "Trace Elements in Human and Animal Nutrition", 4th ed., Academic, New York, 1977, pp. 1-12, 159-169.

106. W. Mertz, *Fed. Proc., 29,* 1482 (1970).

107. M. Anke, B. Groppel, H. Kronemann, and M. Grün, "Nickel in the Human Environment", IARC Sci. Publ. 53 (F. W. Sunderman, Jr., ed.-in-chief), Oxford Univ. Press, Oxford, UK, 1984, pp. 339-365.

108. E. Nieboer and A. A. Jusys, in "Chemical Toxicology and Clinical Chemistry of Metals" (S. S. Brown, and J. Savory, eds.), Academic, London, UK, 1983, pp. 3-16.

109. C. E. Casey and M. F. Robinson, *Br. J. Nutr., 39,* 639 (1978).

110. H. A. Schroeder, M. Mitchener, and A. P. Nason, *J. Nutr., 104,* 239 (1974).

111. M. D. McNeely, F. W. Sunderman, Jr., M. W. Nechay, and H. Levine, *Clin. Chem., 17,* 1123 (1971).

112. C. N. Leach, Jr., J. V. Linden, S. M. Hopfer, M. C. Crisostomo, and F. W. Sunderman, Jr., *Clin. Chem., 31,* 556 (1985).

113. R. K. Thauer, G. Diekert, and P. Schönheit, *Trends Biol. Sci., 5,* 304 (1980).

114. H. L. Drake, *J. Bacteriol., 149,* 561 (1982).

115. J. W. Spears, R. W. Harvey, and L. J. Samsell, *J. Nutr., 116,* 1873 (1986).

5

Nickel Ion Binding to Amino Acids and Peptides

R. Bruce Martin
Chemistry Department
McCormick Road
University of Virginia
Charlottesville, Virginia 22901

1. THE SETTING

Nickel is one of several metal ions in the first transition row that occur commonly in the 2+ state. Others include Mn^{2+}, Co^{2+}, Ni^{2+}, Cu^{2+}, and Zn^{2+}. Comparison of its properties with those of its neighbors deepens our understanding of Ni^{2+} chemistry. Comparison of Ni^{2+} with its neighboring 2+ metal ions and Mg^{2+} appears in Table 1. Although Ni^{2+} exhibits the lowest six-coordinate effective ionic radius, there is little variation in radius among the six metal ions. Thus the capacity of Ni^{2+} to form much stronger complexes with amino acid-type ligands than Mg^{2+} is not explained by the comparable ionic radii. The effective ionic radius decreases with a reduction in coordination number: for Ni^{2+} from 0.69 Å in sixfold coordination, to 0.63 Å in fivefold coordination, and to 0.55 Å in fourfold coordination [1]. Thus there is a pronounced shortening of the Ni^{2+} to donor atom bond lengths upon going from a hexacoordinate geometry found in most complexes to the planar geometry that occurs in some peptide complexes (Sec. 4).

The first deprotonation from a metal ion-bound water on aqueous Ni^{2+} requires quite basic solutions with $pK_a = 10.2$ and does not occur with any of the metal ions in Table 1 (row 2) until pH > 8. On the other hand, this statistic is misleading because hydroxide insolubility causes precipitates to form in neutral solutions (Table 1, row 3) [2]. $Ni(OH)_2$ begins to precipitate near pH 6 in 0.1 M solutions, near pH 7 in 10^{-3} M solutions, and near pH 8 in 10^{-5} M solutions. With normal low concentrations in physiologic fluids, Ni^{2+} is usually sequestered by other ligands so that hydroxide insolubility is not a concern.

A leading distinction between Ni^{2+} and its neighboring 2+ metal ions is the slow rate of ligand exchange in and out of the coordination sphere. Values for the first order water exchange rate constant in the aqueous ion appear in the fourth row of Table 1 [3]. The rate constant for Ni^{2+} is 10^3 times slower than that for Zn^{2+} and 6×10^4 times slower than that for Cu^{2+}. Though rate constants for other ligands differ numerically from that for water, for a single ligand

TABLE 1

Comparisons of First Transition Row 2+ Metal Ions

Characteristic	Mg^{2+}	Mn^{2+}	Co^{2+}	Ni^{2+}	Cu^{2+}	Zn^{2+}
Six-coordinate effective ionic radius, Å[a]	0.72	0.83	0.74	0.69	0.73	0.74
First hydroxo complex formation, pK_a[b]	11.7	10.9	9.9	10.2	8.5	9.3
pH of hydroxide ppt. with 1 mM M^{2+}	9.9	9.1	7.7	6.9	5.3	7.1
Water exchange rate log (sec^{-1})[c]	5.6	7.4	6.3	4.3	9.1	7.4
Stability constant logs[d]						
NH_3	0.2	1.0	2.1	2.8	4.2	2.4
Imidazole			2.4	3.0	4.2	2.5
Diaminoethane	0.4	2.7	5.5	7.3	10.5	5.7
Glycine	2.0	2.8	4.7	5.8	8.1	5.0
Alanine	1.9	2.5	4.3	5.4	8.1	4.6
Histidine		3.3	6.8	8.7	10.2	6.5
Mercaptoethylamine	2.3		7.7	10.0	rx[e]	9.9
Catechol	5.7	7.7	8.6	8.9	13.0	9.9

[a]Ref. 1.
[b]Ref. 2.
[c]Ref. 3.
[d]Ref. 4.
[e]rx = redox reaction.

the relative ordering of the values for the several metal ions remains the same. Chelation slows ligand exchange compared to unidentate ligands. The mechanism of formation and dissociation of amino acid chelate complexes has been controversial [3]. When Ni^{2+} adopts a planar geometry with peptides (Sec. 4), ligand exchange becomes much slower still.

The last portion of Table 1 lists stability constant logarithms for Ni^{2+} and its neighbors with a variety of ligands [4]. Hexacoordinate Ni^{2+} follows in the Irving-Williams stability sequence Mg^{2+} <

$Mn^{2+} < Fe^{2+} < Co^{2+} < Ni^{2+} < Cu^{2+} > Zn^{2+}$. Ni^{2+} binds ligands significantly more weakly than Cu^{2+} and more strongly than Co^{2+}. The variable position of Zn^{2+} in this series has been described in terms of the stability ruler [1,5]. Most often the Zn^{2+} stabilities compare with those for Co^{2+} and are weaker than those for Ni^{2+}, but for hydroxide ion and catechol Zn^{2+} binding is stronger than that for Ni^{2+}. For sulfhydryl-containing ligands the Ni^{2+} and Zn^{2+} binding strengths are often comparable as demonstrated by β-mercaptoethylamine in Table 1. With sulfhydryl-containing ligands Ni^{2+} undergoes a hexacoordinate (octahedral) to planar transformation. In this process the green-blue hexacoordinate Ni^{2+} complex with two unpaired d electrons becomes yellow and planar with all 8 d electrons spin-paired. Once the energy to spin-pair is overcome, the resulting planar complex is stabilized by the planar ligand field compared to the hexacoordinate one.

2. AMINO ACID COMPLEXES

Volume 9 in this series entitled *Amino Acids and Derivatives as Ambivalent Ligands* groups the subject by ligands rather than by metal ion [6]. The index to that volume contains 50 references to Ni^{2+}. Volume 9 is highly recommended for readers interested in amino acid complexes of Ni^{2+}.

2.1. Stability Constant Correlations

Table 2 lists amino group pK_a values and first (K_1) and second (K_2) stability constant logarithms for Ni^{2+} binding to a variety of α-amino acid ligands, including the 20 that occur in proteins when reliable values are available. Since carboxylate group pK_a values exhibit only small variations and are less reliable, they are not listed.

TABLE 2

Ni^{2+}-α-Amino Acid Stability Constant Logs[a]

Amino acid	pK_a	$\log K_1$	$\log K_2$	$\log K_0$	% Closed
1. Glycine	9.67	5.78	4.80	5.33	
2. L-Alanine	9.81	5.40	4.5	5.38	lsl[b]
3. 2-Aminobutanoate	9.74	5.30	4.4	5.36	lsl
4. 2-Aminopentanoate	9.75	5.27	4.38	5.36	
5. 2-Aminohexanoate	9.78	5.40	4.61	5.37	lsl
6. 2-Amino-4-pentenoate	9.38	5.31	4.58	5.23	17
7. 2-Amino-5-hexenoate	9.54	5.38	4.51	5.29	19
8. 2-Amino-6-heptenoate	9.62	5.32	4.40	5.31	lsl
9. L-Leucine	9.66	5.45	4.26	5.33	24
10. L-Phenylalanine	9.20	5.15	4.44	5.17	lsl
11. L-Tryptophan	9.47	5.48	4.92	5.26	40
12. L-Valine	9.60	5.42	4.30	5.31	22
13. 1-NH$_2$ cyclopentane COO$^-$	10.42	5.60	4.63	5.59	
14. 1-NH$_2$ cyclohexane COO$^-$	10.24	5.50	4.55	5.53	
15. 1-NH$_2$ cycloheptane COO$^-$	10.57	5.33	4.5	5.64	
16. Homoserine (HSer)	9.39	5.46	4.55	5.23	41
17. L-Serine	9.16	5.40	4.5	5.15	44
18. L-Threonine	9.07	5.46	4.55	5.12	54
19. L-Methionine	9.16	5.34	4.56	5.15	35
20. L-S-Methylcysteine (SMC)	8.84	5.26	4.56	5.04	40
21. L-Aspartate	9.73	7.15	5.24	5.35	98
22. L-Glutamate	9.70	5.60	4.16	5.34	45
23. L-Asparagine	8.83	5.68	4.55	5.04	77
24. L-Glutamine	9.12	5.16	4.26	5.14	lsl
25. L-Lysine, MLH^0	9.23	4.93	4.22	5.18	
26. L-Arginine, MLH^0	9.13	5.05	4.05	5.14	
27. Sarcosine (Sar)	10.08	5.39	4.36	5.47	
28. L-Proline	10.52	5.95	4.95	5.63	
29. L-Histidine	9.20	8.66	6.86	5.17	100
30. L-Cysteine[c]	10.20	8.7	10.9	5.52	100
31. D-Penicillamine[c]	10.4	10.7	12.2	5.6	100

[a]At 25°C and 0.10 ionic strength from Ref. 4, unless otherwise stated. Values for ligands 4 to 8 and 19 from M. Israeli and L. D. Pettit, *J. Inorg. Nucl. Chem.*, 37, 999 (1975). Values for ligand 11 from J. B. Orenberg, B. E. Fischer, and H. Sigel, *J. Inorg. Nucl. Chem.*, 42, 785 (1980). All pK_a values which are concentration-based in Ref. 4 have been adjusted (as recommended in the Introduction) to the normal activity in hydrogen ion scale by adding 0.11 log unit to the values listed at 0.10 ionic strength.
[b]Signifies point used in establishing least squares line in Fig. 1.
[c]See Sec. 2.5 for sources of values.

   Tyrosine complexes present a special case and are not included
in Table 2. Deprotonation of the ammonium and phenolic groups occurs
in the same pH region. The molar ratio of aminophenol to ammonium
phenolate microforms is 2.2 [7-10]. Since the phenolic group is not
involved in coordination and the metal ions coordinate to the aminophenol form, in evaluating the stability constant it is necessary to
use the pK_a microconstant for the ammonium group deprotonation from
the species with the phenolic group still protonated. Results from
the only study so sophisticated show basicity and stability constants
comparable to those for phenylalanine [11]. The phenolic group depro-
tonates from the phenylalanine-like complex at pH > 9.

   Figure 1 shows a plot of the first stability constant logarithm,
$\log K_1$, vs. the amino group pK_a. Only α-amino acids are listed in
Table 2 and plotted in Fig. 1. Log K_1 values for β-amino acids are

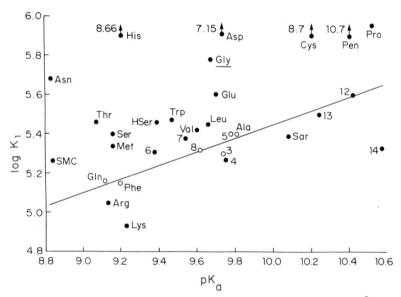

FIG. 1. First stability constant logarithm, $\log K_1$, for Ni^{2+} binding
to α-amino acids vs. amino group pK_a. Symbols and numbers are keyed
to Table 2. Open circles represent points for glutamine, phenylalanine, 2-amino-6-heptenoate (#8), α-aminobutyrate (#3), norleucine
(#5), and alanine that were used to calculate the linear least squares
line. The points for histidine, aspartate, cysteine, and penicillamine are off the scale at the top of the figure.

TABLE 3

Slopes and Intercepts of Log K_1 vs. pK_a
Plots for α-Amino Acids

Metal	Slope[a]	Gly I(9.67)	Log K(Gly) -I(9.67)	Ala I(9.81)	Log K(Ala) -I(9.81)
Co^{2+}	0.38(2)	4.24	0.43	4.29	0.02
Ni^{2+}	0.35(5)	5.33	0.45	5.38	0.02
Cu^{2+}	0.55(6)	8.09	0.04	8.16	-0.01
Zn^{2+}	0.47(5)	4.50	0.46	4.57	-0.01
Cd^{2+}	0.21(9)	3.79	0.45	3.82	-0.02

[a]Number in parentheses represents one standard deviation in last slope digit.

much weaker; for β-alanine of pK_a = 10.21, the Ni^{2+} log K_1 = 4.54 [4], well off the bottom of the Fig. 1 scale. The plot of Fig. 1 shows much scatter, which has dissuaded many investigators from attempting an elaboration. I elect to make a bolder but still conservative approach.

Through points for six ligands designated by open circles in Fig. 1, we calculate the least squares line. The six ligands in order of increasing pK_a are glutamine, phenylalanine, 2-amino-6-heptenoate (#8), α-aminobutyrate (#3), norleucine (#5), and alanine. The side chains of these six ligands are unlikely to interact with the metal ion or provide steric hindrance to chelation through the amino nitrogen and a carboxylate oxygen. The least squares line is completely described by its slope of 0.35 ± 0.05 and an intercept log K_1 value of 5.33 at pK_a 9.67. These values are listed in the second and third columns of Table 3.

The six ligands used in calculating the least squares line were conservatively chosen by simultaneous consideration of results for Co^{2+}, Ni^{2+}, Cu^{2+}, Zn^{2+}, and Cd^{2+}. Only ligands that gave a reasonable fit to a straight line with all five metal ions were included. (Reliable data for binding of glutamine to Zn^{2+} and

Cd^{2+} and of α-aminobutyrate to Cd^{2+} are nonexistent.) For example, slightly low values for ligand #4 appear near the line for Ni^{2+} (Fig. 1) and Cu^{2+} but are more than 0.1 log unit too low for Co^{2+} and Zn^{2+}, and hence ligand #4 is not included in establishing the least squares line. Ligand #13 appears near the least squares line for Co^{2+}, Ni^{2+}, and Zn^{2+} but is too high for Cu^{2+}, and due to its unusual structure and determination at 20° was not included.

For the least squares lines calculated on the above basis for five metal ions, Table 3 lists the slope in column 2 and intercept at pK_a 9.67 designated as I(9.67) in column 3. For all five metal ions the slopes in Table 3 for the binding of bidentate α-amino acids are less than those found for binding of tridentate-substituted iminodiacetates [12], but for Ni^{2+}, Cu^{2+}, and Zn^{2+} the α-amino acid slopes are close to those found for binding of unidentate pyridine and purine N7 or imidazole-type nitrogens [13-15].

The fourth column of Table 2 lists stability constants K_2 for the reaction:

$$ML + L \rightleftharpoons ML_2 \qquad K_2 = \frac{[ML_2]}{[ML][L]}$$

Statistically for octahedral complexes with bidentate ligands the usual argument yields for the ratio $K_1/K_2 = (12/1)/(5/2) = 4.8 = 10^{0.7}$. The first bidentate ligand may bite at any of 12 edges in the aqueous Ni^{2+} octahedron, leaving five edges for the second ligand. The 2 factor in the K_2 denominator arises because there are two identical ligands that may leave the 2:1 complex. The statistical factor applies to symmetrical and unsymmetrical bidentate ligands. For a typical amino acid as alanine the log (K_1/K_2) difference in Table 2 is 0.9, only 0.2 log unit greater than the statistical value of 0.7. Perhaps due to its 7% greater ionic radius (Table 1) reducing steric hindrance, Co^{2+} consistently yields log (K_1/K_2) differences less than those for Ni^{2+} [16]; for Co^{2+} and alanine log $(K_1/K_2) = 0.8$. These nearly statistical differences for Ni^{2+} and Co^{2+} suggest that electrostatic repulsion of the anionic amino acids either does not play a determining role or is offset by other effects. The neutral bidentate ligand 1,2-diaminoethane exhibits $\log (K_1/K_2)$ values of 1.2 and 0.9 for Ni^{2+} and Co^{2+}, respectively, both values being greater than those of 0.9 and 0.8 for alanine. Low $\log (K_1/K_2)$ differences of 0.2-0.5 found for 2,2'-bipyridyl and 9,10-phenanthroline with the two metal ions suggest that π-bonding effects promote addition of the second ligand. A low difference of 0.7 for phenylalanine suggests a favorable interaction between the two aromatic side chains [16]. High $\log (K_1/K_2)$ differences for Ni^{2+} and glutamate (1.4) and aspartate (1.9) derived from Table 2 are ascribed to some chelation of carboxylate side chains giving rise to repulsive interactions. That the side chains are chelated to the metal ion is further supported in Sec. 2.3 by an entirely independent argument.

2.2. Exceptionality of Glycine

To place the least squares lines on the $\log K_1$ axis, the third column of Table 3 lists the calculated $\log K_1$ values for $pK_a = 9.67$, and we designate this intercept on the least squares line as $I(9.67)$. The value of 9.67 on the pK_a axis was chosen because it corresponds to the pK_a of glycine (Table 2). The fourth column of Table 3 lists the difference between the observed $\log K_1$ for glycine and the calculated value based on the pK_a predicted from the least squares line. For Co^{2+}, Ni^{2+}, Zn^{2+}, and Cd^{2+}, this difference is an astonishingly invariant 0.43-0.46 log unit. Note the high position for the glycine point in Fig. 1. In contrast, for Cu^{2+} the difference is an insignificant 0.04 log unit. The atypical values of glycine stability constants have been noted before [16]. For Ni^{2+} the extra stability of the glycine complex resides in a more favorable enthalpy change [17].

What is the explanation for the 0.43-0.46 log unit stronger than predicted stability constants for glycine with Co^{2+}, Ni^{2+}, Zn^{2+}, and Cd^{2+}, but not with Cu^{2+}? Explanations based on five-membered

chelate ring bite angles or other angles in the chelate ring do not
appear to provide an answer since the effective ionic radius of Cd^{2+}
is about 30% greater than those for the other four metal ions [1].
Perhaps one should view the glycine value as "normal" with all the
other C-substituted α-amino acids suffering steric hindrance at the
β carbon with coordinated water molecules. Due to Jahn-Teller dis-
tortions and the weak apical water coordination to tetragonal Cu^{2+},
there is little steric hindrance with this metal ion.

When comparing a property among amino acids, it is common to
employ the value for glycine as the reference to which values of
other amino acids are compared. The results just discussed, reported
in Table 2 and shown in Fig. 1, show that the stability constant of
glycine with most metal ions including Ni^{2+} is exceptional, and misleading impressions arise if glycine is used as the reference. This
unappreciated circumstance has undoubtedly caused consternation among
investigators when comparing stability constants. Alanine should be
used as a reference instead of glycine. The penultimate column of
Table 3 lists the stability constant values for alanine based on its
pK_a of 9.81 and the linear least squares fits. The last column of
Table 3 lists the difference between the observed and calculated
alanine stability constant logarithms. The alanine differences for
all five metal ions are an insignificant $\pm$ 0.02 log unit. Therefore,
when comparing stability constants among amino acids and peptides,
use alanine and not glycine as the reference.

2.3. Percentage Closed Forms

Many of the amino acids plotted in Fig. 1 fall above the linear
least squares line. It is instructive to consider the extent of
side chain-metal ion interaction implied by the vertical distance
between the experimental point and the least squares line.

A metal ion may combine with a deprotonated ligand site such
as an amino acid to give open (ML) and closed (MC) complexes according to

$$M + L \rightleftarrows ML \rightleftarrows MC$$

In this application ML represents a metal ion chelated by an amino acid via its amino and carboxylate groups. MC represents a complex with some degree of interaction between the amino acid side chain and the metal ion. For formation of the open complex we define the equilibrium constant

$$K_0 = \frac{[ML]}{[M][L]}$$

and for isomerization of the open to the closed complex the equilibrium constant or enhancement as

$$E = \frac{[MC]}{[ML]}$$

Over both open and closed forms the observed first stability constant is given by

$$K_1 = \frac{[ML] + [MC]}{[M][L]} = K_0(1 + E)$$

Presence of closed forms leads to a complex stability enhancement compared to that anticipated without closed forms. The stability enhancement becomes

$$E = \frac{K_1 - K_0}{K_0}$$

and the stability enhancement factor

$$1 + E = \frac{K_1}{K_0} = 10^{\log(K_1/K_0)}$$

The enhancement factor is the factor by which the baseline stability constant K_0 should be multiplied to give the observed stability constant K_1. The enhancement factor may be found by comparing the observed stability constant of the complex with that expected for the corresponding complex without a closed form. Usually the comparison is made with logarithms of stability constants [18].

Calculation of an enhancement factor rests on the reliable estimation of a baseline stability constant. With the least squares line in the log K_1 vs. pK_a plot of Fig. 1, we already possess the means to calculate the enhancement factor. We estimate the baseline stability constant log K_0 by finding the ordinate intercept on the least squares line of the ligand pK_a. Baseline stability constants calculated in this way appear in the fifth column of Table 2. The logarithmic difference between the observed stability constant K_1 and the baseline stability constant of the same pK_a becomes log $(K_1/K_0) = \log(1 + E)$. Thus $\log(1 + E)$ is given by the difference between the values in the third and fifth columns of Table 2.

With a potentially chelating side chain, the fraction of interacting side chains may range from zero, where all complexes are open, to nearly unity, where virtually all complexes are closed. The fraction of closed complexes f is given by

$$f = \frac{[MC]}{[ML] + [MC]} = \frac{E}{1 + E} = 1 - 10^{\log(K_0/K_1)}$$

Values for the fraction of closed complexes, converted to percentages, appear in the last column of Table 2. Since the log K_1 values collected in Table 2 were determined over a period of many years by different investigators, $\log(K_1/K_0)$ values of less than 0.12 may not be significant. The 0.12 log unit difference corresponds to 24% closed forms. Values for ligands 6-8 were reported in a single paper so that the suggested 17 and 19% closed forms for ligands 6 and 7 over none for ligand 8 might be significant and imply a weak metal ion interaction with the double bond in the side chain. The remoteness of the double bond from the metal ion chelating center in ligand #8 makes impossible its interaction so that this ligand falls on the least squares line in Fig. 1. The identical pattern for ligands 6, 7, and 8 also occurs with Co^{2+} and Zn^{2+}. The suggested 24 and 22% closed forms for leucine and valine [$\log(K_1/K_0)$ of 0.12 and 0.11] may merely represent experimental differences. Both leucine and valine lie near the least squares line for Zn^{2+}. There is, however, some apparent tendency for hydrophobic side chains to occupy space near metal ions [18,19]. A stronger tendency for aromatic side chains to do so accounts for the 40% closed forms for tryptophan.

For log (K_1/K_0) values greater than 0.12 it is likely that the side chain interacts appreciably with the metal ion. Table 2 lists 35-54% closed forms for methionine, S-methylcysteine, homoserine, serine, glutamate, and threonine. We can visualize metal ion interactions with side-chain donor atoms in these amino acids chelated primarily via the amino and α-carboxylate groups. The 35% closed form calculated for methionine accounts for, but does not demand, the weak stereoselectivity observed in its 2:1 Ni^{2+} complex [20]. Even stronger metal ion-side chain interactions occur with asparagine and aspartate with 77 and 98% closed forms, respectively. With this pair of ligands a β-carboxylate oxygen forms a six-membered chelated ring with a metal ion already chelated at the amino and α-carboxylate groups. A crystal structure determination shows a tridentate aspartate ligand with Ni^{2+} in a mixed complex with three imidazole groups [21].

Finally, the side chains for the last three entries in Table 2 coordinate strongly to Ni^{2+}. The imidazole side chain in histidine coordinates in 99.97% of the complexes, and the sulfhydryl group in cysteine and penicillamine coordinates in 99.93 and 99.9992% of the complexes, respectively. The higher percentage for penicillamine over cysteine arises because the pair of methyl groups at the β-carbon reduces the frequency of nonchelating conformations.

2.4. Histidine Complexes

Histidine complexes of many metal ions, including Ni^{2+}, are unusually strong and often of great importance biologically. Almost all Ni^{2+} in blood plasma travels in histidine complexes. The interaction of histidine with transition metal ions including Ni^{2+} in chemical and biological systems has received an extensive review [22].

2.4.1. *The Imidazole Ring*

Imidazole and derivatives contain two types of ring nitrogens: a
basic pyridine nitrogen with $pK_a \sim 6-7$ and a nonbasic pyrrole nitrogen
that bears a hydrogen. The pyrrole nitrogen does not bind metals;
metal ion binding at the imidazole pyrrole nitrogen would yield a
complex of high energy whose existence has never been established
[22]. Metal ion binding at imidazole derivatives occurs at the pyridine nitrogen or at a deprotonated pyrrole nitrogen which then adopts
pyridine character. Ni^{2+} bound at an imidazole pyridine nitrogen
fails to promote across the ring pyrrole hydrogen deprotonation [23].

There is a rapid tautomeric equilibrium between the two ring
nitrogens for the single hydrogen in a neutral imidazole ring. In
asymmetrically substituted imidazoles such as histidine NMR studies
reveal [24,25] a ratio ≥ 4 in favor of the tautomer that conforms to
the organic chemist's (but not biochemist's) convention of labeling
the pyrrole nitrogen as N(1)-H and describing the imidazole ring as
4-substituted [16]. Thus metal ion chelation between the amino
nitrogen and the imidazole ring pyridine nitrogen occurs in the
favored histidine and histamine tautomers with N(3) the pyridine
nitrogen. Chelation is not possible in the other tautomer.

Complexing of Ni^{2+} by histidine and histidine methyl ester
exhibits biphasic kinetics attributed to Ni^{2+} complexing at both the
pyridine and pyrrole ring nitrogens [26]. But complexing of Ni^{2+} at
a pyrrole nitrogen even as a reaction intermediate is unacceptable,
and another interpretation is required [22]. More recently and sat-
isfactorily biphasic kinetics in Ni^{2+} complexing to histamine have
been explained by initial complexing of Ni^{2+} at pyridine nitrogens
in the tautomeric imidazole ring. In the favored tautomer with the
pyridine nitrogen at N(3) there is rapid chelate ring closure. Minor
tautomer complexation at the pyridine nitrogen gives an unproductive
unidentate complex that cannot undergo chelate ring closure, and thus
dissociation must occur before a chelate ring may form by the first
pathway [27]. The difficulty in making the distinction in biphasic
kinetics between a reaction intermediate on the pathway to products

and a metastable compound made in an unproductive, reversible side
reaction has been detailed [28].

2.4.2. Stereoselectivity

Ni^{2+} complexes of histidine exhibit easily demonstrable stereoselectivity. In principle a complex of a metal ion and an L-ligand interacts differentially with any other L,D ligand pair in forming a 2:1
complex. The formulation is simplified when both ligands are the
same chemical species so that one compares pure MLL or MDD complexes
with a mixed MLD complex. The question is whether a difference in
stabilities or other properties is detectable [22].

Unlike the mixed complex MDL, the pure 2:1 complexes can dis-
sociate in two different ways to yield a specified 1:1 complex. As
a result when a solution contains equal amounts of each enantiomer,
the statistical distribution of complexes is 50% MDL and 25% each
of MLL and MDD. Two potentiometric titration studies agree that,
instead of 50% MDL, solutions of Ni^{2+} and equal amounts of L- and
D-histidine contain 70-72% MDL [29,30]. The corresponding percentages for histidine and other metal ions are as follows: Co^{2+} and
Zn^{2+} 62%; Cd^{2+}, 53%; and Cu^{2+}, 50% [30]. Thus, perhaps due to its
regular octahedral stereochemistry and small size (Table 1), Ni^{2+}
exhibits the greatest stereoselectivity.

These potentiometric results contrast to a structure determination on crystals obtained from a solution containing racemic
histidine. Crystals of the Ni^{2+} complex contained equal amounts of
the two pure complexes MLL and MDD [31]. The crystals contain species that represent only 15% of the complexes in solution.

Despite the absence of an observable thermodynamic stereoselectivity in the mixed complexes of histidine and its methyl ester,
the ester hydrolyzes 40% faster when the histidine is of the opposite
configuration [32]. The metal ion-promoted hydrolysis of amino acid
esters and peptides has received thorough review [33]. Later papers
provide additional information concerning the roles of Ni^{2+} [34,35].

Greater degrees of thermodynamic stereoselectivity are found upon adding bulky groups to histidine. The percentage of mixed MDL species rises to 85% for N-methylhistidine and to 94% for N,N-dimethylhistidine [36]. In contrast, no stereoselectivity was observed in the Ni^{2+} complexes of aspartate, asparagine, glutamate, and glutamine [37].

Stereoselectivity with Ni^{2+} complexes has been used to separate α-amino acid enantiomers by ligand exchange chromatography [38]. Stereoselectivity has also been observed in a variety of diastereomeric dipeptide complexes [39]. This last reference furnishes a general review of stereoselectivity.

Ni^{2+} binding has been considered as a function of a lengthening chain in the multifunctional α,ω-diaminocarboxylates: 2,3-diaminopropanoate, 2,4-diaminobutanoate, ornithine, and lysine [16,40]. With anionic 2,3-diaminopropanoate in a 2:1 molar ratio to Ni^{2+}, the Ni^{2+} is hexacoordinate at high temperature, and as the temperature is lowered two carboxylate groups are released and four nitrogen donors take up positions in a plane about the Ni^{2+} [41]. The planar to octahedral conversion occurs because of the higher negative enthalpy change of the diaminoethane over glycine binding modes of the ligand.

Several studies have been made of mixed complexes containing Ni^{2+}, an amino acid, and another amino acid or other ligand [23,42-47]. Because they contain two different donor atoms, chelated amino acids form internally mixed complexes. More extensive internal mixing occurs with histidine and 3,4-dihydroxyphenylalanine (DOPA) (Sec. 2.6).

2.5. Cysteine and Penicillamine (β,β-Dimethylcysteine)

Since $pK_1 = 1.9$ for the carboxylic acid and ionization in both compounds, over almost the entire pH range they occur as carboxylates. The potentiometrically determined macroconstants (0.16 ionic strength, 25°C) $pK_2 = 8.24$ and $pK_3 = 10.37$ in cysteine [48] and $pK_2 = 7.95$ and $pK_3 = 10.55$ in penicillamine [4,49] refer to a blend of the sulfhydryl and ammonium group deprotonations. For cysteine at 25°C the molar ratio of the two anionic forms $[^-SCH_2CH(COO^-)NH_3^+]/[HSCH_2CH(COO^-)NH_2] =$

2.1 has been consistently found by several investigators over the past 30 years [49-52], and the ratio is nearly independent of ionic strength [53]. Due to the greater positive enthalpy change for ammonium than for sulfhydryl group deprotonations, the molar ratio increases to 3.0 at 5°C and becomes less than unity above 70°C [52]. For penicillamine reanalysis by nonlinear least squares of the earlier data [49] suggests, with greater uncertainty than for cysteine, a corresponding ratio of about 2.5. Thus in weakly basic solutions both cysteine and penicillamine occur predominantly as ammonium thiolates rather than as amino thiols.

In assessing the extent of metal ion chelation in Sec. 2.3, we are interested in the acidity constant for the ammonium group deprotonation from a ligand with an already ionized sulfhydryl group. Complete resolution of the standard microconstant equilibria [7,54] in conjunction with the pK_2 and pK_3 macroscopic acidity constants given above yields (at 0.16 ionic strength and 25°C), where the subscript 2 applies to the sulfhydryl group and the subscript 3 to the ammonium group, for cysteine $pk_{12} = 8.41$, $pk_{13} = 8.73$, $pk_{123} = 10.20$, and $pk_{132} = 9.88$. For penicillamine the estimated values are, respectively, 8.1, 8.5, 10.4, and 10.0. The pk_{123} microconstant value appears for the α-ammonium group deprotonation in the second column of Table 2.

Only a few stability constant studies with Ni^{2+} and cysteine have considered and allowed for polynuclear complex formation through sulfur bridging, signaled by appearance of an absorption band at 380 nm. The trinuclear complex $Ni_3Cys_4^{2-}$ probably consists of two Ni^{2+}, each with two cysteines bound in a cis arrangement with two amino and two anionic sulfhydryl donors, joined by a central Ni^{2+} bound to the four cysteine sulfhydryl groups, each of which bridges the central and one other Ni^{2+}. All three Ni^{2+} are planar and diamagnetic [55,56]. Steric hindrance to bridging by the β-methyl groups reduces polynuclear complex formation in penicillamine complexes [55,57]. Results of three studies confirm that for both cysteine and penicillamine $\log K_1 < \log K_2$ as shown in Table 1 [55,57,58]. This inversion

of the usual stability order results from the additional ligand
field stabilization provided by spin pairing in two sulfur, two
nitrogen donor atom systems and formation of a 2:1 planar, diamagnetic complex from a 1:1 complex with a hexacoordinate, paramagnetic
Ni^{2+}. Little 1:1 complex exists in these solutions.

The additional stability provided by generation of planar,
diamagnetic, mononuclear complexes with penicillamine renders small
the amount of mixed complexes in most systems. For example, a 1:1:1
solution containing penicillamine, glycine, and Ni^{2+} (and three
equivalents of base) yields virtually no mixed complex, but rather
half of the Ni^{2+} occurs as a 2:1 planar, diamagnetic pencillamine
complex and half as a 2:1 hexacoordinate, paramagnetic glycine complex [55]. When histidine is substituted for glycine, a small percentage of diamagnetic mixed complex is formed [55]. With an anionic
sulfhydryl and amine nitrogen donors from cysteine, evidently the
amino and especially the imidazole nitrogen donors from histidine
are sufficient to effect spin pairing and a small amount of mixed,
planar complex.

2.6. Catecholamines

L-3,4-dihydroxyphenylalanine (DOPA) is important as a precursor of
the so-called catecholamines. DOPA presents two separate metal ion
binding sites: a glycine or phenylalanine type site and a catechol
site. At which site a metal ion binds depends on both the metal ion
and the pH [59]. Macroscopic acidity constants in DOPA of pK_1 = 8.76
and pK_2 = 9.84 are blends of catechol and ammonium group deprotonations. The ammonium-catecholate-to-catecholamine ratio is 1.6 in
DOPA with greater values of 2-5 in the epinephrines [60-62]. Thus
these so-called catecholamines are better described as ammonium
catecholates.

A fortunate circumstance allows easy comparison of the relative binding strengths of the phenylalanine and catechol binding

modes in DOPA. The DOPA microconstant from the zwitterion for the ammonium group deprotonation is $pk_2 = 9.17$ and for the first catechol deprotonation is $pk_1 = 8.97$ [60,61,63]. The first value is nearly identical to the $pK_a = 9.20$ value for phenylalanine in Table 2, and the second corresponds closely to the first ionization in catechol, $pK_1 = 9.32$ [4]. The second catechol deprotonation in DOPA also corresponds closely to the same deprotonation in catechol [4]. These comparable acidity constants permit direct use of phenylalanine and catechol stability constants to represent the corresponding binding modes in 1:1 DOPA metal ion complexes.

Since there is only one deprotonation from neutral phenylalanine and two from catechol, the metal ion from the preferred phenylalanine site at low pH switches to the catechol site at high pH. From the phenylalanine and catechol stability constants [4] and the acidity constant analysis given above, I calculate the crossover pH from phenylalanine like to catechol binding in 1:1 complexes of DOPA to increase in the following order (with crossover pH in parentheses): Fe^{3+} (<0), Ga^{3+} (2.5), Pb^{2+} (4.1), Al^{3+} (4.8), Cu^{2+} (7.0), Zn^{2+} (7.5), Mn^{2+} (7.8), Co^{2+} (8.5), Mg^{2+} (8.5), Cd^{2+} (8.6), and Ni^{2+} (9.3). Thus Ni^{2+} ends the list and retains weak amino acid like binding through the neutral pH region.

These conclusions from the model compound analysis receive direct support from experiments performed with DOPA as a ligand. The experiments with excess ligand produces three forms of 2:1 complexes: both ligands bound in a phenylalanine mode, both in a catecholate mode, or one ligand in each binding mode. Due to the tendency to form mixed complexes (especially as electrostatics works against both ligands in a catecholate mode), we expect a strong tendency for 2:1 complexes to be of mixed mode. A study of DOPA complexes concludes that the order of increasing pH to convert 2:1 complexes from mixed to the catecholate mode for both ligands follows the order $Cu^{2+} < Zn^{2+} < Co^{2+} < Mn^{2+}$, Ni^{2+}, the last pair not converting significantly at pH <11 [64]. (The last study is exemplary as it considers microconstant equilibria in the free ligands and in their complexes.) In neutral solutions both DOPA molecules are bound to Ni^{2+} in an amino acid mode, the mixed mode dominating when pH > 8 [63,65].

Without the strong amino acid-type site, binding of Ni^{2+} and other metal ions to dopamine and epinephrine (with a weak ethanolamine site) occurs nearly exclusively at the catecholate locus [59,62, 66,67]. The binding of most metal ions such as Mg^{2+} and Ni^{2+} is so weak in neutral solutions, however, that complexes of these catecholamines are unimportant in living systems [15].

3. CITRATE COMPLEXES

Citrate has been proposed as a ligand for Ni^{2+} in plants, especially nickel accumulator plants [Chap. 3]. At about 0.1 mM concentration in the blood plasma, citrate must be considered as a potential ligand for metal ions. It is the primary nonprotein ligand for Fe^{3+} and Al^{3+} in plasma [68,69].

Potentiometric titrations reveal that in neutral solutions Ni^{2+} forms with citrate a 1:1 complex with log K_1 = 5.5 and a 2:1 complex with log K_2 = 2.3 [70]. A crystal structure determination shows a tridentate citrate chelating one Ni^{2+} by two carboxylate oxygens and the hydroxy oxygen [71]. The chelated hydroxy group does not lose its proton until pH > 8 [72].

The stability constant log K_1 = 5.5 for Ni^{2+} and citrate is in the same range as those for the bidentate amino acids in Table 2. In neutral solutions, however, all three carboxylate groups are in their anionic state while the bidentate amino acids in Table 2 remain as zwitterions. The conditional stability constant [1,69] for Ni^{2+} with alanine at pH 7.4 is given by log K_c = log K_s + pH - pK_a = 5.4 + 7.4 - 9.8 = 3.0, a value 2.5 log units less than that for citrate under the same conditions. Thus under plasma conditions citrate rather than the bidentate amino acids chelate Ni^{2+}.

However, Table 2 indicates that histidine and cysteine bind Ni^{2+} considerably more strongly than typical bidentate amino acids.

The ordering at pH 7.4 of conditional stability constants (logarithms in parentheses) is as follows: Cys, K_2 (7.0) > His, K_1 (6.8) > citrate, K_1 (5.4) > His, K_2 (5.0) > Cys, K_1 (4.8). The stabilities of mixed complexes with two different ligands are uncertain. There is slightly less histidine than citrate in blood plasma and only about one-quarter as much cysteine [73], which may be tied up in other ways as well. With the relative concentrations the above series suggests that a mixed histidine-citrate Ni^{2+} complex may constitute a significant fraction of plasma Ni^{2+}. A computer simulation study proposes that plasma Ni^{2+} is carried as histidine and cysteine complexes [74]; citrate was not considered in the simulation.

Carboxylate and amino groups anchor chelation of metal ions to sugars. The amino group of glucosamine participates in chelation of Ni^{2+} [75,76]. The reaction of 1,2-diaminoethane with glucosamine yields an N-glycoside that according to a crystal structure forms a bistridentate complex in which there are six nitrogen donors to Ni^{2+} [77].

4. PEPTIDE COMPLEXES

4.1. Hexacoordinate and Planar Complexes

The coordinating properties of the peptide (amide) bond have been thoroughly reviewed [78], and the review should be consulted as an entry point to the subject. This much shorter section deals with selected aspects and provides references to more recent articles.

Ni^{2+} forms more interesting complexes with peptides than other metal ions. It is necessary to distinguish sharply Ni^{2+} interactions with dipeptides on one hand and tri- and higher peptides including proteins on the other. In both cases the mutual interaction in acidic solutions is chelation at an amino nitrogen and the carbonyl oxygen of the first peptide bond, the nitrogen of which remains protonated. Metal ion interaction at the peptide nitrogen occurs only upon its deprotonation [78]. Substitution of the peptide hydrogen by the metal ion results in a tridentate chelate involving the amino nitrogen, a deprotonated peptide nitrogen, and a carboxylate oxygen in dipeptides, and the second peptide carbonyl oxygen in tri- and higher peptides. The process does not end there in tri- and higher peptides because Ni^{2+} substitution for the second peptide hydrogen occurs in the same slightly basic pH region as the first due to a cooperative transition from hexacoordinate (octahedral) to planar geometry about Ni^{2+} giving rise to yellow complexes [78,79]. Dipeptides are unable to effect the hexacoordinate to planar transformation and instead form bis complexes in basic solutions in which the two amide deprotonated tridentate ligands take up positions at right angles to each other. Peptide bond planarity enforces a planar Ni^{2+} geometry with tri- and higher peptides and a near octahedral geometry in the bis dipeptide complexes. (Glycylglycinamide should also initially adopt the bis hexacoordinate structure with loss of two interior peptide protons, followed by conversion to a planar structure upon loss of one of the two ligands from the complex and in the other ligand loss of the terminal amide proton with a switch from amide nitrogen coordination. This possibility may have been overlooked in [80]).

Table 4 summarizes equilibrium constants obtained in glycyl peptide systems. The second column lists the amino group pK_a values.

TABLE 4

Stability Constants for Glycyl Peptides of Ni^{2+}[a]

	pK_a	log K_s	pK_1	pK_2	pK_3
Diglycine	8.15	4.0/3.2	9.3	9.9	
Triglycine	7.96	3.70	8.7[b]	8.0[b]	12.8[c]
Tetraglycine	7.97	3.65	9.3[b]	7.4[b]	7.8[b]

[a]At 25° and 0.1-0.16 ionic strength. From Ref. 78 unless otherwise stated.
[b]Original values; see Sec. 4.3.
[c]Displacement of bound carboxylate by hydroxide from Ref. 91.

The third column tabulates the stability constants for Ni^{2+} chelation at the amino terminal nitrogen and first carbonyl oxygen. Two stability constant values are listed for Gly-Gly because two ligands bind at right angles in a hexacoordinate complex. The pK_1 and pK_2 values for Gly-Gly refer to loss of a peptide proton from the two ligands in the bis complex. Due to steric inhibition of deprotonation from the planar peptide bond [78,81], higher pK_a values obtain with bulky noncoordinating side chains in the carboxylate terminal dipeptide residue [82,83]. Derivation of the pK_a values for the cooperative deprotonations from tri- and tetraglycine complexes are discussed in Sec. 4.3.

No reliable stability constant values appear to exist for Ni^{2+} association at the carboxylate terminus of a peptide when the ammonium group remains protonated. We take as a guide the stability constant log K_s = 0.58 for Ni^{2+} binding at acetylglycine [84]. This very weak constant implies an insignificant buildup of a $NiHL^{2+}$ complex under any set of conditions (see Fig. 3).

4.2. Optical Properties

Hexacoordinate (octahedral) Ni^{2+} complexes exhibit only three weak absorption bands in the visible and near-infrared regions due to electric dipole forbidden d-d transitions [85]. The two bands in the visible near 380 nm and 620 nm are also magnetic dipole forbidden, but the one in the near-infrared is magnetic dipole allowed and is a candidate for circular dichroism (CD) studies only on instruments that can reach 1,000 nm. Deviations from strict octahedral symmetry result in weak observable CD in the visible region [86].

Due to reduced symmetry, greater intensity is associated with the absorption bands in planar, yellow Ni^{2+} complexes that absorb in the 410- to 450-nm region. Of three transitions appearing in this region two are magnetic dipole allowed [85]. As a result CD studies have been usefully employed in investigating planar complexes of tri- and higher peptides and of cysteine. Typically tripeptide complexes with the triglycine-type structure that contain one or more L-amino acid residues display an easily measurable negative CD at about 470 nm. The total CD magnitude in a tri- or tetrapeptide complex is an additive function of contributions from optically active centers acting independently [78,85,87]. This regularity led to the development of the hexadecant rule for CD in tetragonal transition metal ion complexes [78,85,87,88]. In contrast to the simple negative CD provided by most L-amino acid residues, histidyl and cysteinyl residues produce CD with both positive and negative regions through the absorption band. These differences have been put to practical use in determining whether histidine or cysteine side chains are involved in Ni^{2+} binding to proteins [22,85,89].

Both hexacoordinate and planar Ni^{2+} complexes display strong charge transfer absorption bands in the ultraviolet [90] that move into the visible when a sulfhydryl group is involved [87].

Lack of space prevents citing the many ways in which both the absorption and circular dichroism properties of Ni^{2+} complexes have been put to use. The reader will notice many examples throughout the references.

4.3. Cooperative Deprotonations in Tri- and Tetraglycine Complexes

The degree of cooperativity of peptide deprotonations in Ni^{2+} complexes has provoked controversy. The issue is whether the degree of cooperativity is too high to permit resolution of successive polypeptide complex amide deprotonations, or what is equivalent, whether there is a significant buildup of complexes with intermediate degrees of deprotonation. The following analysis examines the criterion for irresolvability and indicates that for both the triglycine and tetraglycine complexes of Ni^{2+} it is possible either to resolve the successive individual acidity constant values or provide an excellent estimate of likely values.

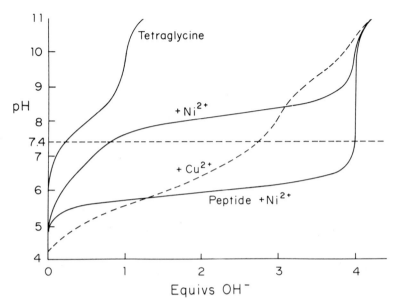

FIG. 2. Titration curves. Top, left, curve of tetraglycine alone. The curves labeled $+Ni^{2+}$ and $+Cu^{2+}$ are for tetraglycine with an equimolar amount of the metal ion. Peptide represents the amino terminal model peptide for human plasma albumin: L-Asp-L-Ala-L-His-N-methyl-amide [122]. Curves constructed at 5 mM each component from equilibrium constants in Table 4 and Refs. 78 and 122.

Titration curves for tetraglycine and Ni^{2+} are drawn in Fig. 2. The curve to the top and left of tetraglycine alone shows an endpoint after addition of one equivalent of base and a midpoint pH = 7.97 at 0.50 equivs added OH^- corresponding to the amino group pK_a. The next curve contains equimolar amounts of Ni^{2+} and tetraglycine and shows consumption of 4 equivs of OH^- corresponding to loss of one amino group and three peptide hydrogens. The flatness of the curve from 1 to 4 equivs OH^- is due to the cooperative hexacoordinate to planar transition. The flat titration curve actually crosses that of an equimolar solution of Cu^{2+} and tetraglycine at 3.1 equivs and pH 8.4, above which tetraglycine binds Ni^{2+} more strongly than it binds Cu^{2+}.

The data obtained in 1959 [79] in equimolar solutions of Ni^{2+} and triglycine or tetraglycine were analyzed for the successive

deprotonations from the complexes by a nonlinear least squares computer program. The results are for two peptide deprotonations from
triglycine: pK_1 = 8.7 ± 0.1, pK_2 = 8.0 ± 0.1, and the sum pK_1 +
pK_2 = 16.72 ± 0.02. These pK_1 and pK_2 values are listed in Table 4.
They are intermediate between two pairs of values 8.8 and 7.7 [91],
and 8.55 and 8.04 [82], both pairs of which are on a H^+ concentration
scale that gives lower pK_a values. This difference is confirmed by
the lesser pK_1 + pK_2 sums, 16.5 and 16.6, for the last two sets of
values. For the three peptide deprotonations from tetraglycine the
program yields from the 1959 data [79], pK_1 + pK_2 = 16.7 ± 0.2,
pK_3 = 7.8 ± 0.2, and pK_1 + pK_2 + pK_3 = 24.55 ± 0.04. For complexes
of both peptides, the pK_a sums are more accurately known than the
individual values. This circumstance arises because of the cooperative nature of the deprotonations leading to similar pK_a values that
in some cases even lead to an inversion of the normal order, e.g.,
for the triglycine complex pK_1 > pK_2. The cooperativity reaches an
extreme in the tetraglycine complex where pK_1 for the first peptide
deprotonation cannot be determined at all because very little of the
$Ni(LH_{-1})$ complex exists. Once produced, it efficiently undergoes a
subsequent deprotonation with pK_2 < pK_1 as the hexacoordinate, paramagnetic Ni^{2+} complex spin pairs to yield a planar, diamagnetic,
yellow complex followed by the third peptide deprotonation with
pK_3 = 7.8 (Fig. 2).

Two independent lines of argument furnish a resolution of the
sum pK_1 + pK_2 = 16.7 for the tetraglycine complex into individual
pK_a values.

In the limit of $K_2/K_1 \to \infty$ for two highly cooperative deprotonations, the ratio of H^+ concentration at 0.5 and 1.5 equivs of added
strong base is 3, corresponding to 0.48 log unit. To resolve definitively the individual pK_1 and pK_2 values we require that the difference $\Delta pH = pH_{3/2} - pH_{1/2} > 0.48$. For K_2/K_1 = 50 or pK_1 - pK_2 = 1.7
(the second deprotonation occurs with a lower pK_a than the first),
ΔpH = 0.51, sufficiently different from 0.48 to begin to permit resolution of pK_1 and pK_2. Since the pK_1 + pK_2 = 16.7 sum is not resolv-

able in the Ni^{2+}-tetraglycine complex, the difference $pK_1 - pK_2 \geq$ 1.7. Thus we obtain $pK_1 \geq 9.2$ and $pK_2 \leq 7.5$.

In the second argument I contend that the pK_1 value for the tetraglycine complex should be significantly greater than that of the triglycine complex for which $pK_1 = 8.7$ (Table 4). In these peptide complexes the last bound nitrogen is succeeded by a bound peptide carbonyl oxygen on the same amino acid residue. The main determinant of the strength of the metal ion carbonyl oxygen interaction is the basicity of the nitrogen to which the carbonyl group joins in a peptide bond (Sec. 4.4). Just before the first peptide deprotonation, the chelated carbonyl group is joined to diglycine in the triglycine complex and to triglycine in the tetraglycine complex. As Table 4 indicates, the pK_a values for the amino groups in diglycine and triglycine are close. After the first peptide nitrogen deprotonation, the newly chelated carbonyl group is joined to glycine in the triglycine complex and to diglycine in the tetraglycine complex. Now the amino group of glycine ($pK_a = 9.7$) is substantially more basic than that of diglycine ($pK_a = 8.2$). Thus the metal ion-peptide carbonyl oxygen interaction after the first peptide deprotonation should be significantly stronger in the triglycine than in the tetraglycine complex, producing a more acidic pK_1 in the triglycine complex. Of the difference in pK_a values $9.7 - 8.2 = 1.5$ log units for the amino groups on the dangling portion of the peptides we take 40% (see Table 3) and add the 0.6 log unit to the $pK_1 = 8.7$ for the triglycine complex and estimate $pK_1 = 8.7 + 0.6 = 9.3$ for the tetraglycine complex. This value accords with the limit inferred above in the first argument. From the known sum $pK_1 + pK_2 = 16.7$, we also estimate that $pK_2 = 7.4$. These estimated values appear in Table 4.

From the constants in Table 4 it may be calculated that the maximum amount of the $Ni(LH_{-1})$ complex of tetraglycine that occurs is only 2.2% of total metal ion at pH 8.1 in solutions 5 mM in both Ni^{2+} and tetraglycine. (For the corresponding solutions with triglycine, there is a maximum 14% $Ni(LH_{-1})$.) The relative magnitudes of the pK_1, pK_2, and pK_3 values for tetraglycine in Table 4 assure that in the cooperative transition the intermediate complexes $Ni(LH_{-1})$

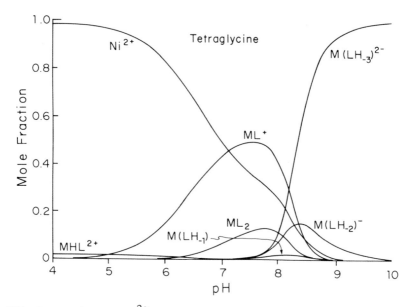

FIG. 3. Equimolar Ni^{2+}-tetraglycine species distribution curves of mole fraction Ni^{2+} species vs. pH. Curves constructed at 5 mM each component from equilibrium constants in Table 4 and $pK_a = 3.3$ for carboxylic acid ionization from tetraglycine, $\log K_2 = 2.9$ [79] for $ML^+ + L^- \rightarrow ML_2$, and $\log K_S = 0.58$ [84] for coordination of Ni^{2+} at carboxylate group with protonated ammonium group, $M^{2+} + HL \rightarrow MHL^{2+}$. This complex is always less than 2% of total Ni^{2+}. At lower total concentrations there is less complex formation, especially of ML_2 species where two ligands are chelated by the amino group and the first peptide oxygen.

and $Ni(LH_{-2})^-$ are squeezed out between the dominant NiL^+ at pH <8 and $Ni(LH_{-3})^{2-}$ at pH >8. The results are plotted in Fig. 3, which shows the distribution of complexes in an equimolar mixture of Ni^{2+} and tetraglycine.

4.4. Chelation at the Amino Terminus

The reaction for association of metal ions at the amino terminus of peptides involves chelation between the amino group and the carbonyl oxygen of the first peptide bond.

$$M + L^- \rightleftharpoons ML \qquad K_1 = \frac{[ML]}{[M][L^-]}$$

Table 5 shows the results for log K_1 for a variety of glycyl peptides. Figure 4 displays the results in the form of a plot of log K_1 vs. pK_a of the amino terminus (solid circles, scale at top of plot). For the linear least squares line through the 12 solid circles (exluding Gly-Sar) the Ni^{2+} slope is 1.9 ± 0.2. A virtually identical steep slope of 2.0 ± 0.1 was found from an identical plot with Cu^{2+} and a closely similar group of peptides (Fig. 4 in Ref. 78). Such steep slopes are inconsistent with the much lower values in Table 3, and we gain insights by considering the sources of the steepness.

In seeking the source of the steep slope in Figure 4, we first note the narrow pH range of 7.9–8.6 for the peptide pK_a values in

TABLE 5

Stability Constants for Glycyl-L-amino Acid Peptides[a]

Peptide	Peptide pK_a	Ni^{2+} log K_1	Deglypeptide pK_a
Gly-Gly (GG)	8.15	4.0	9.67
Gly-β-Ala (GβA)	8.20	4.19	10.21
Gly-Ala (GA)	8.25	4.23	9.81
Gly-Leu (GL)	8.28	4.26	9.66
Gly-Phe (GP)	8.22	4.03	9.20
Gly-Val (GV)	8.20	4.25	9.60
Gly-Met (GM)	8.19	4.15	9.16
Gly-Sar (GSar)	8.59	4.44	10.08
Gly-Pro (GPro)	8.55	4.69	10.52
Glycinamide (GN)	8.04	3.8	9.38
Gly-Gly-amide (GGN)	7.91	3.4	8.04
Triglycine (G_3)	7.96	3.70	8.15
Tetraglycine (G_4)	7.97	3.65	7.96

[a]At 25° and 0.10 ionic strength. From Refs. 78 and 4, which misinterprets the scale of some pK_a values in the literature.

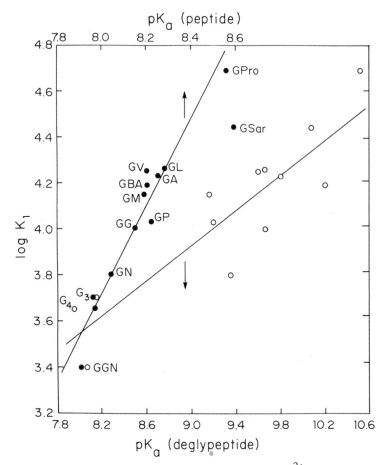

FIG. 4. First stability constant logarithm for Ni^{2+} binding to
Table 5 glycyl peptides vs. peptide amino terminal pK$_a$ (solid circles
and upper scale) and vs. deglycinated peptide pK$_a$ (open circles and
lower scale). For each peptide there is one solid and one open point
on a horizontal line. See Table 5 for meaning of peptide abbreviations. An interval on the lower pK$_a$ scale contains twice as many log
units as the same interval on the upper scale. Straight lines are
least squares fits to solid and open points. The point for Gly-Sar
was excluded from the least squares fit to the line through the solid
points.

Table 5. The narrow 0.7 log unit range arises because all the peptides possess the same glycyl amino terminus. That there is any
range at all in the glycyl amino pK_a is due to a remote reflection
of differences in the second (and subsequent) residues among the
peptides.

Since in the reaction the metal ion forms a five-membered
chelate ring with the amino terminal nitrogen and the carbonyl oxygen
of the first peptide bond, it is necessary to consider the role of
the latter in determining complex stability. In amino acids the pK_a
of the carboxylate group spans an even narrower range than the pK_a
of the amino group. Furthermore, all the peptides of Table 5 possess
an amino terminal glycyl residue so that no differences in peptide
carbonyl oxygen basicity can arise from differences in the amino
terminal residue. It is the primary thesis of this section that the
major determinant of variations in peptide carbonyl oxygen basicity
is the basicity of the amino group that joins to make up the peptide
bond.

To analyze the role of the basicity of amino groups of the
second residue on the basicity of the carbonyl oxygen in the first
peptide bond, we consider the species that results from removal of
the amino terminal glycyl residue from the peptides in Table 5. The
last column in Table 5 lists the amino group pK_a values of the deglycinated peptides; for the first nine entries the value corresponds
to that for an amino acid. The open circles in Fig. 4 result from
a plot of log K_1 vs. the deglycinated peptide pK_a values appearing
in the last column of Table 5. Though there is considerable scatter
a strong trend is apparent. The slope of the least squares line
through all the open circles for Ni^{2+} in Fig. 2 is 0.38 ± 0.05. A
similar treatment for a similar set of Cu^{2+} peptide complexes in
Fig. 4 of Ref. 78 yields a slope of 0.43 ± 0.07. These slopes are
within one standard deviation of those in Table 3.

The close correspondence of the slope through the open circles
for Ni^{2+} complexes of deglycinated peptides in Fig. 4 and the slope
in Table 3 for amino acids (and a similar result for Cu^{2+} complexes)
strongly suggests that peptide oxygen basicity is an important

factor in determining the $\log K_1$ stability of peptide complexes.
The basicity of this group is only remotely reflected in the original
peptide amino group basicity accounting for the steep slope of 1.9.
Furthermore, these results provide support for the view that the primary determinant of variations in amide carbonyl group basicity is
the basicity of the amino group that participates in the amide bond.
That the open circles in Fig. 4 exhibit some scatter just implies
that the amino group basicity is an imperfect indicator of associated
carbonyl group metal ion binding capability. For both the Ni^{2+} and
Cu^{2+} plots, the scatter would be reduced markedly by the elimination
of three points, those for glycinamide, Gly-Gly, and Gly-β-Ala, all
three of which lie below the least squares line in Fig. 4. The
respective deglycinated species are NH_3, Gly^-, and $β-Ala^-$, and there
are reasons for suspecting that their amino groups contribute in a
somewhat different way to amide carbonyl oxygen basicities than do
the amino groups of most other deglycinated peptides in Table 5 and
Fig. 4. Nonetheless the correlation shown by the open circles in
Fig. 4 for Ni^{2+} and found also for Cu^{2+} indicates a primary role for
amino group basicity in determining peptide carbonyl oxygen basicity
for all peptide bonds.

It has been previously suggested that the high occurrence of
peptide carbonyl oxygens bound to a proline nitrogen in providing an
oxygen donor ligand to Ca^{2+} in proteins results from the high basicity
of this oxygen [92]. Proline never furnishes the carbonyl oxygen
coordinated to Ca^{2+}. The proline amino group is 0.8 log unit more
basic than the next most basic α-amino acid amino group. Thus the
carbonyl oxygen bound to a proline nitrogen in an amide bond should
be the most basic amide oxygen and also the strongest metal ion
binder. That this is the case for Ni^{2+} is proved by the solid circle
point for Gly-Pro atop Fig. 4. A similar top position occurs for
Cu^{2+} (Fig. 4 in Ref. 78).

4.5. Kinetics

Acid-induced dissociation of Ni^{2+}-peptide complexes may occur by two pathways: peptide oxygen protonation with metal ion-nitrogen bond breaking, and general acid catalysis directly at the peptide nitrogen forcing metal ion-nitrogen bond breaking [91,93-98]. Addition of nucleophilic amines to the planar Ni^{2+}-triglycine complex speeds the peptide group protonation reactions and conversion to hexacoordinate complexes [99].

The tripeptide composed of three residues of α-aminoisobutyric acid forms a Ni^{2+} complex with two deprotonated peptide nitrogens that is much slower to form and much more stable than the corresponding complex with triglycine [100]. The former tripeptide bears no hydrogens and two methyl groups on each α carbon. The corresponding trivalent Ni(III) complex is also relatively stable compared to other tripeptide complexes [101,102].

4.6. Trivalent Ni(III) Complexes

Compared to other donor atoms, deprotonated peptide nitrogens aid formation of relatively stable complexes of Ni(III) and Cu(III). The reduction potential for the Ni(III) to Ni^{2+} complex of tetra-glycine with three deprotonated peptide nitrogens is $E^{\circ} = 0.79$ V [103]. Since the Ni(III) complexes tend to be hexacoordinate [104] and the Ni^{2+} complexes planar, introduction of a solvent that reduces coordination in apical positions favors Ni^{2+} and increases the reduction potential [105]. Conversely, addition of unidentate donor ligands such as imidazole and ammonia favors the Ni(III) state and decreases the reduction potential [106]. The ESR spectra of Ni(III) complexes of histidine-containing tripeptides and bleomycin have been described [107]. Stability of Ni(III)-peptide complexes is limited by protonation and loss of coordination of the amino terminal nitrogen in acidic solutions and by decomposition in basic solutions [108].

Margerum and coworkers continue to report studies of Ni(III)-peptide complexes. These include properties of bis dipeptide [109] and bis tripeptide [110] complexes and mixed tripeptide and amine complexes [111]. Also described are electron transfer kinetics between Ni(III)- and Ni^{2+}-deprotonated peptide complexes [112] and with the corresponding Cu^{2+} and Cu(III) complexes [113].

Though Ni(III) has been identified in some hydrogenase enzymes (Chap. 7), it appears unlikely that a peptide-deprotonated complex would be involved.

4.7. Histidine- and Cysteine-Containing Peptides

Recent papers strengthen principles for the modes of interaction of metal ions with peptides. Interaction at the amino terminus with chelation at successive peptide nitrogens occurs with noncoordinating side chains in the amino terminal residue. Presence of a chelating side chain in the amino terminal position hinders peptide deprotonations, which require planarity about the metal ion. Examples of chelating side chains which strongly hinder peptide deprotonations occur when histidine [114,115] and cysteine [116,117] residues occur in the amino terminal position. The ligands then function as substituted histidine and cysteine. An amino terminal methionine is less effective in hindering peptide deprotonations [118].

When histidine and cysteine appear further along the peptide chain they may facilitate peptide deprotonation by providing a strong anchor for a metal ion [78]. Especially strong anchors exist when cysteine or histidine occur as the second or third residue along a peptide chain.

Glycylcysteine serves as a tridentate ligand with an amino nitrogen, a deprotonated peptide nitrogen, and the sulfhydryl groups in a binuclear sulfur-bridged complex with Ni^{2+} [117]. Presumably this structure does not occur with glutathione [116] because of the seven-membered ring demanded by the amino terminal γ-glutamyl residue. One wonders whether glutathione did not evolve with the γ-glutamyl residue to avoid being a strongly tridentate chelate like

glycylcysteine. Glycylglycylcysteine forms a straightforward mononuclear Ni^{2+} complex with amino, two deprotonated peptide nitrogen
and sulfhydryl donors about planar, diamagnetic Ni^{2+} [117].

In strong contrast to histidylglycine chelating as a substituted histidine, glycylhistidine serves as a tridentate ligand with
amino, deprotonated peptide, and imidazole donor atoms [114,119].
This complex dominating at neutral pH is succeeded at pH >9 by formation of a tetramer in which an imidazole pyrrole hydrogen is substituted by the fourth coordination point from planar Ni^{2+} in another
complex [120]. Ni^{2+} is ineffective in promoting across the ring
pyrrole deprotonations without pyrrole hydrogen substitution [22,23].

Interest in the behavior of the tripeptide Gly-Gly-His was
heightened because histidine also appears as the third residue in
bovine and human serum albumins. The tripeptide serves as a quadridentate ligand with amino, two deprotonated amide, and imidazole
nitrogen donors completed by pH 7 in a planar, yellow Ni^{2+} complex
[121]. The amino terminal peptide model for human albumin, L-Asp-L-
Ala-L-His-N-methylamide, also yields a Ni^{2+} complex with the same
donors completed by pH 7 [122]. The titration curve appears in
Fig. 2.

More than a quarter of a century ago it was reported that
equimolar solutions of Ni^{2+} and either bovine or human albumin yields
in neutral and weakly basic solutions a yellow color akin to peptide
deprotonations in small peptides, and it was suggested that Ni^{2+}
interacts at the amino terminus, two deprotonated peptide nitrogens,
and the imidazole ring of the histidyl residue in the third position
[123]. As more recent studies have verified [122,124] in contrast
to the model peptide, the midpoint for the transition to planar,
yellow Ni^{2+} occurs at about pH 7 in albumin with complete conversion
delayed to pH >9. Cu^{2+} binds more strongly at the same albumin site.
The importance of the histidyl residue in the third position is
emphasized by parallel experiments on dog albumin, which contains a
tyrosine residue in the third position. The interaction with both
Cu^{2+} and Ni^{2+} is much weaker than with human albumin and is similar
to the chelates formed by tetraglycine [125]. The phenolic function

on the tyrosine residue in dog albumin is not involved in Ni^{2+}
binding [126]. Even though the Ni^{2+}-albumin complex is only about
2/3 formed at pH 7.4, this fraction is sufficient to furnish an
antigenic determinant in occupational nickel sensitivity [127, and
Chap. 4]. Most Ni^{2+} in plasma is bound by histidine in a 2:1 complex, and exchange of Ni^{2+} between albumin and histidine is slow
[124]. For Ni^{2+} there does not seem to be a physiologic role for
the amino terminal metal ion binding site of albumin. The strongly
chelating quadridentate metal ion binding site at the amino terminus
of bovine and human albumins is specific only for the tetragonal,
covalent metal ions Ni^{2+}, Cu^{2+}, and Pd^{2+}.

4.8. Summary and Applications to Proteins

We exclude nickel-containing metalloproteins, discussed in Chap. 6
and 7 of this volume. The potential reaction of Ni^{2+} with proteins
at the amino terminus is modeled by tetraglycine in Fig. 2; by pH
7.4 only reaction at the amino terminus has occurred, and no peptide
deprotonations have taken place. At pH 7.4 the binding at the amino
terminus is weak. Binding at isolated histidine or cysteinyl side
chains is almost as likely. With a properly placed cysteinyl or
histidyl side chain, as the histidyl residue in position three in
human plasma albumin, the binding becomes much stronger. Figure 2
shows that for Ni^{2+} interaction at the albumin model peptide, the
highly cooperative reaction is complete by pH 7 with loss of three
peptide protons and generation of planar Ni^{2+}. In albumin itself,
however, peptide deprotonations take place with a midpoint only near
pH 7. As a result the free amino acids histidine and cysteine, with
possibly citrate, are the major carriers of Ni^{2+} in the plasma.
More specific binding of Ni^{2+} to proteins receives review in Chap. 6.

REFERENCES

1. R. B. Martin, in "Metal Ions in Biological Systems", Vol. 20
    (H. Sigel, ed.), Marcel Dekker, New York, 1986, p. 21.

2. C. F. Baes, Jr., and R. E. Mesmer, "The Hydrolysis of Cations",
    John Wiley, New York, 1976.

3. D. W. Margerum, G. R. Cayley, D. C. Weatherburn, and G. K.
    Pagenkopf, in "Coordination Chemistry", Volume 2 (A. E. Martell,
    ed.), ACS monograph 174, American Chemical Society, Washington,
    D.C., 1978.

4. A. E. Martell and R. M. Smith, "Critical Stability Constants",
    Vols. 1-5, Plenum, New York, 1974-82.

5. R. B. Martin, *J. Chem. Educ.*, *64*, 402 (1987).

6. "Metal Ions in Biological Systems", Vol. 9 (H. Sigel, ed.),
    Marcel Dekker, New York, 1979.

7. J. T. Edsall, R. B. Martin, and B. R. Hollingworth, *Proc. Natl.
    Acad. Sci. USA*, *44*, 505 (1958).

8. R. B. Martin, J. T. Edsall, D. B. Wetlaufer, and B. R. Holling-
    worth, *J. Biol. Chem.*, *233*, 1429 (1958).

9. E. Coates, P. G. Gardam, and B. Rigg, *Trans. Farad. Soc.*, *62*,
    2577 (1966).

10. R. F. Jameson, G. Hunter, and T. Kiss, *J. Chem. Soc. Perkin II*,
    1105 (1980).

11. T. Kiss and A. Gergely, *J. Chem. Soc. Dalton*, 1951 (1984).

12. R. B. Martin and H. Sigel, *Comments Inorg. Chem.*, in press (1988).

13. S.-H. Kim and R. B. Martin, *Inorg. Chim. Acta*, *91*, 19 (1984).

14. R. B. Martin, *Accts. Chem. Res.*, *18*, 32 (1985).

15. R. B. Martin, *Met. Ions Biol. Syst.*, *23*, 315 (1988).

16. R. B. Martin, *Met. Ions Biol. Syst.*, *9*, 1 (1979).

17. I. Sóvágó, A. Gergely, and J. Posta, *Acta Chim. Acad. Sci.
    Hung.*, *85*, 153 (1975).

18. S.-H. Kim and R. B. Martin, *J. Am. Chem. Soc.*, *106*, 1707 (1984).

19. P. I. Vestues and R. B. Martin, *J. Am. Chem. Soc.*, *102*, 7906
    (1980).

20. J. L. M. Swash and L. D. Pettit, *Inorg. Chim. Acta*, *19*, 19
    (1976).

21. L. P. Battaglia, A. B. Corradi, L. Antolini, G. Marcotrigiano,
    L. Menabue, and G. C. Pellacani, *J. Am. Chem. Soc.*, *104*, 2407
    (1982).

22. R. J. Sundberg and R. B. Martin, *Chem. Rev.*, *74*, 471 (1974).

23. I. Sóvágó, T. Kiss, and A. Gergely, *J. Chem. Soc. Dalton*, 964 (1978).

24. W. F. Reynolds, I. R. Peat, M. H. Freedman, and J. R. Lyerla, Jr., *J. Am. Chem. Soc.*, *95*, 328 (1973).

25. F. Blomberg, W. Maurer, and H. Ruterjans, *J. Am. Chem. Soc.*, *99*, 8149 (1977).

26. J. E. Letter, Jr., and R. B. Jordan, *Inorg. Chem.*, *10*, 2692 (1971).

27. P. Dasgupta and R. B. Jordan, *Inorg. Chem.*, *24*, 2721 (1985).

28. R. B. Martin, *J. Chem. Educ.*, *62*, 789 (1985).

29. J. H. Ritsma, J. C. Van de Grampel, and F. Jellinek, *Recl. Trav. Chim. Pays-Bas*, *88*, 411 (1969).

30. P. J. Morris and R. B. Martin, *J. Inorg. Nucl. Chem.*, *32*, 2891 (1970).

31. K. A. Fraser and M. M. Harding, *J. Chem. Soc. A*, 415 (1967).

32. J. E. Hix and M. M. Jones, *J. Am. Chem. Soc.*, *90*, 1732 (1968).

33. R. W. Hay and P. J. Morris, *Met. Ions Biol. Syst.*, *5*, 173 (1976).

34. D. E. Newlin, M. A. Pellack, and R. Nakon, *J. Am. Chem. Soc.*, *99*, 1078 (1977).

35. S. A. Bedell and R. Nakon, *Inorg. Chem.*, *16*, 3055 (1977).

36. J. H. Ritsma, *J. Inorg. Nucl. Chem.*, *38*, 907 (1976).

37. J. H. Ritsma, G. A. Wiegers, and F. Jellinek, *Rec. Trav. Chim. Pays-Bas*, *84*, 1577 (1965).

38. F. Lafuma, J. Boue, R. Audebert, and C. Quivoron, *Inorg. Chim. Acta*, *66*, 167 (1982).

39. L. D. Pettit and R. J. W. Hefford, *Met. Ions Biol. Syst.*, *9*, 173 (1979).

40. E. Farkas, A. Gergely, and E. Kas, *J. Inorg. Nucl. Chem.*, *43*, 1591 (1981).

41. E. Farkas, E. Homoki, and A. Gergely, *J. Inorg. Nucl. Chem.*, *43*, 624 (1981).

42. A. Gergely, I. Sóvágó, I. Nagypál, and R. Kiraly, *Inorg. Chim. Acta*, *6*, 435 (1972).

43. I. Sóvágó and A. Gergely, *Inorg. Chim. Acta*, *37*, 233 (1979).

44. R. Griesser and H. Sigel, *Inorg. Chem.*, *10*, 2229 (1971).

45. A. Gergely and T. Kiss, *J. Inorg. Nucl. Chem.*, *39*, 109 (1977).

46. M. Tanaka, *J. Inorg. Nucl. Chem.*, *35*, 965 (1973).

47. A. Gergely and I. Sóvágó, *Inorg. Chim. Acta*, *20*, 19 (1976).

48. H. L. Conley, Jr., and R. B. Martin, *J. Phys. Chem.*, *69*, 2923 (1965).

49. E. W. Wilson, Jr., and R. B. Martin, *Arch. Biochem. Biophys.*, *142*, 445 (1971).

50. R. E. Benesch and R. Benesch, *J. Am. Chem. Soc.*, *77*, 5877 (1955).

51. E. L. Elson and J. T. Edsall, *Biochemistry*, *1*, 1 (1962).

52. E. Coates, C. G. Marsden, and B. Rigg, *Trans. Farad. Soc.*, *65*, 3032 (1969).

53. G. Gorin and C. W. Clary, *Arch. Biochem. Biophys.*, *90*, 40 (1960).

54. J. T. Edsall and J. Wyman, "Biophysical Chemistry", Vol. 1, Academic, New York, 1958, pp. 496-503.

55. I. Sóvágó, A. Gergely, B. Harman, and T. Kiss, *J. Inorg. Nucl. Chem.*, *41*, 1629 (1979).

56. A. Gergely and I. Sóvágó, *Met. Ions Biol. Syst.*, *9*, 77 (1979).

57. D. D. Perrin and I. G. Sayce, *J. Chem. Soc. A*, 53 (1968).

58. J. H. Ritsma and F. Jellinek, *Recl. Trav. Chim. Pays-Bas*, *91*, 923 (1972).

59. R. K. Boggess and R. B. Martin, *J. Am. Chem. Soc.*, *97*, 3076 (1975).

60. R. B. Martin, *J. Phys. Chem.*, *75*, 2657 (1971).

61. R. F. Jameson, G. Hunter, and T. Kiss, *J. Chem. Soc. Perkin II*, 1105 (1980).

62. A. Gergely, T. Kiss, G. Deák, and I. Sóvágó, *Inorg. Chim. Acta*, *56*, 35 (1981).

63. A. Gergely, T. Kiss, and G. Deák, *Inorg. Chim. Acta*, *36*, 113 (1979).

64. T. Kiss and A. Gergely, *Inorg. Chim. Acta*, *78*, 247 (1983).

65. A. Gergely and T. Kiss, *Met. Ions Biol. Syst.*, *9*, 143 (1979).

66. T. Kiss and A. Gergely, *Inorg. Chim. Acta*, *36*, 31 (1979).

67. T. Kiss, G. Deák, and A. Gergely, *Inorg. Chim. Acta*, *91*, 269 (1984).

68. R. B. Martin, *J. Inorg. Biochem.*, *28*, 181 (1986).

69. R. B. Martin, *Clin. Chem.*, *32*, 1797 (1986).

70. G. R. Hedwig, J. R. Liddle, and R. D. Reeves, *Aust. J. Chem.*, *33*, 1685 (1980).

71. E. N. Baker, H. M. Baker, D. F. Anderson, and R. D. Reeves, *Inorg. Chim. Acta*, *78*, 281 (1983).

72. F. R. Still and P. Wikberg, *Inorg. Chim. Acta*, *46*, 153 (1980).

73. P. M. May, P. W. Linder, and D. R. Williams, *J. Chem. Soc. Dalton*, 588 (1977).

74. A. Cole, C. Furnival, Z.-X. Huang, D. C. Jones, P. M. May, G. L. Smith, J. Whittaker, and D. R. Williams, *Inorg. Chim. Acta, 108*, 165 (1985).

75. J. Lerivrey, B. Dubois, P. Decock, G. Micera, J. Urbanska, and H. Kozlowski, *Inorg. Chim. Acta, 125*, 187 (1986).

76. E. B. V. Appelman-Lippens, M. W. G. DeBolster, D. N. Tiemersma, and G. Visser-Luirink, *Inorg. Chim. Acta, 108*, 209 (1985).

77. S. Yano, Y. Sakai, K. Toriumi, T. Ito, H. Ito, and S. Yoshikawa, *Inorg. Chem., 24*, 498 (1985).

78. H. Sigel and R. B. Martin, *Chem. Rev., 82*, 385 (1982).

79. R. B. Martin, M. Chamberlin, and J. T. Edsall, *J. Am. Chem. Soc., 82*, 495 (1960).

80. T. F. Dorigatti and E. J. Billo, *J. Inorg. Nucl. Chem., 37*, 1515 (1975).

81. P. J. Morris and R. B. Martin, *Inorg. Chem., 10*, 964 (1971).

82. G. Brookes and L. D. Pettit, *J. Chem. Soc. Dalton*, 2106 (1975).

83. W. S. Kittl and B. M. Rode, *Inorg. Chim. Acta, 66*, 105 (1982).

84. J. W. Bunting and K. M. Thong, *Can. J. Chem., 48*, 1654 (1970).

85. R. B. Martin, *Met. Ions Biol. Syst., 1*, 129 (1974).

86. R. A. Haines and M. Reimer, *Inorg. Chem., 12*, 1482 (1973).

87. J. W. Chang and R. B. Martin, *J. Phys. Chem., 73*, 4277 (1969).

88. R. B. Martin, J. M. Tsangaris, and J. W. Chang, *J. Am. Chem. Soc., 90*, 821 (1968).

89. J. M. Tsangaris, J. W. Chang, and R. B. Martin, *Arch. Biochem. Biophys., 130*, 53 (1969).

90. J. M. Tsangaris, J. W. Chang, and R. B. Martin, *J. Am. Chem. Soc., 91*, 726 (1969).

91. E. J. Billo and D. W. Margerum, *J. Am. Chem. Soc., 92*, 6811 (1970).

92. R. B. Martin, *Met. Ions Biol. Syst., 17*, 1 (1984).

93. C. F. V. Mason, P. I. Chamberlain, and R. G. Wilkins, *Inorg. Chem., 10*, 2345 (1971).

94. E. B. Paniago and D. W. Margerum, *J. Am. Chem. Soc., 94*, 6704 (1972).

95. G. K. Pagenkopf and V. T. Brice, *Inorg. Chem., 14*, 3118 (1975).

96. C. E. Bannister and D. W. Margerum, *Inorg. Chem., 20*, 3149 (1981).

97. J. M. T. Raycheba and D. W. Margerum, *Inorg. Chem.*, *19*, 497 (1980).

98. C. E. Bannister, J. M. T. Raycheba, and D. W. Margerum, *Inorg. Chem.*, *21*, 1106 (1982).

99. E. J. Billo, G. F. Smith, and D. W. Margerum, *J. Am. Chem. Soc.*, *93*, 2635 (1971).

100. W. R. Kennedy and D. W. Margerum, *Inorg. Chem.*, *24*, 2490 (1985).

101. S. T. Kirksey, Jr., T. A. Neubecker, and D. W. Margerum, *J. Am. Chem. Soc.*, *101*, 1631 (1979).

102. J. M. T. Raycheba and D. W. Margerum, *Inorg. Chem.*, *20*, 1441 (1981).

103. F. P. Bossu and D. W. Margerum, *Inorg. Chem.*, *16*, 1210 (1977).

104. A. G. Lappin, C. K. Murray, and D. W. Margerum, *Inorg. Chem.*, *17*, 1630 (1978).

105. M. P. Youngblood and D. W. Margerum, *Inorg. Chem.*, *19*, 3068 (1980).

106. C. K. Murray and D. W. Margerum, *Inorg. Chem.*, *21*, 3501 (1982).

107. Y. Sugiura and Y. Mino, *Inorg. Chem.*, *18*, 1336 (1979).

108. E. J. Subak, Jr., V. M. Loyola, and D. W. Margerum, *Inorg. Chem.*, *24*, 4350 (1985).

109. S. A. Jacobs and D. W. Margerum, *Inorg. Chem.*, *23*, 1195 (1984).

110. G. E. Kirvan and D. W. Margerum, *Inorg. Chem.*, *24*, 3245 (1985).

111. T. L. Pappenhagen, W. R. Kennedy, C. P. Bowers, and D. W. Margerum, *Inorg. Chem.*, *24*, 4356 (1985).

112. C. K. Murray and D. W. Margerum, *Inorg. Chem.*, *22*, 463 (1983).

113. G. D. Owens, D. A. Phillips, J. J. Czarnecki, J. M. T. Raycheba, and D. W. Margerum, *Inorg. Chem.*, *23*, 1345 (1984).

114. E. Farkas and A. Gergely, *J. Chem. Soc. Dalton*, 1545 (1983).

115. B. Radomska, T. Kiss, and I. Sóvágó, in prep.

116. I. Sóvágó and R. B. Martin, *J. Inorg. Nucl. Chem.*, *43*, 425 (1981).

117. H. Kozlowski, B. Decock-LeReverend, D. Ficheux, C. Loucheux, and I. Sóvágó, *J. Inorg. Biochem.*, *29*, 187 (1987).

118. I. Sóvágó and G. Petocz, *J. Chem. Soc. Dalton*, 1717 (1987).

119. R. B. Martin and J. T. Edsall, *J. Am. Chem. Soc.*, *82*, 1107 (1960).

120. P. J. Morris and R. B. Martin, *J. Inorg. Nucl. Chem.*, *33*, 2913 (1971).

121. T. Sakurai and A. Nakahara, *Inorg. Chem.*, *19*, 847 (1980).

122. J. D. Glennon and B. Sarkar, *Biochem. J.*, *203*,15 (1982).

123. R. B. Martin, *Fed. Proc. 20*, Supplement 10, 54 (1961).

124. S. H. Laurie and D. E. Pratt, *J. Inorg. Biochem.*, *28*, 431 (1986).

125. J. D. Glennon and B. Sarkar, *Biochem. J.*, *203*, 25 (1982).

126. J. D. Glennon, D. W. Hughes, and B. Sarkar, *J. Inorg. Biochem.*, *19*, 281 (1983).

127. J. Dolovich, S. L. Evans, and E. Nieboer, *Br. J. Indust. Med.*, *41*, 51 (1984).

6

Nickel in Proteins and Enzymes

Robert K. Andrews, Robert L. Blakeley, and Burt Zerner
Department of Biochemistry
University of Queensland
St. Lucia, Queensland 4067 Australia

> From Beowulf's door
> To the strumpets of yore,
> From small ugly horses
> To goats' disease courses,
> From netherworld jives
> To the clicking of knives,
> Nickel has always been part of our lives.
>
> ANON.

1. INTRODUCTION

Nickel has undoubtedly been of great value to mankind from the
earliest times: it has been found in significant (2-5%), but probably fortuitous concentration in bronze and copper artifacts from
Mesopotamia, Egypt, South Africa, Syria, India, and China, the oldest
of which date from 3500 B.C. [1]. Bactrian coins which have survived
from the reigns of Euthydemus II, Agathocles, and Pantaleon (ca. 200-
165 B.C.) are remarkable alloys containing approximately 19% Ni, 77%
Cu, and 2% Fe [2]. *Pai thung* (modern romanization: *baitong*), the
copper-nickel-zinc alloy of the Chinese known as *white copper* which
dates securely from 945 A.D. [3], found its way to Europe in the
sixteenth century.

Two millenia later, copper-nickel alloys were reintroduced into
coinage, and *pai thung* was reinvented as German silver (*argentan* or
nickel silver).

But nickel has always been a rogue! When the silver miners of
Saxony in the middle ages first came across *niccolite* (NiAs), a
heavy reddish brown ore frequently stained green, they called it
Kupfernickel (rogue copper) because all attempts at smelting it

failed to yield copper (or silver): the ore had been jinxed by the
netherdwelling Nicks and Kobolds [4].

The pure element was isolated by the Swede A. F. Cronstedt in
1751, some 13 years after his mentor Georg Brandt had isolated
cobalt, and was given its present name in 1754. The third Swedish
metal, manganese, was isolated 20 years later [5].

Two hundred years on, the behavior of nickel was true to form.
When Walter Mertz reviewed nutritional trace element research in
1969, nickel was unique among the elements of the first transition
series, since all its neighbors [vanadium, chromium, manganese, iron,
cobalt, copper (and zinc)] had been reported to be essential or as
having a particular biological function [6]. He commented with acute
farsight: "This element is particularly interesting chemically
because of its easy transition between several coordination structures".

2. NICKEL IN BIOLOGY

2.1. Availability of Metals

With the benefit of hindsight, the absence of nickel from the foregoing group of biologically significant elements was somewhat anomalous from the point of view of availability of metal ions to life
evolving on earth, because its abundance in the earth's crust and
in seawater is comparable to that of many other transition metal
ions with long established biological importance (Table 1).

Much of the nickel available to living systems derives ultimately from the soil. The world mean for nickel in soils is reported
to be about 20 µg/g, and examples are commonly in the range 1-100
µg/g [13]. Variations in soil content of nickel partially explain
the range of nickel concentrations among individuals of the same
plant species as well as the range (typically 0.05-5 µg/g dry weight)
displayed by plants in general including those used as food for animals and humans [13-16]. The occurrence of metal ions in living

TABLE 1

Mean Abundance of Some Metallic Elements

At. no.	Element	Earth's crust[a] (μg/g)	Ocean water[b,c] (μg/liter)	Human plasma or serum[c,d] (μg/liter)
3	Lithium	20	180	31
11	Sodium	24×10^3	10.77×10^6	3.2×10^6
12	Magnesium	20×10^3	1.29×10^6	0.04×10^6
13	Aluminum	82×10^3	2	10
19	Potassium	24×10^3	0.399×10^6	0.17×10^6
20	Calcium	42×10^3	0.412×10^6	0.20×10^6
23	Vanadium	135	2.5	0.03
24	Chromium	100	0.3	0.15
25	Manganese	950	0.2	0.5
26	Iron	56×10^3	2	1200
27	Cobalt	25	0.02	0.1
28	Nickel	75	0.56	0.3
29	Copper	55	0.25	1000
30	Zinc	70	4.9	1000
42	Molybdenum	1.5	10	0.6

[a]Adapted from Ref. 7.
[b]Adapted from Refs. 8 and 9. Normalized to a chloride content of 19.35 g/liter.
[c]Considerable variations and uncertainties occur for most elements which are present at low concentrations.
[d]Ref. 10 except for Li, Na, Mg, K, and Ca (Ref. 11) and Ni (Ref. 12).

systems has been expressed in terms of the enrichment factor of an element: the ratio of the concentration of the element in an organism to its concentration in the earth's crust [17]. The enrichment factor for nickel in fungi, plants, and land animals is within the broad range observed for other transition metal ions. Despite many uncertainties about the actual concentration of nickel in many living systems, the common presence of nickel in terrestrial and aquatic

biota [13,16,18,19] ensures that this element is potentially available for a suitable biological function in virtually all forms of
life.

2.2. Nickel in Biological Material

The history of analysis of nickel in human serum serves to illustrate
a major limitation in any proposition based on the concentration of
nickel in biological material. For the decade prior to 1970, the
generally accepted mean concentration of nickel in plasma or serum of
healthy human beings was about 21 μg/liter, since similar values were
measured in four different laboratories [10]. During the 14-year
period which began in 1970, independently measured values were all
about 2.6 ± 1 μg/liter [10]. In 1984, the mean nickel concentration
in serum of healthy adult humans was reported to be 0.46 ± 0.26 μg/
liter (39 subjects) [20], and this lower level was soon confirmed in
the same laboratory (0.28 ± 0.24 μg/liter in 30 subjects) [12] and
elsewhere (0.44 ± 0.18 μg/liter in 18 subjects) [21]. The decrease
from 2.6 to 0.46 μg/liter for the mean serum nickel concentration of
healthy individuals as measured in the same laboratory was ascribed
to refinement of techniques and equipment [20]. It is clear that
accurate measurement of low concentrations of nickel in biological
specimens is very difficult, and much of the literature on nickel
content of tissues may be unreliable.

    An argument sometimes offered in support of homeostatic control
of serum nickel concentration in vertebrates is that the mean serum
concentrations of nickel measured in one laboratory are rather similar (2.0-9.3 μg/liter) across 15 different species including human
beings [22]. Since the mean value for human serum (2.6 μg/liter) in
this work is apparently in error by a factor of ~9 [12], these data
cannot comment critically on the question of nickel homeostasis in
vertebrates.

    The mean concentration of nickel in serum or plasma of healthy
humans is compared with that of other metals in Table 1. The serum

concentration of nickel is very low compared to that of the trace
elements iron, copper, and zinc, but is comparable to that of the
ultratrace elements chromium, manganese, cobalt, and molybdenum (the
latter three of which have known enzymatic or coenzymatic roles).
Further, the whole-body load of nickel in human beings appears to be
similar to or somewhat higher than that of these last four essential
metals [23]. While it may be tempting to interpret the presence of
nickel in a biological system as evidence for a specific role, such a
conclusion would be unwarranted. For example, human serum contains
much more aluminum than several of the transition metal ions with
known enzymatic roles (Table 1), but there does not appear to be any
evidence that aluminum has a fundamental biochemical role in humans.

2.3. The Functions of Nickel in Biology

One of the simplest definitions of essentiality is that "an element
is essential if its deficiency reproducibly results in impairment of
function from optimal to suboptimal" [6]. Evidence for the essentiality of nickel generally [24], and in three nonruminant vertebrates (chickens, pigs, and rats), has been recently reviewed [25]
(see Chap. 4), and Nielsen has shown that inappropriate diets in some
of the early nutritional studies with nickel may account for the lack
of reproducibility of certain results [26]. In nutritional studies,
there is the further problem at the molecular level that many enzymes
may be activated by any of several metal ions to a high degree of
catalytic efficiency (Sec. 8). Thus a readily available zinc ion
(say) might well exert a sparing effect by replacing a withheld
nickel ion in the latter's normal functional role in a metalloenzyme. *It follows that an externally experimentally valid nutritional study may fail to identify the existence of a normal functional role for a trace metal ion.*

Because of the variability in nickel concentration between
different individuals in any given species [13,15,18,19], it may be
surmised that biological systems have flexibility in dealing with

nickel ions. However, it is not necessarily easy to decide whether
a given component binds nickel ion because (1) nickel ion happens to
be there, (2) the component has a purpose such as storage, transport,
or excretion of nickel ion (and perhaps other metal ions), or (3)
the nickel ion has a functional catalytic or binding role in the
component. Nickel toxicity is a well-known phenomenon in microorganisms, plants, and animals (see Chap. 10), and it poses a whole
new set of questions about interactions of nickel ion with biological
molecules. Moreover, hyperaccumulation of nickel ion occurs in certain plant species [15] (see Chap. 3), and this phenomenon apparently
involves complexes of Ni(II) with relatively simple chelating agents
such as malonate, malate, and citrate.

Consequently, if the role of a metal ion in electron transfer
is excluded, its function in a biological system may be considered
in two categories, viz., *catalysis* and *binding*. In *catalytic systems*,
the metal ion takes part in the active site chemistry of enzymes and,
potentially, in other reactions such as the transesterification and
hydrolysis reactions of RNA [27]. *Binding* includes all other interactions of the metal ion with biological molecules (for nickel, see
Sec. 3 and Chaps. 5, 8, and 9). Detailed chemical studies may be
required in order to decide which of these two essential categories
applies to a newly recognized system, but the possible complications
do not alter the essential question.

3. NICKEL-BINDING PROTEINS

3.1. Isolation of Nickel-Binding Proteins

Proteins which bind Ni^{2+} (Chap. 5) are potentially of considerable
interest because of the light they may shed on the enzymology and
toxicity of Ni^{2+}. A brief summary is included here. Some enzymes
which are activated by Ni^{2+} are described in Sec. 8.

Identification of molecules which bind to Ni^{2+} in animals has
been based largely on treatment of serum or intact animals with

$^{63}Ni^{2+}$ which is a soft β emitter (0.067 MeV) with a half-life of
92 years. After fractionation by chromatographic or electrophoretic
methods, Ni^{2+}-containing components have been located by liquid
scintillation counting or autoradiography. Such experiments require
careful consideration of experimental conditions to ensure that the
rate and position of equilibria of Ni^{2+} complexation reactions do
not lead to artifacts because of the time scale and conditions of
equilibration and separation procedures. For example, Tris buffers
bind to Ni^{2+} at neutral pH [28,29], and they form polynuclear complexes with Ni^{2+} ion at alkaline pH [28a].

   Some tracer experiments in the literature have used $^{63}Ni^{2+}$ of
low specific activity. For comparison, the specific activity of
carrier-free ^{63}Ni may readily be calculated to be 66.2 Ci/g, and the
specific activity of typical commercially available ^{63}Ni is about
10 Ci/g. Tracer experiments to identify natural nickel-binding proteins in vivo require the highest possible specific activity in
order to avoid swamping the system with levels of nickel substantially higher than the normal ones.

3.2. Nickel-Binding Proteins in Serum

3.2.1. Serum Albumin

This 69-kDa protein is present at ~42 mg/ml (0.6 mM) in human serum
and accounts for ~60% by weight of the protein in serum [30]. Serum
albumin has a single site to which most divalent metal ions bind
reversibly, but it also binds additional metal ions with ill-defined
equilibria. The majority of Ni^{2+} in human serum is associated with
albumin, and the stability constant of the 1:1 Ni^{2+}-albumin complex
is $10^{9.57}$ M^{-1} at 25°C in 0.1 M N-ethylmorpholinium chloride buffer
(pH 7.53, 0.06 M in NaCl) [31]. The metal ion binding site has been
characterized in terms of the effect of pH on the electronic absorption spectrum of Ni^{2+} and Cu^{2+} complexes of albumin and of synthetic
N-terminal model peptides together with NMR studies on the Cu^{2+} and
Ni^{2+} complexes of the 24-residue N-terminal fragment of albumin.

Results suggest that Ni^{2+} in its complex with the large N-terminal
fragment is diamagnetic at pH 8.0 and has five ligands (the α-amino
group and the side chain of the N-terminal Asp residue, the deprotonated amide nitrogens of Ala-2 and His-3, and the side-chain
imidazole of His-3) [32], and that the same structure occurs in the
Ni^{2+}-albumin complex (see Sec. 6.4.) [31]. Resonances of the Ni^{2+}
ligands of the peptide are reported to be invariant between pH 6.0
and 11.0 [32], whereas the large variations which occur in the visible absorption spectrum of 1 mM Ni^{2+}-albumin between pH 6.36 and
10.32 [31] indicate that the Ni^{2+}-albumin system may not yet be
completely described.

3.2.2. "Nickeloplasmin" and α_2-Macroglobulin

The term "nickeloplasmin" was coined by Sunderman in 1971 to identify
a putative large Ni^{2+}-containing protein in rabbit and human serum.
Upon gel filtration of serum on Sephadex G-200 (0.1 M $Tris.H^+Cl^-$
buffer, pH 8.6), a peak of Ni^{2+} eluted near the void volume [33,34].
The pooled nickel peak was submitted to DEAE-cellulose chromatography
at pH 8.6 using a linear gradient of Tris buffer at the same pH for
elution. This resulted in a peak of Ni^{2+} which was "distinctly separated" from α_2-macroglobulin (see below) but was located among
unresolved protein peaks [33,34]. On agarose gels, the major protein component of the nickel peak behaved electrophoretically as an
α_2-globulin, but no evidence was adduced to indicate that Ni^{2+} was
or had been associated with the major protein [34]. In subsequent
experiments involving $^{63}Ni^{2+}$, the major Ni^{2+}-binding macroglobulin in
rabbit serum was described as an α_1-macroglobulin but stoichiometry
of nickel binding was not reported [35]. The existence of nickeloplasmin appears to be unsubstantiated.

Human α_2-macroglobulin is a glycoprotein composed of four
identical ~185-kDa subunits which are covalently linked in pairs by
disulfide bonds. It contains 8-11% carbohydrate and is present in
serum at a concentration of about 2.5 mg/ml. Its primary physiologic function is thought to be the rapid clearance of toxic

proteinases from serum via a proteolytically inactive macroglobulin-proteinase complex [36]. The tetramer appears to bind ~20 Zn^{2+} ions with an apparent dissociation constant of 0.8 μM in 20 mM $Tris.H^+Cl^-$ buffer (pH 7.4, 150 mM in NaCl) [37], and this may be compared to an early report of three to seven Zn^{2+} ions per tetramer in the purified protein [38]. Native α_2-macroglobulin and the α_2-macroglobulin-trypsin complex bind comparable amounts of Zn^{2+}, and the proteinase inhibitory activity of macroglobulin is not affected by zinc binding at physiologic concentrations of Zn^{2+} (10-15 μM) nor by the removal of Zn^{2+} by EDTA [37]. The interaction of α_2-macroglobulin with Cd^{2+} and Ni^{2+} appears to be different from that with Zn^{2+}, but the former two interactions have not been thoroughly investigated [37]. There appears to be no evidence which would enable critical comment on whether or not Ni^{2+} in serum is significantly complexed by α_2-macroglobulin.

3.2.3. Histidine-Rich Glycoprotein

A glycoprotein which reversibly binds heme and several divalent metal ions has been isolated from rabbit and human serum [39]. The size of the protein, the number of metal ions bound, and the dissociation constants of the complexes all appear to be uncertain [39-41]. It has been suggested that a small percentage of the Ni^{2+} in human serum may be associated with this glycoprotein [41].

3.3. Nickel-Binding Proteins in Kidney, Lung, and Liver

When rats are treated parenterally with small doses of soluble $^{63}Ni^{2+}$ salts, most of the radioactivity is excreted via the urine within 3 days. The organ with the highest concentration of nickel after 3 days is the kidney [42]. In homogenates of the whole kidney of $^{63}Ni^{2+}$-treated rats, the majority of the radioactivity is in the supernatant after the microsomal particles have been removed by ultracentrifugation [43,44]. An acidic Ni^{2+}-binding oligopeptide

(estimated molecular weight, ~4 kDa) has been isolated from the
supernatant of human kidney [44], and the kidney supernatant of
$^{63}Ni^{2+}$-treated rats has been fractionated by gel filtration into
several partially resolved but as yet unidentified peaks of radioactivity [43,44].

It was reported in 1983 that most of the radioactivity in the
kidney supernatant of $^{63}Ni^{2+}$-treated rats was associated with a ~15-
kDa glycoprotein [45], but comparison with subsequent results from
the same laboratory [44] indicates that the suggested protein is an
artifact, possibly associated with effects of a Tris buffer. Tris
buffers were also used in a study of incorporation of $^{63}Ni^{2+}$ into
lung and liver tissues of mice [46]. A large number of radioactive
protein bands were resolved electrophoretically under denaturing
conditions using sodium dodecyl sulfate-polyacrylamide gels [46].
It seems unlikely that there will be any simple relationship between
the denatured proteins which bind $^{63}Ni^{2+}$ and the Ni^{2+}-binding proteins
in vivo.

3.4. Other Nickel-Binding Proteins

Ni^{2+} ion binds to several proteins at sites which are normally
occupied by other divalent metal ions, and a few examples are
summarized below.

3.4.1. Lectins (Saccharide-Binding Proteins)

Lentil lectin, pea lectin, and jack bean concanavalin A are D-glucose/
D-mannose binding proteins with very similar physicochemical, structural, and biological properties [47]. In the native state, these
proteins contain one Mn^{2+} and one Ca^{2+} per subunit. However, the
metal ions are readily exchanged for several divalent transition
metal ions under appropriate conditions [47,48]. The Ca^{2+}-Co^{2+},
Ca^{2+}-Ni^{2+}, Ca^{2+}-Zn^{2+}, and also the Cd^{2+}-Cd^{2+} forms of the lentil and
pea lectins are all as active as the Ca^{2+}-Mn^{2+} form with respect to

reversible binding to saccharides, and a Ca^{2+}-Ca^{2+} form also exists
[47]. In the (25.5-kDa) subunit of concanavalin A, Mn^{2+} and Ca^{2+}
bind at adjoining sites to form a binuclear complex with two octa-
hedra sharing a common edge. The six ligands of Mn^{2+} are two water
molecules and the side chains of Glu-8, Asp-10, Asp-19, and His-24.
The six ligands of Ca^{2+} are two water molecules, the side chains of
Asp-10, Asn-14, and Asp-19, and the peptide carbonyl oxygen of Tyr-12
[49]. Ni^{2+} presumably binds at the Mn^{2+} site.

3.4.2. Azurin

Azurin is a 14-kDa type I, blue copper protein from *Pseudomonas*
aeruginosa. It normally contains a single Cu^{2+} with distorted tetrahedral coordination by the side chains of His-46, Cys-112, His-117,
and Met-121 [50]. After the Cu^{2+} has been removed, the apoprotein
will slowly sequester a single Ni^{2+} at the Cu^{2+} site. The magnetic
moment (μ_{eff} = 3.2 BM) [51] and the ligand-to-metal charge transfer
transitions [52] of Ni^{2+}-azurin are consistent with a (distorted)
tetrahedral coordination for the metal ion.

3.4.3. Metallothionein

Metallothioneins are sulfhydryl-rich highly conserved proteins
present in all vertebrates, invertebrates, and fungi. Mammalian
forms are found in many tissues, particularly kidney and liver.
They contain 20 cysteine residues among a total of 61 or 62 amino
acid residues (6.1 kDa), and they are isolated as a complex with up
to seven Zn^{2+} and/or Cd^{2+} ions. All 20 cysteine residues participate in binding metal ions [53]. Metallothionein complexes with
seven Co^{2+}, Ni^{2+}, Zn^{2+}, Cd^{2+}, or Hg^{2+} ions have been investigated,
and the seven metal ions are each tetrahedrally coordinated in one
trinuclear and one tetranuclear cluster with a mixture of bridging
and nonbridging cysteine side-chain ligands (see Sec. 9) [54]. The
complex of metallothionein with Cd^{2+} is chemically and metabolically
relatively stable, and detoxification of Cd^{2+} appears to be an important function of this protein [55]. It does not appear to be known

whether metallothionein has a significant role in either the normal
or abnormal metabolism of nickel.

3.4.4. Hemoproteins

Hemoglobin (Hb) and myoglobin (Mb) have been reconstituted with
Ni^{2+}-protoporphyrin IX in place of the normal Fe^{2+}-protoporphyrin IX
(heme) to give ^{Ni}Hb and ^{Ni}Mb, respectively. In ^{Ni}Mb, the Ni^{2+} is
five-coordinate and appears to have the imidazole side chain of a
histidine residue as the axial ligand. In ^{Ni}Hb, there are two distinct types of Ni^{2+}. One type has four-coordinate Ni^{2+}-protoporphyrin
IX and the other is equivalent to that in ^{Ni}Mb. In contrast to normal
hemoglobin, ^{Ni}Hb does not bind oxygen [56].

3.4.5. Conalbumin (Ovotransferrin)

This protein is a transferrin found in the white of chickens' eggs,
and it has two binding sites for multivalent metal ions. An anion
such as bicarbonate is bound concomitantly with each metal ion. The
function of conalbumin is apparently the storage of metal ions, and
the relative stability of the transition metal complexes is in the
following order: $Fe^{3+} > Cr^{3+}$, $Cu^{2+} > Mn^{2+}$, Co^{2+}, $Cd^{2+} > Zn^{2+} > Ni^{2+}$
[57]. The side chains of two or three tyrosine residues and one or
two histidine residues are involved in the binding of each metal ion
in serum transferrin with which conalbumin is highly homologous [58].

3.4.6. G-Actin

At low ionic strength, the protein actin from rabbit muscle exists
as a globular 42-kDa monomer called G-actin. In the presence of a
divalent metal ion (e.g., Mg^{2+}, Ca^{2+}, Mn^{2+}, Co^{2+}, Ni^{2+}), ATP binds
very strongly to G-actin. It has been suggested that the divalent
metal ion is complexed by the two terminal phosphate residues of ATP
in the ternary complex [59,60]. Magnesium ion is present at high
concentration in muscle cells, and it is presumably the normal metal
ion in the ternary complex.

3.4.7. *T4D Bacteriophage Baseplate Protein*

The baseplate of T4D bacteriophage is a multiprotein complex with an aggregate molecular mass of ~8 MDa. The normal five or six Zn^{2+} ions in the baseplate proteins may be largely replaced by Co^{2+} and apparently by Ni^{2+} in vivo to give a baseplate with enhanced stability [61].

4. ENZYMES AND METAL IONS

4.1. Classification of Enzymes

In considering the role of nickel and other metals in biology, it is useful to provide a general perspective on metalloproteins and metalloenzymes. Ideally, this would be done through the system of enzyme classification promulgated by the Nomenclature Commission (formerly the Commission on Enzymes) of the International Union of Biochemistry (IUB) [62]. This system recognizes six classes of enzymes according to the type of reaction that is catalyzed:

Class 1:	oxidoreductases	Class 2:	transferases
Class 3:	hydrolases	Class 4:	lyases
Class 5:	isomerases	Class 6:	ligases

While the enormous diversity of enzymes and their sources necessarily leads to some ambiguities in any classification system, the IUB system has assisted the enzymologist by providing a unique four-part descriptor for each enzyme (e.g., EC 3.5.1.5 for urease) or in many cases for an arbitrary group of enzymes (e.g., EC 3.1.3.2 for all types of acid phosphatase). One of the strengths of the IUB system is its potential for computerized retrieval of references to a particular enzyme. The EC designations are now listed in the Chemical Abstracts Index Guide under "E.C." (Enzyme Commission) leading to the preferred name and the registry number of the enzyme as a chemical substance. These may be used for computer searches of the Chemical Abstracts files. Further, in the Registry file on STN International, EC numbers may be searched in the complete name field.

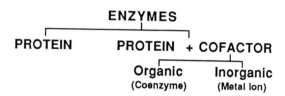

FIG. 1. Classification of enzymes according to the chemical nature
of their active site constituents.

One of the limitations of the IUB and the Chemical Abstracts
systems is that they generally provide *little or no guide to the*
essential components of the active site of an enzyme. It is therefore useful for the enzymologist to think in terms of the classification scheme in Fig. 1. Many enzymes effect catalysis utilizing
nothing but protein (i.e., the amino acid side chains and the polypeptide backbone) for catalytic machinery. Many other enzymes
utilize also a cofactor (a nonprotein component) to provide an
essential part of the chemistry of catalysis. This cofactor may
be either a metal ion or a small organic molecule called a coenzyme.

4.2. Vitamins, Coenzymes, and Epigenetic
Covalent Modification of Enzymes

By definition, a vitamin is an organic molecule which is required
in trace amounts for effective growth and reproduction of a given
species, and which that species cannot synthesize and must obtain
from external sources. Vitamin K, possibly vitamin C, and all of
the B vitamins are either coenzymes or precursors of coenzymes.
New coenzymes are discovered every so often by enzymologists, and
more vitamins for human beings may yet await discovery. Although
a few coenzymes (e.g., protoporphyrin IX for heme) may be synthe-
sized de novo by human beings, many cannot.

The relationship between an enzyme and a coenzyme may be
described thus:

Apoenzyme + coenzyme = holoenzyme

Here, the *apoenzyme* is the "protein-only" component whereas the *holoenzyme* is the complete, functional enzyme. In general, the coenzyme is responsible for the essential chemistry of enzymatic catalysis. Often, the coenzyme without any protein catalyzes the same reaction as the holoenzyme, but at a rate many orders of magnitude smaller and with no substantial selectivity for the structure of the substrate (i.e., little specificity). The apoenzyme by itself is devoid of catalytic activity, but it contributes substrate selectivity and high catalytic efficiency to the holoenzyme.

Coenzymes always undergo covalent bond changes and/or redox changes during enzymatic catalysis, and therefore are necessarily recyclable substrates for their associated enzyme. In some holoenzymes, the coenzyme (e.g., biotin [63,64] and sometimes flavins [65,66]) is covalently attached to the protein, generally via the side chain of an amino acid residue. In other holoenzymes the coenzyme is not covalently attached but instead binds reversibly using physical interactions (hydrophobic and electrostatic interactions and hydrogen bonds).

Many enzymes undergo *epigenetic* covalent modification, i.e., modification during or after synthesis of the polypeptide backbone on the ribosomes [67,68]. While some of the covalent modifications involve attachment of a coenzyme, others alter the physical and/or catalytic properties of the enzyme without accompanying covalent bond changes in the modifying agent during catalysis. An organic molecule which modifies the catalytic efficiency of an enzyme without undergoing covalent bond changes during catalysis is often referred to as a *modifier*, even if the same molecule may be a coenzyme in a different enzymatic system.

4.3. Metalloenzymes and Metalloproteins

The general relationship between a cofactor and a protein may be expressed thus:

Apoprotein + cofactor = holoprotein

where the *apoprotein* is the protein-only component and the *holo-protein* is the complete, functional protein. If the cofactor is a metal ion, the holoprotein is a *metalloprotein*. The term *cofactor* may also be used for any small molecule which has an essential function in any holoprotein which serves a binding, storage, or transport function even though the holoprotein is not an enzyme. In this sense, heme is a cofactor in hemoglobin. An organic cofactor or coenzyme which is either covalently attached or otherwise essentially nonexchanging with free cofactor has sometimes been called a *prosthetic group* [69].

The relationship between an enzyme and a metal ion may often be described as follows:

Apoenzyme + metal ion = holoenzyme

Such holoenzymes have been called *metalloenzymes* [70] even though the metal ion sometimes takes no part in the catalytic reaction. Metal ions clearly may serve either directly in catalysis or as a *modifier* which is concerned with conformation or stability of the enzyme, and we suggest that the term metalloenzyme should denote the former function. After the stoichiometry of a metalloenzyme has been established (in terms of moles of metal ion per mole of protein), the next critical question is whether the functional role of the metal ion is (1) structural or (2) primarily catalytic. These questions also apply to those holoenzymes which contain both a coenzyme and a metal ion and to those which contain multiple metal ions.

An enzyme is often said to be *metal ion-activated* if addition of a metal ion to the assay system results in an increased catalytic rate. This is a useful but crude phenomenological description which by itself clearly cannot carry any implication about the detailed chemical role of the metal ion.

The trace and ultratrace metal requirements of an organism are often understood in terms of a cofactor relationship between the metal ions and certain enzymes or proteins. However, several metal ions are essential in animals in terms of a "dietary requirement" but do not yet have a clearly defined chemical role. Such metal ions include V, Cr, and Ni [23].

4.4. Catalytically Essential Metal Ions

Two limiting categories of interaction between an enzyme and a
catalytically essential metal ion may be defined. These categories
need to be examined when dealing with any newly isolated or otherwise ill-characterized metalloenzyme. To the extent that the chemistry becomes understood, it speaks for itself and the need for
categorization recedes.

(1) *An individual metal ion involved in enzymatic catalysis
interacts primarily with the substrate.* The metal ion (M^{n+}) may
never actually coordinate to the side chains of the enzyme protein,
or it may coordinate only transiently during catalysis. A M^{n+}-
enzyme-substrate complex of the latter type would at best be too
unstable to allow its isolation free of excess metal ions. Data on
labile M^{n+}-enzyme, M^{n+}-substrate, and M^{n+}-enzyme-substrate complexes
have been gathered by magnetic resonance techniques which are sensitive to individual species or to the statistical behavior of several
equilibrating species [71]. Sometimes a substrate-M^{n+} complex is
more stable than the corresponding enzyme-M^{n+} complex [59]. For
example, virtually all enzymatic reactions of ATP involve the Mg^{2+}
or Mn^{2+} complex of ATP as substrate [72].

(2) *An individual metal ion involved in enzymatic catalysis
forms a definable complex with the enzyme.* Such a metal ion is
often called an *active site metal ion,* and it does not necessarily
exchange with free metal ion under catalytic conditions. This type
of metalloenzyme may have considerable kinetic (and/or thermodynamic)
stability with respect to dissociation of the metal ion near neutral
pH. However, the metal ion may often be reversibly removed from
such a metalloenzyme on a suitable time scale by one of several
procedures: (a) treatment with a chelating agent, (b) dialysis or
gel filtration, (c) decreasing the pH partially to protonate the
side-chain ligands of the metal ion, (d) partial denaturation of
the protein with urea, guanidinium chloride, organic solvents, or
a detergent, (e) alteration of the oxidation state of the metal ion,
or (f) combinations of the above or related procedures. Nonintegral

stoichiometry may be found when the M^{n+}-enzyme complex undergoes
partial dissociation under the normal conditions of isolation of
the enzyme, and variable stoichiometry in different batches of a
purified enzyme may reflect competition between metal ions in vivo
or in the purification procedures.

4.5. Replacement of Active Site Metal Ions

Active site metal ions may often be replaced by another metal ion of
similar size and redox properties (Sec. 8) with retention of a high
degree of enzymatic activity. Further, the substrate may promote
interchange of an active site metal ion with its free counterpart
[73] or with other metal ions in solution [74]. There is accord-
ingly a risk in the uncritical use in enzymology of "cocktails"
which routinely include, for example, Mg^{2+} or Ca^{2+}.

It is conceivable that it might be impossible for a metal ion
to be reversibly removed from a metalloenzyme or metalloprotein.
This could occur if conditions forcing enough to remove the metal
ion resulted in concomitant irreversible denaturation of the enzyme.
Such a result could clearly be caused by experimentation that is
insufficiently careful and imaginative. However, there is a reasonable but currently hypothetical circumstance that could account for
such behavior. Such behavior would be explained if a precursor of
the enzyme (a *zymogen* or *proenzyme*) folded into its final tertiary
structure around a metal ion and then underwent proteolytic cleavage
or other covalent modification such that the polypeptide chain of the
active enzyme was not in its thermodynamically most stable form [75].

The principle that "the three-dimensional structure of a native
protein in its normal milieu is determined by the amino acid sequence"
was established by Anfinsen with bovine ribonuclease A, which has a
single 13.7-kDa polypeptide chain and no zymogen [76]. This principle has been strongly supported by the poor recovery of native conformation from unfolded single-subunit globular nonmetalloproteins
whose multiple chains were originally formed by proteolytic nicking
of a single-chain precursor [77].

The importance of an active site metal ion in the refolding of a denatured enzyme has been demonstrated with the 29-kDa single-subunit zinc metalloenzyme, bovine carbonic anhydrase. Zn^{2+} is bound in the initial steps of renaturation, and the metal ion markedly influences the path and the rate of reactivation even though in this case it doesn't affect the final conformational state [78, 79]. The loss of a metal ion from a proteolytically nicked or otherwise epigenetically modified metalloenzyme could therefore lead to the inability to recover the active metalloenzyme from the partially unfolded apoenzyme. Further, incorrect conformation of a refolded subunit could lead to incorrect assembly of an oligomeric enzyme [80].

Jack bean urease is a possible candidate for such a metalloenzyme, since its active site nickel has resisted many attempts at reversible removal [74,75,81-83]. The nickel ion is readily accessible to several inhibitors which bind reversibly to it and to a variety of substrates (Secs. 6.5, 6.6, and 11).

Another potential candidate is pyruvate carboxylase (EC 6.4.1.1) from chicken liver. The Mn^{2+} in each subunit of this enzyme has resisted all attempts to reversibly remove it. However, active enzyme containing Mg^{2+} was achieved by isolating the enzyme from the liver of Mn^{2+}-deficient chickens [84, cf. 74]. The active site Mg^{2+} could be partially replaced with Mn^{2+}, Ni^{2+}, or Co^{2+} with little effect on the specific activity of the enzyme. Unfortunately, the enzyme partially reconstituted with Mn^{2+} was not further studied to see if it behaved as the native Mn^{2+} enzyme. If the inability to reversibly remove Mn^{2+} from the native enzyme is due to epigenetic modification or proteolytic nicking of pyruvate carboxylase, then the enzyme isolated from Mn^{2+}-deficient chickens must be modified differently.

5. METAL ION CATALYSIS

5.1. Catalysis and Promotion

The first isolation of an enzyme preceded by four years the invention
of the word *catalysis*. In 1833, Payen and Persoz partially purified
from malt a heat-sensitive mixture that could convert starch into
sugar: they called the mixture *diastase* (from the Greek word for
separation, διάστασις) because it effected the separation of soluble
sugar from the insoluble covering of the starch grain [85]. In 1836,
Schwann coined the name *pepsin* for the digestive principle of gastric
juice. In 1837, Berzelius drew on these and other examples to present the concept of catalysis in the third edition of his great textbook of chemistry [86]: "Catalytic power seems really to consist in
this, that bodies can, by their mere presence, and not by their
affinity, arouse affinities slumbering at this temperature, and as
a result the elements in a complex body become rearranged" [87].

The concept that a catalyst enhances the rate of a chemical
reaction but has no effect on the position of equilibrium was first
formulated by Wilhelm Ostwald: "A catalyst is a substance which
alters the velocity of a chemical reaction without appearing in the
final products" [88]. An agent which enhances the rate of a reaction
but is itself consumed is called a *promoter*.

The difference between catalysis and promotion may be exemplified by the hydrolysis of a carboxylic ester. The stoichiometry of
this reaction is described by Eq. (1), (2), or (3), depending on the
pH:

$$R\text{-}C(O)\text{-}OR' + H_2O \rightarrow R\text{-}COOH + HOR' \tag{1}$$

$$R\text{-}C(O)\text{-}OR' + {}^-OH \rightarrow R\text{-}COO^- + HOR' \tag{2}$$

$$R\text{-}C(O)\text{-}OR' + 2\,{}^-OH \rightarrow R\text{-}COO^- + {}^-OR' + H_2O \tag{3}$$

The lowest rate of hydrolysis generally occurs at neutral or somewhat acidic pH, where Eq. (1) or (2) describes the stoichiometry.
The rate of disappearance of esters is markedly enhanced at high pH,
the phenomenon being universally known as specific base catalysis.

The stoichiometry corresponding to this situation may be described by either Eq. (2) or (3), depending on whether pH < $pK_a^{HOR'}$ or pH > $pK_a^{HOR'}$, respectively. In Eq. (2) and (3), the agent (^-OH) which produces the enhancement of rate is consumed, and the effect of hydroxide ion on the rate of hydrolysis is therefore properly described as promotion rather than catalysis.

The majority of the literature on metal ion catalysis of hydrolytic reactions is actually concerned with metal ion promotion rather than catalysis: the metal ion is present in excess or at least in stoichiometric quantities, and remains complexed to the products of the reaction. The difference between catalysis and promotion lies in the rates of establishment of, and the positions of equilibria involving the products and the rate-enhancing agent.

5.2. Superacid Catalysis by Metal Ions

The ability of a metal ion to provide positive charge to the seat of reaction at neutral pH is the essential aspect of Westheimer's proposal [89] that "metal-ion catalysis can perhaps be described as superacid catalysis in neutral solution", since acid catalysis is not generally observed here. This concept has not always been very well understood, perhaps because catalysis by acid may be significantly more efficient than the corresponding catalysis by a metal ion. An obvious difference between a proton and a metal ion is the size: the small proton, because of its far greater field, can better effect polarization of the substrate, i.e., withdraw electrons from the reaction center [75]. However, proton catalysis is seldom available under cellular conditions, and it is therefore not surprising that metal ions play an essential Lewis acid role in many enzymatic reactions [74].

The catalytic roles of metal ions as Lewis acids may be grouped into four categories [cf. 90-92]:

    (I) Attachment of the metal ion to a suitable site on an electrophile enhances the reactivity of the electrophile.

(II) Attachment of the metal ion to a departing nucleophile
      ("leaving group", nucleofuge) increases its leaving
      ability [74].

(III) The metal ion acts as a center for the simultaneous
      attachment of both an electrophile and the attacking
      nucleophile, thus converting an intermolecular into an
      intramolecular process (template effects).

(IV) The metal ion increases the acidity of a proton on a
      coordinated species effectively to provide a better
      general acid catalyst [93] or to increase the concentra-
      tion of the more nucleophilic conjugate base of that
      species.

Generalized examples of categories I to IV for metal ion promotion are given in Fig. 2. An example of category I is the alkaline
hydrolysis of N,N-dimethylformamide: complexation of $(H_3N)_5Co(III)$
to the carbonyl oxygen provides an enhancement of greater than 10^4
in the second order rate constant for alkaline hydrolysis (k_{-OH}) as
compared with that for the uncomplexed amide [94].

Category II (Fig. 2) is exemplified by the Hg^{2+}-promoted
hydrolysis of thiol esters [e.g., R-C(O)-S-Et], in which coordination of the sulfur to Hg^{2+} activates the carbonyl group and stabilizes the leaving group [95]. In the presence of ~0.01 M Hg^{2+} in
aqueous solution at low pH, a small proportion of the thiol ester
exists as the reactive thiol ester-Hg^{2+} complex present in a rapid
equilibrium. When R is 4-nitrophenyl, the carbonyl group is efficiently attacked by water, whereas when R is 4-methoxyphenyl, the
rate limiting step is formation of an acylium ion which subsequently
reacts with water.

Potential examples of the template effect (category III) are
generally complicated by one or more effects in the other categories,
as well as by the lability of the complexes with which it has been
studied. Promotion of the formation of a Schiff base from 2-hydroxy-
benzaldehyde (salicylaldehyde) and glycine by several divalent metal
ions may be an example [96]. A reasonable mechanism involves minor

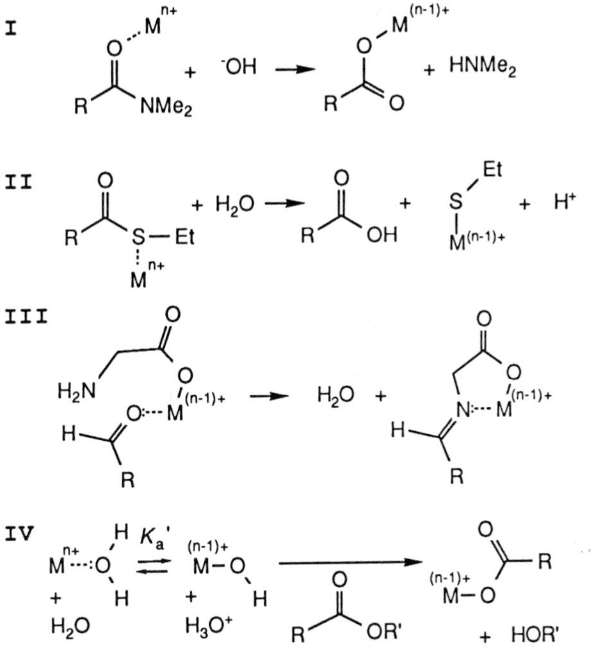

I

II

III

IV

FIG. 2. Generalized examples of promotion by metal ions. Category I: alkaline hydrolysis of a carboxamide bond in a Co(III) complex of N,N-dimethylformamide [94]. Category II: Hg^{2+}-promoted hydrolysis of a thiol ester [95]. Category III: template effect in the divalent metal ion promoted formation of a Schiff base from salicylaldehyde and glycine [96,97]. Category IV: nucleophilic attack by Co(III)-coordinated $^-$OH ion on 4-nitrophenyl acetate [98].

complexes in which the reactive glycine is coordinated only by its carboxylate group, leaving the amino group free to attack the coordinated carbonyl group of salicylaldehyde (Fig. 2). The kinetically equivalent attack of free glycine on the activated carbonyl group of complexed salicylaldehyde could not be rigorously excluded [97]. This and related systems are not simple, and the detailed mechanism of Schiff base formation has not been considered.

An example of category IV is the nucleophilic reaction of a Co(III)-coordinated hydroxide ion with 4-nitrophenyl acetate (Fig. 2). The reactive hydroxo complex is the conjugate base of the corresponding aquo ion whose pK_a' is 6.4 at 25°C. The Co(III)-coordinated hydroxide ion is actually only about 10^{-4} times as reactive as free $^-$OH in terms of a second order rate constant for reaction with 4-nitrophenyl acetate [98]. However, it is available in high concentration at neutral pH due to the low pK_a' of the coordinated water molecule (cf. 15.7 for H_2O itself), and the markedly higher concentration of the reactive conjugate base at neutral pH markedly outweighs its decreased reactivity. In general, the magnitude of the pK_a' of molecules coordinated to a divalent or trivalent metal ion will depend on the particular metal ion, its charge, its other ligands, and the polarity of its environment.

The four categories of catalytic roles for metal ions constitute a summary of chemical principles gleaned from many studies which have been presented elsewhere [74,93,99-103]. Roles I and II for a metal ion are directly analogous to the corresponding catalytic roles for protons, whereas roles III and IV represent additional capabilities of metal ions which are not available to protons. A fifth role for metal ions which is of potential importance has been recently demonstrated by Suh and Chun [92], and this role may best be presented in terms of its relationship to protonic equilibria in unstable intermediates.

5.3. Effects of pH on the Partitioning of Unstable Intermediates

The effect of pH on simple reactions may generally be understood in terms of the relative reactivity of a base and its conjugate acid in the reaction in question. For example, stabilization of a departing weak nucleophile by its protonation often accounts for specific acid catalysis (proton catalysis) of a reaction. However, protonation of an attacking nucleophile will clearly diminish its reactivity. This may be illustrated by the reaction of methyl methanethiolsulfonate $[H_3C-S(O_2)-S-CH_3]$ with 2-mercaptoethanol to form the mixed disulfide and methanesulfinic acid (H_3C-SO_2H), in which the nucleophilicity of the thiolate anion of 2-mercaptoethanol is at least 5×10^9 fold higher than that of the neutral thiol [104].

If a reaction involves a reversibly formed, unstable intermediate, the partitioning of the intermediate between products and reactants will depend on pH if the forward and back reactions involve different ionization states of the intermediate. This was first established in an investigation of the effect of pH on k_{obs} for the hydrolysis of o-carboxyphthalimide (1) [105]. In this system, the

1

carboxyl group catalyzes the attack of water on a carbonyl group to form an unstable tetrahedral intermediate. The bell-shaped log k_{obs} vs. pH profile requires that the forward and back reactions of the reversibly formed steady-state intermediate involve different ionization states of the intermediate. This phenomenon has subsequently been observed in a variety of systems which involve unstable intermediates [90,106,107].

The acid-base properties of unstable intermediates have generally been treated as simple equilibria, but it has been suggested that proton transfer may be rate limiting in some systems [108].

5.4. Reverse Attack by a Departed Intramolecular Nucleophile: Inhibition by Protons and by Metal Ions

The carboxylate group in 2-carboxyphenyl acetate (aspirin) and its derivatives serves as an intramolecular catalyst of the hydrolysis of the ester linkage (Scheme 1) [92,109]. Two catalytic mechanisms have been established:

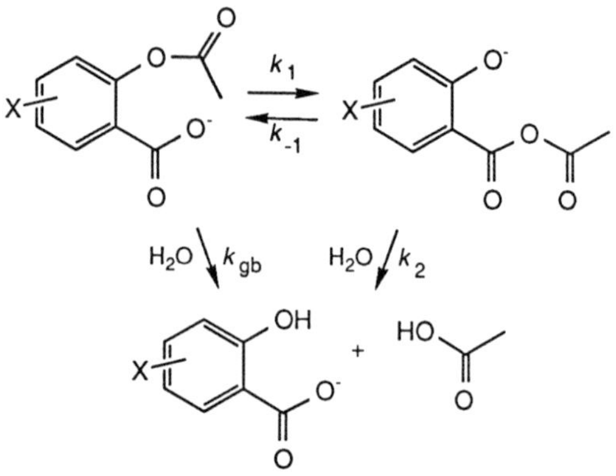

SCHEME 1. Competing mechanisms in the hydrolysis of monoanionic derivatives of aspirin. Acid-base equilibria involving the carboxylate and phenoxide ions have been omitted for simplicity. In the k_{gb} reaction, the carboxylate ion acts as a general base to abstract a proton from a water molecule as the latter attacks the carbonyl group. (Adapted from Ref. 92.)

(1) A general base mechanism in which the 2-carboxylate group abstracts a proton from a water molecule as it adds to the ester carbonyl group, leading directly to product (k_{gb} in Scheme 1), and

(2) A nucleophilic mechanism in which the 2-carboxylate group reversibly attacks the ester carbonyl group (the k_1 step) to give the steady-state intermediate (I) which is a mixed anhydride. I may be either hydrolyzed (the k_2 step) to form the products or nucleophilicly attacked by the phenoxide ion (the k_{-1} step) to regenerate the starting material. The phenoxide ion in I may serve as a general base to catalyze hydrolysis of the mixed anhydride in the k_2 step [92].

The carboxylate form of aspirin itself undergoes hydrolysis at neutral pH via the general base mechanism (k_{gb}) since even though the mixed anhydride is formed very rapidly (i.e., $k_1 \gg k_{gb}$ in Scheme 1), its formation is completely reversible (i.e., $k_{-1} \gg k_2$) [92,109]. However, derivatives of aspirin are hydrolyzed via the mixed anhydride intermediate in two circumstances: (1) when breakdown of the mixed anhydride (the k_2 step) to products is greatly catalyzed either by

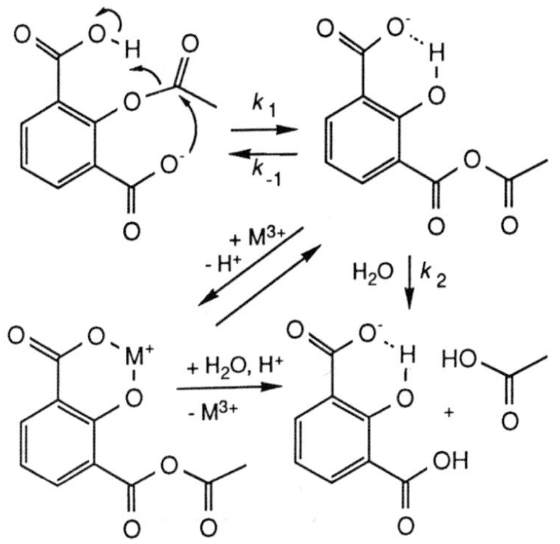

SCHEME 2.   Catalysis by Fe^{3+} and Al^{3+} ions in the hydrolysis of the monoanion of 2,6-dicarboxyphenyl acetate.   The competing general base mechanism has been omitted, as have acid-base equilibria involving the carboxylate and phenoxide ions.   (Adapted from Ref. 92.)

general base catalysis (proton abstraction), by the phenoxide ion, or by nucleophilic intervention by an additional group, and/or (2) when the reverse attack of the phenoxide ion on the mixed anhydride (the k_{-1} step) is suppressed by protonation of the phenoxide nucleophile [92]. In the limit of the latter situation, equilibrium protonation of the departed nucleophile (i.e., the phenoxide ion) serves to alter the rate-limiting stage of the nucleophilic pathway from the k_2 step to the k_1 step [92].

Suh and Chun recently studied the effect of pH and metal ions on the hydrolysis of 2,6-dicarboxyphenyl acetate (Scheme 2) [92]. Although the monoanion interacts only weakly with metal ions, its corresponding mixed anhydride has a good chelation site. Rapid coordination of the intermediate with a metal ion (Fe^{3+}, Al^{3+}) serves the same function as simple protonation of the phenoxide ion in preventing the reverse reaction and thereby altering the rate-limiting step (Scheme 2). The net result is pronounced metal ion promotion of the hydrolysis of 2,6-dicarboxyphenyl acetate via the

nucleophilic pathway. On the basis of their results, Suh and Chun
recognized a fifth general catalytic role for metal ions as Lewis
acids which may be formulated as follows:

(V) The metal ion coordinates to and deactivates a departed
nucleophile which by virtue of proximity to an intermediate would otherwise bring about the reversal of the
reaction in which it was formed.

This newly established role is different from role II above
with respect to the timing of interaction of the metal ion with the
departing nucleophile. Borderline cases may be envisaged.

5.5. Metal Ion Catalysis of the Hydrolysis of Carboxamides

An important new insight into metal ion catalysis of the hydrolysis
of amides of carboxylic acids was recently published by Sayre [93].
As a point of reference, the rate-limiting step in the hydrolysis
of simple carboxylic acid amides [R-C(O)-NHR'] at neutral pH is the
breakdown of the tetrahedral intermediate to form products. In contrast to the facile elimination of an alkoxide ($^-$OR) leaving group
in the hydrolysis of an ester, the expulsion of a highly basic $^-$NHR'
anion from a tetrahedral intermediate is very unfavorable, and protonation of the nitrogen must take place prior to, or in concert
with, C-N bond cleavage. In suitable model systems, intermolecular
or intramolecular general acid catalysis greatly facilitates amide
hydrolysis [93].

Metalloenzyme catalysis of the hydrolysis of amides has often
been discussed in terms of two alternative but kinetically equivalent
pathways leading to a tetrahedral intermediate, as shown in Scheme 3.
In pathway A the coordinated metal ion polarizes the carbonyl group
of the amide thereby increasing its susceptibility to nucleophilic
attack, whereas in pathway B the metal ion provides a coordinated
hydroxide ion which is highly nucleophilic at neutral pH. Both

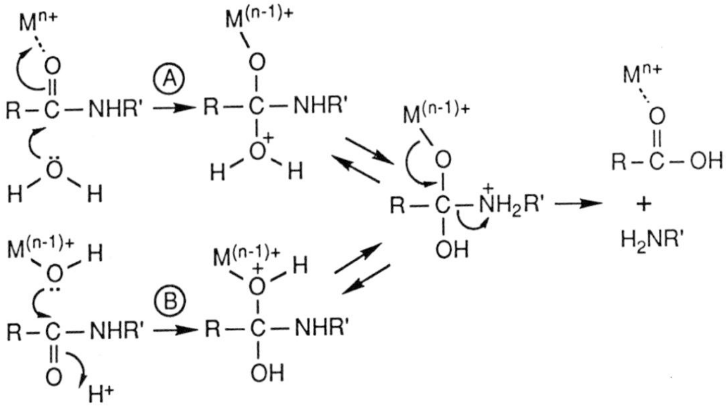

M^{n+} $M^{(n-1)+}$

(A)

$R-C-NHR'$ → $R-C-NHR'$ $M^{(n-1)+}$ M^{n+}

H   O   H   H   O   H   O
                                                        $R-C-OH$

                                        $R-C-NH_2R'$ →   +

$M^{(n-1)+}$ H $M^{(n-1)+}$ H   OH H_2NR'

(B)

$R-C-NHR'$ → $R-C-NHR'$

O   H⁺   OH

SCHEME 3. Two kinetically equivalent pathways for the formation of a tetrahedral intermediate in the metal ion-catalyzed hydrolysis of a carboxamide at neutral pH. (Adapted from Ref. 93.)

pathways lead to the same O-coordinated tetrahedral intermediate(s) when prototropic equilibria are taken into account. Regardless of how the coordinated tetrahedral intermediate is formed, the essential question is how the involvement of the metal ion facilitates the otherwise rate-limiting breakdown of the tetrahedral intermediate at neutral pH [93].

The electron source for cleavage of the C-N bond is a pair of electrons on the oxygen anion which is coordinated to the metal ion in the tetrahedral intermediate (Scheme 3). Even though these electrons are not as readily available as those of a free alkoxide anion (⁻OR), they are far more available than those of the corresponding alcohol (HOR). They are therefore substantially available to serve as an electron source in the product-forming fragmentation step. The hitherto disregarded essence of metal ion catalysis by either pathway A or pathway B in Scheme 3 is that the metal ion facilitates both the formation and the breakdown of the tetrahedral intermediate [93].

In Sayre's perspective, the metal ion remains stoichiometrically coordinated throughout the hydrolysis reaction as shown in Scheme 3. It does not interact with the leaving nitrogen in the tetrahedral intermediate, since this would prevent N-protonation and the $M^{(n-1)+}$-NHR' complex cannot be as good a leaving group as H_2NR'. Acid-base and prototropic equilibria in the tetrahedral intermediate are complicated and will have a major effect on the efficiency of catalysis by metal ions [93 and cf. 29,75].

The proclivity of metal ions to form stable complexes with nonregular geometry and nonstandard coordination number is well known (Sec. 6) [74,93,110]. This raises the possibility that the same metal ion may serve both to activate a carbonyl group and to provide a coordinated hydroxide ion as a nucleophile in the catalysis of amide hydrolysis [93].

6. SOME PROPERTIES OF NICKEL COMPLEXES

Although nickel compounds with formal oxidation states from 1- to 4+ have been characterized, Ni(II) is generally the most stable oxidation state in aqueous solution and its literature is correspondingly extensive. However, Ni(III) is clearly important in enzymology and, further, any intellectual barrier between the ionic and the organometallic chemistry of nickel was securely breached in 1975 when the reaction shown in Fig. 3 was found to occur under mild conditions

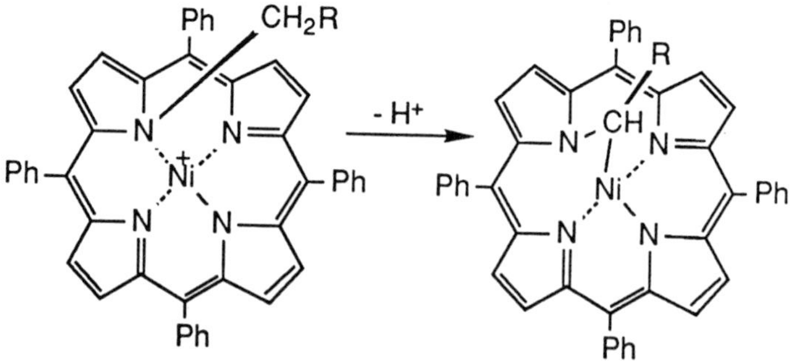

FIG. 3. Formation of an organonickel compound by treatment of N-ethoxycarbonylmethylene-meso-tetraphenylporphinatonickel(II) with triethylamine in CH_2Cl_2 at 25°C [R- = -C(O)OEt]. (Adapted from Ref. 111.)

[111]. In the product, the nickel is four-coordinate with normal Ni-C (1.905 Å) and Ni-N (1.916 Å) bond lengths. The tetrapyrrole ring is substantially nonplanar, as might be expected.

Some aspects of Ni(II) and Ni(III) complexes which are particularly relevant to nickel metalloenzymes and metalloproteins are discussed below in light of the basic properties of nickel complexes as described by Cotton and Wilkinson [112].

6.1. Nickel(II) Complexes: Geometry and Magnetic Susceptibility

Nickel(II) is a d^8 ion and its simple complexes commonly have coordination numbers of 4, 5, or 6. The geometry of simple Ni(II) complexes may be basically square-planar, square-pyramidal, tetrahedral, trigonal-bipyramidal, or octahedral. Moreover, equilibria between square and octahedral or square and tetrahedral forms sometimes exist in solution, and the difference in stability between different forms may be fairly small and is dependent on factors such as the chemical nature of the ligands, steric effects, and the solvent [112].

The adaptability of Ni(II) and other first row transition metal ions to the geometric restrictions of their ligands is dramatically illustrated by the unexceptional behavior of the seven-coordinate pentagonal-bipyramidal complex of Ni(II) [as well as V(III), Fe(II), Co(II), Cu(II), and Zn(II)] with 2,6-diacetylpyridinebissemicarbazone [110]. In the basal plane (2) of the Ni(II) complex, three nitrogens

2

and two carbonyl oxygens of the chelon coordinate to the metal ion. In addition, water molecules (not shown) are coordinated above and below the plane of the chelon. The Ni(II)-O bond distances are 2.05 and 2.09 Å for the two axial water molecules, and 2.22 and 2.48 Å for the Ni(II)-carbonyl oxygen bonds. The Ni(II)-N bond distances are 2.06 Å (pyridine), and 2.11 and 2.22 Å (Schiff bases).

Nickel(II) complexes have either two unpaired electrons or none, and their magnetic properties are sometimes diagnostically useful. Magnetic moments of 3-4 BM at 25°C have generally been observed for paramagnetic ("high spin") forms, and the value may depend on geometric perturbation and on temperature-dependent conformational or ligand equilibria. The magnetic properties of Ni(II) complexes are as follows [112]:

      Paramagnetic: tetrahedral, five-coordinate (sometimes), octahedral

      Diamagnetic: square-planar, five-coordinate (sometimes)

Weak magnetic interactions between paramagnetic Ni(II) ions in binuclear complexes have been identified by magnetic susceptibility measurements over a wide range of temperature approaching 0 K at the lower limit. Antiferromagnetic coupling has been observed in binuclear complexes of the type $(NiLCl_2)_2$ in which L is 2,9-dimethyl-1,10-phenanthroline or 2,2'-biquinoline and each pentacoordinate Ni(II) is square-pyramidal with a shared basal edge consisting of two bridging chloride ions [113]. Ferromagnetic coupling between Ni(II) ions has been observed in the binuclear complex $[Ni(en)_2Cl]_2$ where en is ethylenediamine [114]. Each Ni(II) in this complex is octahedrally coordinated and the shared edge consists of two bridging chloride ions. Ferromagnetic coupling has also been found in the tetranuclear species $[Ni^{2+}(^-O\text{-}CH_3)(HO\text{-}CH_3)(acac^-)]_4$, where $acac^-$ is the conjugate base of acetylacetone. Each Ni(II) ion is octahedrally coordinated and the tetramer has a distorted cubane structure with four bridging methoxide ions [115].

6.2. Octahedral Ni(II) Complexes: Ligand Exchange Reactions

In dilute aqueous solutions of simple Ni(II) salts, the dominant nickel species is the octahedral hexaaquo ion, $Ni(H_2O)_6^{2+}$ [116]. Acetate, pyridine, and ammonia are simple models of individual protein side chains, and their 1:1 complexes with Ni(II) have the structure $Ni(H_2O)_5L^{(2-n)+}$ where the ligand is L^{n-}. The stability constants of 1:1 complexes of Ni(II) with amines are between 100 M^{-1} and 1,000 M^{-1} while that with acetate is only 2 M^{-1} (Table 2). In general, the stability constant for a particular ligand is affected by the chemical nature of the other ligands, e.g., that for binding of NH_3 to $Ni(NH_3)_5(H_2O)^{2+}$ to form $Ni(NH_3)_6^{2+}$ is only 1.4 M^{-1} even without correction for the number of equivalent ammonia ligands [117].

The chelate effect [118] is seen in the enhanced stability of the 1:1 bidentate glycine complex and the 1:1 tridentate histidine complex as compared with monodentate models (Table 1).

Substitution reactions of $Ni(H_2O)_6^{2+}$ are thought to obey Eqs. (4) and (5), where $(H_2O)_5Ni(H_2O)L^{(2-n)+}$ is a reversibly and very

$$(H_2O)_5Ni(H_2O)^{2+} + L^{n-} \xrightleftharpoons{K_0} (H_2O)_5Ni(H_2O)L^{(2-n)+} \qquad (4)$$

$$(H_2O)_5Ni(H_2O)L^{(2-n)+} \xrightleftharpoons[k_{-0}]{k_0} (H_2O)_5NiL^{(2-n)+} + H_2O \qquad (5)$$

rapidly formed *outer sphere complex* in which the original six ligands of the Ni^{2+} are intact [116]. Outer sphere complexes involve weak physical interactions (electrostatic and hydrophobic forces and hydrogen bonding), and the stability constant K_0 is seldom greater than 1 M^{-1}. If the incoming ligand is negatively charged, an outer sphere complex is an *ion pair*. The slow stage in the replacement of coordinated H_2O is the k_0 step in which L^{n-} replaces one H_2O ligand, possibly by way of a seven-coordinate transition state.

TABLE 2

Stability Constants for 1:1 Complexes of Ni(II)
with Simple Ligands in Aqueous Solution

Ligand	Structure	$\log K_1{}^a$ (M^{-1})	Ionic strength	Temp. (°C)
Acetate	$H_3C\text{-}COO^-$	0.28	1 ($NaClO_4$)	25
Pyridine	C_5H_5N	2.13	1 ($NaClO_4$)	25
Ammonia	NH_3	2.72	2 (NH_4NO_3)	30
Glycine monoanion	$H_2N\text{-}CH_2\text{-}COO^-$	5.70	1 ($NaClO_4$)	25
Histidine monoanion		8.50	0.25 (KCl)	25

$^a K_1 = [NiL^{(2-n)+}]/[Ni^{2+}][L^{n-}]$ where all sites not occupied by the ligand (L^{n-}) are occupied by water.
Source: Selected from Ref. 117.

Because K_0 is generally so small that no appreciable proportion of the outer sphere complex is present under accessible conditions, substitution of a water ligand of $Ni(H_2O)_6^{2+}$ generally obeys second order kinetics with an observed formation rate constant $k_f = K_0 k_0$. The measured value of k_f is between $\sim 10^3$ M^{-1} sec^{-1} and $\sim 10^5$ M^{-1} sec^{-1} at 25°C for a series of neutral and charged incoming ligands [116,119].

The measured value of k_0 is 3×10^4 sec^{-1} at 25°C for replacement of a particular water ligand in $Ni(H_2O)_6^{2+}$. Very similar values have been estimated for a variety of incoming anionic and neutral ligands on the basis of calculated values of K_0 [116].

The first bond between $Ni(H_2O)_6^{2+}$ and several chelating agents is formed via the mechanism described above and with rate constants similar to those displayed by simple ligands. Wilkins has discussed factors which affect the rate of formation and breaking of subsequent bonds between chelating agents and Ni(II) ions [116].

The rate constant [corresponding to k_0 in sec^{-1} in Eq. (5)] for replacement of H_2O in aquo ions $[M(H_2O)_m^{n+}]$ by solvent water has been measured for many elements, and a self-consistent set of rate constants is as follows [M^{n+}, approx. $\log k_0$]: Al^{3+}, 1.0; V^{2+}, 2.0; Ni^{2+}, 4.1; Mg^{2+}, 5.1; Co^{2+}, 5.4; Fe^{2+}, 6.1; Mn^{2+}, 6.6; Zn^{2+}, 7.4;

Cu^{2+}, 8.5; Ca^{2+}, 8.5; Na^{+}, 8.9; K^{+}, 9.1 [120,121]. The rate constant for water exchange in $Ni(H_2O)_6^{2+}$ is relatively small compared with those for other transition metal ions commonly found at the active site of enzymes. Further, this rate constant is not much larger than k_{cat} for an efficient metalloenzyme. Active site metal ions are mixed-ligand complexes [122], and rate constants for their ligand substitution reactions will presumably be different from those of the corresponding aquo ions. However, it is possible that ligand substitution could be a rate-limiting step in the interaction of substrates, products, or inhibitors with an active site Ni(II) in a metalloenzyme. Further, the very low water exchange rate for Al^{3+} may be sufficient to account for the lack of any known catalytic role for this ion in enzymology.

6.3. Octahedral Ni(II) Complexes: Electronic Absorption Spectra

Absorption spectra from the near-ultraviolet through the visible and into the near-infrared region are frequently useful guides to the geometry of Ni(II) complexes [112,123]. However, the molar absorption coefficients (ε) associated with ligand field transitions of octahedral Ni(II) are generally less than 50 M^{-1} cm^{-1} (often less than 10 M^{-1} cm^{-1}), and the associated absorption peaks may readily be obscured by (1) other chromophores, (2) light scattering associated with very large protein molecules [124], or (3) charge transfer transitions of thiolate or other ligands (Sec. 6.5). Octahedral Ni(II) complexes generally display at least three distinct absorption maxima.

In Table 3 are listed some of the spectral characteristics of jack bean urease together with those of three known Ni(II) enzymes and of several simple octahedral complexes of Ni(II). In Ni(II)-phosphoglucomutase, Ni(II)-carboxypeptidase A, and Ni(II)-carbonic anhydrase, the normal active site metal ion has been replaced by Ni(II) to produce a nickel metalloenzyme (Sec. 8) whose electronic

TABLE 3

Electronic Absorption Spectra of Nickel(II) Enzymes and of Simple Octahedral Nickel(II) Complexes

No.	Complex	Donor atom set	λ_{max} (nm), ε_{max} (M^{-1} cm^{-1})			Ref.
			($^3A_{2g} \rightarrow ^3T_{1g}(P)$)	($^3A_{2g} \rightarrow ^3T_{1g}(F)$)	($^3A_{2g} \rightarrow ^3T_{2g}$)	
1	Ni$_2$-urease[a]	?	407,[b] —	745, 46	910, 14	1060, 10	124
2	Ni-phosphoglucomutase	O$_6$?	410, 23	760, 7	700, 6	1300, 5	125,126
3	Ni-carboxypeptidase A	N$_2$O$_3$-4?	412, 24	685, 7	770sh, 5[b]	1060, 3	126,127
4	Ni-carbonic anhydrase	N$_3$O$_3$?		625, 10			126,128
5	Ni(NH$_3$)$_6^{2+}$	N$_6$	355, 6.3	571, 4.8	760, 0.5	930, 4.0	129
6	Ni(histidinate)$_2$	N$_4$O$_2$	357, 10	556, 7.3		935, 6.7	130
7	Ni(glycinate)$_3^-$	N$_3$O$_3$	362, 14.4	602, 8.2	763, 2.0	990, 9.9	128
8	Ni(MTDPC)$_2$(pyridine)$_2^{c,d}$	N$_2$O$_2$S$_2$	395sh, —	641, 30	870sh, 10[b]	1075, 20	131,132
9	Ni(MTB)$_2$(pyridine)$_2^{d,e}$	N$_2$O$_2$S$_2$	380sh, —	635, 35.5	665sh, —	1080, 13.5	133,134
10	Ni(BCEE)$_2$Cl$_2^f$	N$_2$Cl$_2$S$_2$	392, 20	634, 9.7		1106, 6.8	135
11	Ni(H$_2$O)$_6^{2+}$	O$_6$	395, 5.2	725, 2.1	658, 1.9	1180, 2.0	128,136
12	Ni(dimethylacetamide)$_6^g$	O$_6$	418, —	775, —	680sh, —	1300, —	137

[a]Based on the 96.6-kDa subunit and corrected for light scattering of the hexameric enzyme [29].
[b]Approximate value. sh = shoulder.
[c]Solvent: CH$_2$Cl$_2$. MTDPC = monothiodi(n-propyl)carbamate ion.
[d]Bidentate -C(O)S$^-$ coordination.
[e]Solvent: acetonitrile. MTB = monothiobenzoate ion.
[f]Solvent: CHCl$_3$. BCEE = S-benzylcysteine ethyl ester. Bidentate N,S(thioether) coordination.
[g]Solvent: N,N-dimethylacetamide.

absorption spectrum is consistent with Ni(II) in an octahedral environment [126]. Moreover, as Hardman and Lipscomb [127] point out,
x-ray diffraction results for Ni(II)-carboxypeptidase A approximate
an octahedral site in which the sixth position is vacant. The
absorption peaks in jack bean urease (Table 3) are also consistent
with Ni(II) in an octahedral environment, although this description
may well be incomplete because the two nickel ions in each subunit
may have different environments (Sec. 11) [75,124].

The simple octahedral Ni(II) complexes in Table 3 contain
ligands typical of those available from the side chains and peptide
bonds of proteins. These examples provide an idea of the range of
spectral properties that may be expected for normal octahedral
Ni(II) in metalloenzymes.

6.4. Ni(II)-Oligopeptide Complexes

Complexes of Ni(II) with simple oligoglycine peptides $[(Gly)_n$ and
their derivatives, where $n \geq 3$] in alkaline solution are square-
planar with the α-amino group and up to three deprotonated peptide
nitrogens as ligands as in 3 [138,139]. However, with a suitable

3   4

chelating agent, N-coordination of a deprotonated peptide bond may
also occur in an octahedral Ni(II) complex as in the 1:2 complex
(4) with diglycine [138].

Near neutral pH, a moderately concentrated 1:2 mixture of Ni(II) and glycinamide exists largely as an octahedral complex $Ni(H_2O)_4L_2^{2+}$ in which each of two glycinamide molecules $[L = H_2N-CH_2-C(O)-NH_2]$ is coordinated simply by its α-amino group. At much higher pH, the dominant species at equilibrium is a square-planar complex $Ni(H_{-1}L)_2$ in which each of two molecules of glycinamide anion $[H_{-1}L = H_2N-CH_2-C(O)-N(H)^-]$ is chelated to Ni(II) via its α-amino group and its anionic deprotonated peptide nitrogen. The apparent pK_a' associated with *each* cyclization/proton loss is ~10. The 1:1 complex of Ni(II) with tetraglycine $[(Gly)_4]$ has a square-planar structure analogous to that with tetraglycinamide (3), and in its formation the three peptide NH protons titrate cooperatively with apparent pK_a' values of ~8.2 [140].

Reversible transformations between octahedral and square planar oligopeptide complexes of Ni(II) at alkaline pH require several minutes at 25°C for equilibration to take place [140].

The decomposition of the Ni(II)-tetraglycinamide complex is specific acid-catalyzed below pH ~7, and the mechanism apparently involves equilibrium protonation on oxygen of N-coordinated peptide bonds. The pK_a' values for successive protonations on oxygen are 2.4, 1.3, and <1 [139].

The N-terminal sequence of human serum albumin is modeled by the acidic tripeptide N-methylamide L-Asp-L-Ala-L-His-NHMe whose interaction with Ni(II) has been investigated [31]. Although several species were present between pH 5.1 and 9.5 in a 1:1 mixture, the species which is dominant above 6 and quantitative at pH 9.5 is $Ni(H_{-2}L)^-$, where $H_{-2}L$ represents the trianionic peptide which has lost two protons from peptide nitrogens as well as the third from the side chain of aspartic acid. This species has λ_{max} = 420 nm (ε = 135 M^{-1} cm^{-1}) with a weak shoulder near 450-480 nm, consistently with a square-planar or square-pyramidal structure. A minor octahedral Ni(II)-peptide complex is also evident in the spectra at low pH. The effects of pH on the electronic absorption spectra in the Ni(II)-tripeptide system as well as the spectra themselves are very similar to those for the Ni(II)-albumin system. The structure proposed for the alkaline form of each involves a five-coordinate

square-pyramidal structure: the α-amino group of Asp-1, the deprotonated peptide nitrogens of Ala-2 and His-3, and the side-chain
imidazole of His-3 are coordinated in a plane, and the side-chain
carboxylate of Asp-1 is coordinated axially (see Sec. 3.2.1) [31].

The square-planar complex of Ni(II) with deprotonated digly-
cinamide [Ni(H$_{-2}$L)] forms a ternary complex with 2,6-lutidine (2,6-
dimethylpyridine) with the composition Ni(H$_{-2}$L)(2,6-lutidine) and
stability constant 112 M^{-1} at 25°C for replacement of the tertiary
amine by water. By way of contrast, the 2,6-lutidine complex of
Ni(H$_2$O)$_6^{2+}$ had previously been reported as too weak to measure, and
2,6-lutidine had therefore been regarded as a "noncomplexing" buffer
for use at neutral pH [141].

6.5. Spectra of Ni(II) Complexes with Thiolate Ligands

6.5.1. *Model Complexes*

Thiolate ligand-to-metal charge transfer transitions have been identified in the electronic absorption spectra of many complexes of
Ni(II) [124]. Table 4 summarizes spectral data for *mononuclear*
thiolate complexes of Ni(II) with a variety of geometries at the
metal ion. For comparison, free thiolate ions have an intense
absorption at low wavelength (thioglycolate, $^-SCH_2CO_2^-$, λ_{max} = 245 nm,
ε = 4,000 M^{-1} cm^{-1} [149]; HS$^-$, λ_{max} = 228.5 nm, ε = 7,140 M^{-1} cm^{-1}
[153]). The absorption peak of HS$^-$ at 228.5 nm has been ascribed to
an HS$^-$ → H$_2$O charge transfer transition [153].

Complexes 1-3 in Table 4 are Ni(II) proteins in which the
Ni(II) ion has a distorted tetrahedral coordination and one, two, or
four RS$^-$ ligands. The d-d ligand field transitions are thought to
occur at wavelengths greater than 600 nm [54,143]. Each of these
Ni(II) proteins has a prominent absorption peak at 355-360 nm with
an ε of ~1,500 M^{-1} cm^{-1} per RS$^-$ ligand. Although this peak was
originally assigned to an S(Met) → Ni(II) charge transfer transition
in Ni(II)-azurin [52], its presence in the other Ni(II) proteins
which do not have methionine ligands suggests that it actually arises

TABLE 4

Electronic Absorption Spectra of Nickel(II) Complexes Which Contain Thiolate Ligands

No.	Complexing agent (X)	Ligand types[a]	Complex	Possible RS⁻ → Ni(II) charge transfer transitions		Other transitions		Ref.
				λ_{max} (nm)	ε_{max}[b] (M^{-1} cm^{-1})	λ_{max} (nm)	ε_{max}[b] (M^{-1} cm^{-1})	
1	Azurin	S⁻(Cys), S(Met), [N(im)]₂	NiX[c,d]	355 439 562	1570 3300 130			52
2	Aspartate transcarbamylase	[S⁻(Cys)]₄	Ni₆c₆r₆[c,e]	360 440	5500 2800	665 720	250[f] 330[f]	142
3	Alcohol dehydrogenase	[S⁻(Cys)]₂, N(im), O(H₂O)	Ni₁.₃Zn₂X₂[c,g]	357 407 505 570	2900 3500 300 130	680	80	143
4	2-Mercaptopropionyl-L-cysteine	S⁻, N⁻(am), S⁻(Cys), O(H₂O)	NiX[h,i]	410	1200	567	450	144
5	2-Mercaptoacetyl-L-histidine	S⁻, N⁻(am), N(im), O(H₂O)	NiX[h]	382	770	543	240	144
6	3-Mercaptopropionyl-L-histidine	S⁻, N⁻(am), O(-CO₂⁻), O(H₂O)	NiX[h]	405	850	515	250	144
7	3-Mercaptopropionylglycine	S⁻, N⁻(am), O(-CO₂⁻), O(H₂O)	NiX[h]	400	1020	510	290	144
8	2-Mercaptopropionylglycine	S⁻, N⁻(am), O(-CO₂⁻), O(H₂O)	NiX[h,i]	375	2750	475	410	144
9	2,3-Dimercaptopropionylglycine	[S⁻]₂, N⁻(am)?, O	NiX[h]	440	1300	575	440	144
10	D-Penicillamine	[S⁻]₂, [N(alk)]₂	NiX₂[h]	<300	Large	463 560sh[f]	127[f] 40[f]	145

No.	Compound	Donor atoms	Complex					Ref.
11	2-(N,N-Dimethylamino)ethanethiol	$[S^-]_2$, $[N(alk)]_2$	NiX_2[i,j]	328	11700	505 / 676	89 / 38	146
12	Cysteine ethyl ester	$[S^-]_2$, $[N(alk)]_2$	NiX_2[h,k]	318	8230	470 / 631	128 / 35	135
13	Cysteine	$[S^-]_2$, $[N(alk)]_2$	NiX_2[h]	272 / 310sh[f]	11600 / 3080[f]	476 / 575sh[f]	100[f] / 35[f]	147,148
14	Thioglycolic acid	$[S^-]_2$, $[O(-CO_2^-)]_2$	NiX_2[h]	285[f]	11500[f]	515 / 640[f]	100[f] / 47[f]	147,149 150,151
15	2,3-Dimercaptopropanol	$[S^-]_4$	NiX_2[h]	307[f]	16000[f]	475 / 610[f]	120[f] / 74[f]	147,152
16	$(n\text{-Propyl})_2NC(O)S^-$	$[S^{-1/2}]_4$, $[O(MTC)]_2$	$(NiX_2)_n$[l]	360sh	430	429sh / 714 / 1200	250[f] / 37 / 18	131

[a]Cys = cysteine; Met = methionine; im = imidazole moiety of histidine; am = N-deprotonated carboxamide; alk = mono-, di- or tri-alkylamine; MTC = monothiocarbamate.
[b]Based on concentration of nickel.
[c]Distorted tetrahedral geometry at the nickel ion.
[d]The normal Cu(II) ion in Pseudomonas aeruginosa azurin has been quantitatively replaced by Ni(II) ion.
[e]The normal Zn(II) ion in the regulatory polypeptide chain (r) of the Escherichia coli enzyme was replaced by Ni(II) ion. The catalytic polypeptide chain (c) and the Ni(II) regulatory polypeptide chain aggregate normally to form an enzyme with specific activity similar to that of the native enzyme but displaying somewhat less cooperativity.
[f]Approximate values.
[g]The catalytic Zn(II) ion in each of the two identical subunits of the horse liver enzyme was totally removed and partially replaced by Ni(II) ion, while the noncatalytic Zn(II) ion in each subunit remained intact. The resulting enzyme had 12% of the specific activity of the native enzyme.
[h]Presumably square-planar geometry at the Ni(II) ion.
[i]Diamagnetic complex.
[j]A trans square-planar structure exists in benzene solution and in crystals.
[k]Solvent: pyridine.
[l]N,N-Dialkylmonothiocarbamate complexes of Ni(II) are paramagnetic oligomers with two sulfur bridges between adjacent Ni(II) ions. Aggregation is concentration dependent and the Ni(II) ions are octahedrally coordinated with bidentate -C(O)S⁻ coordination. Solvent: benzene.

from an RS$^-$ → Ni(II) charge transfer transition [143]. Another
RS$^-$ → Ni(II) charge transfer absorption peak occurs between 407 and
440 nm with a ligand ε of ~3,000 M^{-1} cm^{-1} independently of the number
of RS$^-$ ligands. Weaker RS$^-$ → Ni(II) charge transfer absorption peaks
occur at wavelengths as high as 570 nm in Ni(II)-metallothionein [54],
Ni(II)-azurin, and Ni(II)-alcohol dehydrogenase. S(Met) → Ni(II) and
additional RS$^-$ → Ni(II) charge transfer transitions have been identified in circular dichroism spectra of these nickel-containing proteins [52,54,143]. The absorption spectrum of Ni(II)-metallothionein
is particularly complicated because of multiple types of thiolate
ligands (Secs. 3.4.3 and 9) [54].

In Table 4, complexes 4-9 have (distorted?) square-planar
geometry and one or two RS$^-$ ligands in a tridentate chelate. In
many simpler square-planar complexes of Ni(II), the d-d ligand field
transitions occur at 450-600 nm with an ε of ~60 M^{-1} cm^{-1} [112],
although this description may well require substantial amendment
when oligopeptide complexes [31,139] are considered. Complexes 4-9
display an absorption peak in this region. In addition, each of
them has a strong absorption peak at 375-440 nm (ε = 850-2,750 M^{-1}
cm^{-1}), and it is likely that this peak corresponds to an RS$^-$ → Ni(II)
charge transfer transition.

Complexes 10-15 in Table 4 also have (distorted) square-planar
geometry. Complexes 10-13 contain derivatives of the bidentate
chelon, 2-mercaptoethylamine (HS-CH$_2$CH$_2$-NH$_2$), and each complex has
two RS$^-$ ligands. All of these complexes have an intense absorption
peak at 300 ± 30 nm whose ε of 8,230-11,700 M^{-1} cm^{-1} is equivalent
to 5,000 ± 1,000 M^{-1} cm^{-1} per RS$^-$ ligand, and which is clearly an
RS$^-$ → Ni(II) charge transfer transition. The broad absorption peaks
at wavelengths greater than 450 nm are grossly consistent with d-d
ligand field transitions. However, both λ_{max} and ε_{max} are affected
by overlap with adjacent peaks.

Complex 16 in Table 4 is an octahedral complex of Ni(II) with
a relatively simple anionic sulfur ligand. This complex contains
two bidentate thiocarbamate [R$_2$N-C(O)S$^-$] chelons per nickel center,

and it exists in multinuclear oligomers with a pair of sulfur bridges
between adjacent nickel ions. Strong absorption in the ultraviolet
region is consistent with an $RS^- \rightarrow Ni(II)$ charge transfer transition.

From the examples in Table 4 plus complexes 8 and 9 in Table 3,
it appears that the intensity and wavelength of $RS^- \rightarrow Ni(II)$ charge
transfer transitions are quite sensitive to the particular coordination environment of the Ni(II) center. It may also be concluded that
complexes of Ni(II) with RS^- anions will in general have multiple
charge transfer peaks which are partially overlapping.

6.5.2. The 2-Mercaptoethanol-Urease Complex

2-Mercaptoethanol ($HS-CH_2CH_2-OH$) is a competitive inhibitor of the
urease-catalyzed hydrolysis of urea at pH 7.1, with an apparent dissociation constant (K_i) of 0.72 ± 0.26 mM at 25°C [154]. Inhibition
is associated with rapidly reversible changes in the electronic
absorption spectrum of urease (Fig. 4) [124]. The spectrophoto-
metrically evaluated dissociation constant of 0.95 ± 0.05 mM is in
excellent agreement with K_i, considering the enormous difference in
enzyme concentration between the initial rate experiments (10^{-9} M)
and the spectral studies (10^{-4} M). Spectral parameters for the 2-
mercaptoethanol-urease complex are listed in Table 5. The absorption peaks in the 2-mercaptoethanol-urease complex are clearly charge
transfer transitions of a thiolate anion coordinated to Ni(II) in
urease. This coupled with competitive inhibition of urea hydrolysis
by 2-mercaptoethanol provides secure evidence of an active site role
for Ni(II) in urease [124,154].

6.6. Ni(II)-Acetohydroxamate Complexes

Acetohydroxamate ion [$H_3C-C(O)-NH-O^-$] forms 1:1 and 2:1 complexes
with divalent transition metal ions. The stability constant for
binding the first acetohydroxamate residue is in the range 10^4-10^5
M^{-1} for Mn^{2+}, Fe^{2+}, Co^{2+}, Ni^{2+}, Zn^{2+}, and Cd^{2+}, while that for the

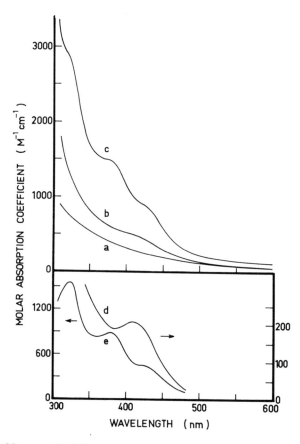

FIG. 4. Effects of light scattering and 2-mercaptoethanol on the electronic absorption spectrum of jack bean urease (14.5 mg/ml, 0.15 mM in subunits) in oxygen-free 0.05 M N-ethylmorpholinium chloride buffer (pH 7.12, 1 mM in EDTA) at 25°C [29,124]. (a) Molar scattering coefficient $\varepsilon_T^{subunit}$ of urease expressed in terms of the 96.6-kDa subunit and evaluated as $\varepsilon_T^{subunit} = 1.45023 \times 10^{-26} \, M^2/q\lambda^4$ where M is the molecular weight of the protein (6 x 96.6 kDa), q is the number of identical subunits (6), and λ is the wavelength in cm. This equation provides an approximation of the apparent absorbance due to Rayleigh scattering by a typical globular protein. (b) Observed ε of urease in the absence of 2-mercaptoethanol, based on the 96.6-kDa subunit. (c) ε of the 2-mercaptoethanol-urease complex. (d) Calculated difference spectrum (b - a), showing one of the absorption peaks associated with Ni(II). (e) Calculated difference spectrum ($\Delta\varepsilon$) of the 2-mercaptoethanol-urease complex (c - b). (Reproduced from *Biochim. Biophys. Acta* [124] by permission of Elsevier Biomedical.)

TABLE 5

Electronic Absorption Spectrum of Urease: Effect of
Light Scattering and 2-Mercaptoethanol[a]

Wavelength (nm)	ε_τ[b] $(M^{-1}\ cm^{-1})$	ε_{urease}[c] $(M^{-1}\ cm^{-1})$	$\Delta\varepsilon_{ME}$[d] $(M^{-1}\ cm^{-1})$
324	737	1270	1550 (max)
380	389	580	890 (max)
420	261	460	460 (sh)

[a]Molar absorption coefficients are based on the 96.6-kDa subunit
[124]; from *Biochim. Biophys. Acta* by permission of Elsevier Biomedical.
[b]Molar scattering coefficient (Fig. 4).
[c]Observed ε of urease in the absence of 2-mercaptoethanol (Fig. 4b).
[d]Calculated $\Delta\varepsilon$ of the 2-mercaptoethanol-urease complex (Fig. 4e).

second residue is about an order of magnitude smaller than that for
the first for any particular ion [155]. The metal ions are presumably chelated to Ni(II) through the two oxygen atoms of the ligand,
by analogy with known five-membered chelate ring structures of
several other hydroxamate complexes [29,75]. This is supported by
the O,S-coordination in cis- (Z-) and trans- (E-) 2:1 (distorted)
square-planar crystalline complexes of thionoacetohydroxamate
[H_3C-C(S)-NH-O$^-$] with Ni(II) [156]. Further, N-hydroxyurea forms
soluble 1:1 and 2:1 complexes with Mn(II), Co(II), and Ni(II) for
which five-membered chelate rings are proposed on the basis of infra-
red studies [157].

Acetohydroxamic acid reversibly forms a 1:1 complex with jack
bean urease, thereby inhibiting the enzyme. The release of acetohydroxamate from this complex is very slow at 4°C and this allowed
[^{14}C]acetohydroxamate to be used to evaluate the first secure equivalent weight for this enzyme (96.6 kDa) in 1975 [158-160]. The electronic absorption spectrum of the urease-acetohydroxamate complex
and competition studies with 2-mercaptoethanol establish that acetohydroxamate coordinates to Ni(II) in urease [29,75,160,161]. The

apparent stability constant of the enzyme-acetohydroxamate complex
is 2.5×10^5 M^{-1} in 0.05 M N-ethylmorpholinium chloride buffer at
25°C (pH 7.12, 1 mM in EDTA), and a chelate structure has been proposed for this complex [160,161].

N-Hydroxyurea is both an inhibitor and a substrate of urease
[162-164]. Isomeric N-hydroxyurea-urease complexes are responsible
for inhibition and substrate activity, respectively [165].

The observed second order rate constant for the reaction of
acetohydroxamate ion with $Ni(H_2O)_6^{2+}$ to form the 1:1 complex is only
five times larger than that for the reaction of acetohydroxamic acid
itself (3.2×10^4 M^{-1} sec^{-1} vs. 6.5×10^3 M^{-1} sec^{-1}, respectively,
at 20°C) [166]. The mechanism of formation of the chelate ring from
the monodentate acetohydroxamic acid complex with loss of a proton
has been discussed [166,167]. The observed second order rate constant for binding of acetohydroxamic acid to active site Ni(II) in
jack bean urease is 17 M^{-1} sec^{-1} at 26°C in 0.02 M phosphate buffer
(pH 7.00, 1 mM in EDTA) [160]. The behavior of acetohydroxamic acid
with Ni(II) in urease is not remarkably different from its behavior
with $Ni(H_2O)_6^{2+}$.

Hydroxamic acids reversibly inhibit several enzymes, and such
inhibition may be indicative of an essential metal ion [29].

6.7. Ni(III) Complexes

Nickel(III) is a d^7 species and its electron spin resonance spectrum
is predicted to display anisotropic resonances corresponding to a
single unpaired electron associated mainly with the d_z^2 orbital [168].
Although some (bis)dithiolato complexes of nickel were once thought
to contain Ni(III), it has been shown by EPR spectroscopy that the
unpaired electron is largely delocalized on the ligands and that the
system may be better described as a Ni(II) complex of a dithiolate
radical anion [169,170].

The first successful EPR proof of a Ni(III) center involved a
series of $NiLX_2^+$ complexes, where L is the tetradentate macrocycle

5   6

2,3-dimethyl-1,4,8,11-tetraazacyclotetradecane (5) and X^- is Br^-, Cl^-, NCO^-, NO_3^-, or 1/2 SO_4^{2-} [171,172]. The square-planar precursor Ni(II) complex (NiL^{2+}) as the perchlorate salt was oxidized very rapidly in water by an equivalent amount of ammonium persulfate, and the axial sulfates of the resulting crystalline Ni(III) complex were readily exchanged with other anions in acid solution. Axial coordination of simple anions stabilizes these cationic Ni(III) complexes markedly, and the tetragonally distorted octahedral complexes are very stable when dry but decompose slowly in water [171]. Redox potentials in acetonitrile have been evaluated for the Ni(III)/Ni(II) couple of a large series of Ni complexes with varying degrees of unsaturation in tetraazamacrocyclic ligands [172].

1,4,7-Triazacyclononane (6) forms an octahedral 2:1 crystalline complex with Ni(II) perchlorate which may be oxidized quantitatively to the corresponding octahedral Ni(III) complex by Co(III) in 1 M $HClO_4$. The Ni(III) complex is moderately stable in water, and its electron transfer reactions with a variety of Fe, Co, and Ni complexes occur via the outer sphere route. Unlike most other Ni(III)/Ni(II) couples, this system maintains octahedral symmetry around the metal center upon electron transfer [173].

The square-planar Ni(II) complex of triglycinamide [$Ni(H_{-3}L)^-$] may be quantitatively oxidized, either electrochemically or with $IrCl_6^{2-}$, to the corresponding Ni(III) species [$Ni(H_{-3}L)$]. The Ni(III) complex is reported to decompose over a period of several hours at pH 5.5 at 25°C and in a few minutes at pH 9.6 [174,175]. It has been shown by EPR studies that the Ni(III) has a tetragonally distorted octahedral geometry with four donor groups (the α-amino group, two deprotonated peptide nitrogens, and the deprotonated N-terminal amide nitrogen) from the oligopeptide in the equatorial plane and two water molecules coordinated axially [176]. Ammonia will replace the axial water molecules, with stability constants of $\sim 2 \times 10^7$ M^{-1} and ~ 60 M^{-1} at 23°C for binding of the first and second ammonia molecules, respectively.

For Ni(III) complexes of simple oligopeptides, species with three-coordinated peptide amide nitrogen anions display an intense charge transfer peak near 325 nm (ε, $\sim 5,500$ M^{-1} cm^{-1}) and near 240 nm (ε, $\sim 11,000$ M^{-1} cm^{-1}) while those with only two have corresponding peaks near 350 and 255 nm [158]. Electronic absorption and/or EPR spectral studies with histidine- and thiol-containing oligopeptides have been reported [175,177], and the redox and spectral characteristics of a wide variety of Ni(III) complexes have been summarized [168].

The complex of Ni(II) with tetraglycine has a structure analogous to that of the tetraglycinamide complex (3). It reacts with oxygen at pH 8 in a light-insensitive reaction to produce (in part) triglycine N-hydroxymethylamide and carbon dioxide [178,179]. The maximum rate of uptake of oxygen at pH 8.2 occurs after an induction period and corresponds to the time when the concentration of Ni(III) is at a maximum. It has been suggested that a ligand hydroperoxide is an intermediate in the oxidation of the peptide [178], but the detailed chemistry is not clear.

The tripeptide derived from α-aminoisobutyric acid forms a very stable Ni(III) complex [Ni(H$_{-2}$L), where L is the tripeptide with an anionic α-carboxylate group] which apparently involves the α-amino group, two deprotonated peptide nitrogens, and the α-carboxylate group in a square-planar conformation [180]. This Ni(III) complex is remarkably stable toward substitution, decomposition in acid, and internal oxidation-reduction in aqueous solution in the dark. However, it is very sensitive to photochemically induced self oxidation-reduction in which Ni(III) disappears and the C-terminal α-aminoisobutyric acid residue is oxidized to acetone and carbon dioxide [161]. The stability of the Ni(III) complex of the tripeptide derived from α-aminoisobutyric acid in the absence of light suggests the possibility that ill-defined *spontaneous* decomposition of the Ni(III) complex of tetraglycine may involve a self oxidation-reduction in which the terminal glycine residue decarboxylates to form an N-coordinated $R-C(O)-N=CH_2$ moiety concurrently with the (transient) reduction of Ni(III) to Ni(I). This amounts to dehydrogenation [181] at the α-carbon of the coordinated peptide.

6.8. Reversible Oxygenation of Ni(II) Complexes

The first example of reversible binding of oxygen to a Ni(II) complex was adduced in 1982 by Kimura and coworkers [181,182]. The complex of Ni(II) with 15-benzyl-1,4,7,10,13-pentaazacyclohexadecane-14,16-dione (7) has a magnetic susceptibility of 2.8 BM at 35°C and an

R = CH₂-Ph

7

8

absorption spectrum consistent with octahedral Ni(II). The structure is described as square-pyramidal although the possibility that H_2O is a sixth ligand is not discussed. In aqueous solution at pH 9.5, 7 reversibly binds one molecule of dioxygen at 35°C to form an octahedral complex (8) with unchanged magnetic susceptibility. Complex 8 displays a Ni → O_2 charge transfer absorption peak at 290 nm (ε = 4,700 M^{-1} cm^{-1}), and this is the first example of a 1:1 Ni(II)-dioxygen complex. Imidazole competes with dioxygen for the binding site, and the stability constant of the imidazole-7 complex is 6.7 M^{-1} [182]. Further, complex 7 has a superoxide dismutase activity [183].

The complex 8 is an hydroxylating agent! In the presence of a reducing agent such as borohydride ion, 8 converts benzene in aqueous solution into phenol, and toluene into a mixture of o- and p-cresol [181].

Complex 7 may be reversibly oxidized electrochemically to the corresponding Ni(III) complex which has the expected anisotropic EPR signal and a magnetic susceptibility of 1.7 BM. This Ni(III) complex is quite stable at pH 7 but decomposes immediately at pH 10 [181,182]. The electrode potential of the Ni(III)/Ni(II) couple is +0.24 V vs. SCE (25°C), and this is the lowest value ever reported for high-spin Ni(III)/Ni(II) couples [182].

7. NICKEL(II) CATALYSIS IN MODEL SYSTEMS

As with the invention of the word catalysis, so it was that the enzymology of the exopeptidases which required a metal ion cofactor [184,185] led to model studies of metal ion catalysis. The early work of Kroll [186], while not without some experimental defects [187,188], clearly indicated that a 1:1 complex of an amino acid ester and metal ion was the species involved in the catalyzed hydrolysis of amino acid esters. Moreover, he clearly showed that the product amino acid is also complexed with the metal ion [189]. These findings stimulated a good deal of speculation on the mechanism of action of metal ion-activated enzymes [190].

Nickel has not featured in systematic studies, presumably because its biological coming of age has been of recent advent. However, the early studies which were undertaken reveal that nickel(II) is a competent promoter in a variety of systems such as the decarboxylation of β-keto acids [191,192] and the hydrolysis of amino acid amides [193] and esters [187].

In spite of the fact that a great deal of effort has now been put into the investigation of metal ion-promoted reactions, the field still abounds with speculation, and speculation upon speculation. Some mechanisms and propositions appear to gain credibility largely by dint of usage (e.g., the "mechanism" of carboxypeptidase A [194], and entasis [195-197, but cf. 29,74]). It is perhaps not difficult to see how this situation arises. In model systems, the metal ion complex is frequently labile, and this fact makes it difficult to draw secure structural inferences. Angelici's investigation of the metal ion-promoted alkaline hydrolysis of ethyl oxalate anion [EtOx$^-$] serves to illustrate the point. Of the three possibilities for the structure of the reactive MEtOx$^{(n-1)+}$ species (9-11), only 10 could

9   10   11

be ruled out by substitution of sulfur for the ether oxygen. The authors reasonably concluded that 9 was the most likely possibility [198]. The relative catalytic efficiencies of Pb^{2+}, Ni^{2+}, and Mg^{2+} were $10^4:10^2:1$. In spite of the deceptive simplicity of the system, few other models have been analyzed with this degree of sophistication.

In 1976, we suggested that metal ion stabilization of a leaving group, as in -NH$_2$···M^{n+}, could be significant in certain enzymatic catalyses [74], and work from Fife's laboratory [199,200] on the Co^{2+}-, Ni^{2+}-, and Zn^{2+}-promoted hydrolyses of acetals (12) and esters (13) has provided a measure of support for the proposal [cf. 201-204].

12   13

A systematic and detailed study of the metal ion-promoted hydrolysis of a series of esters (14) of 2-(2'-hydroxyphenyl)-4(5)-methyl-5(4)-2'',2''-dimethylacetic acid)imidazole has provided good support for the nucleophilic addition of metal-bound $^-$OH in the significant catalysis (10^3-10^5) by Ni^{2+} and Co^{2+} [205].

14   15

Groves and Chambers [206] reported significant rate enhancement in the Cu^{2+}-, Ni^{2+}-, and Zn^{2+}-promoted hydrolysis of 1-[(6-((dimethylamino)methyl)-2-pyridyl)methyl]hexahydro-1,4-diazepin-5-one (15). They purport to have demonstrated the precise geometric orientation of a metal ion in metal ion-promoted amide hydrolysis. However, their arguments in favor of an intramolecular metal-hydroxo path rather than external attack of hydroxide ion are not compelling; while the data derived from entropies of activation may support an intramolecular reaction, the comparison systems are not strictly relevant. Further, D_2O effects ($k_{H_2O}/k_{D_2O} = 1.3$) for the basic reaction of the Cu^{2+} complex are not definitive [106] and, as Sayre [93] points out, their arguments on stereoelectronic control [207] are no more convincing.

We recently reported the first successful catalysis of degradation of a urea by Ni^{2+}, namely, the concurrent ethanolysis and hydrolysis of N-(2-pyridylmethyl)urea is promoted by Ni^{2+} [Eqs. (6) and (7)] [208]. The rate enhancement of the promoted ethanolysis

$$RNH\text{-}C(O)\text{-}NH_2 + EtOH \longrightarrow RNH\text{-}C(O)\text{-}OEt + NH_3 \qquad (6)$$

$$RNH\text{-}C(O)\text{-}NH_2 + 2H_2O \longrightarrow RNH_2 + H_2CO_3 + NH_3 \qquad (7)$$

R = 2-pyridylmethyl

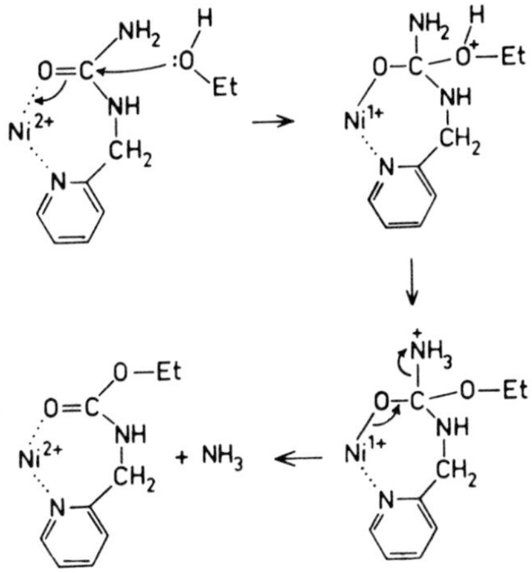

SCHEME 4. Proposed mechanism for the Ni(II)-promoted ethanolysis of N-(2-pyridylmethyl)urea. (Reproduced from Ref. 208 by permission of the American Chemical Society.)

is greater than 7×10^4. Again, the mechanism of ethanolysis has not been unequivocally established, but activation by O-coordination of the urea constitutes a reasonable possibility (Scheme 4).

All of the foregoing examples involve promotion rather than catalysis by the metal ion (Sec. 5.1). However, Breslow and Overman [209] reported the use of Ni^{2+} in an artificial hydrolase which uses cyclohexaamylose as a binding cavity for the substrate, 4-nitrophenyl acetate (cf. Cramer and Mackensen [210]).

Finally, it should be noted that the work of Breslow and Mehta on the catalytic directed chlorination of the nicotinate ester of 3-α-cholestanol using Ni^{2+} as part of a template [211-213] has been retracted [214].

8. NICKEL(II)-ACTIVATED ENZYMES

8.1. Metal Ion Substitution in Metalloenzymes

Replacement of a metal ion cofactor by an alternative metal ion is potentially a powerful tool in the investigation of metalloproteins and metalloenzymes. An alternative metal ion may serve as a *spectroscopic probe* where a spectroscopic technique can provide information about the chemical nature of the ligands or the properties of the complex. For example, cobalt(II) has often been used as a spectrophotometric probe of the metal ion binding site in zinc(II) metalloenzymes [215]. Some metalloproteins in which Ni(II) serves as a spectroscopic probe are described in Secs. 3, 6.3, and 6.5.

An additional function may be served by an alternative metal ion in a metalloenzyme: if the metal-substituted enzyme is catalytically active, then a new range of questions about the mechanism of action may be investigated [74]. This approach is now widely used, and it is appropriate to draw attention to several caveats which are important to an interpretation of the literature (see also Secs. 4.3 and 4.5):

1. Too often the relative enzymatic activities with different metal ions refer to initial rates at single substrate concentrations. Differences in listed activities may therefore reflect variation in k_{cat} or K_m (or both) for the substrate(s) with different metal ions.

2. Initial enzymatic rates may have been determined from the concentration of product formed during a fixed time period without evidence that the reaction is zero order in the presence of all metal ions investigated.

3. The enzymatic rates may have been determined at a single concentration of metal ion, without regard to (generally unknown) rates and equilibria for formation of the enzyme-metal ion complex.

4. The relative activity with different metal ions will often depend on the particular substrate, and nonphysiologic substrates properly addressed can provide useful information.

5. The relative activities with different metal ions will often show differing pH effects.

6. The enzyme preparation may be substantially impure, so that various added metal ions may be differentially available because of binding to the impurities.

7. The metal content of apoenzymes and reconstituted enzymes is frequently not reported, and a relatively low activity with a particular metal ion may be an artifact due to a few percent of another highly effective metal ion present in the system.

8. In a mixed metal form (e.g., Zn-Fe) of a two-metal (e.g., Fe-Fe) enzyme, slow equilibration to produce a mixture containing all forms (e.g., Zn-Zn, Zn-Fe, and Fe-Fe) will undoubtedly occur under some conditions.

It follows that meaningful information on the enzymatic activity of metal-substituted metalloenzymes requires meticulous work, and careful consideration must be given to such fundamental details as the purity and concentration of the metalloenzyme and to the time scale of experiments. Finally, some published results may be erroneous for reasons that are not readily deduced, and some results may therefore have to be at least temporarily discarded.

8.2. Nickel(II)-Substituted and Activated Enzymes

For a growing list of enzymes, nickel is one of several divalent metal ions which provides substantial catalytic activity. Some metalloenzymes for which several metal ions including Ni(II) provide activation are listed in Table 6. Most of the systems in this table are reasonably well characterized, with the probable exception of DNA-directed DNA polymerases from sea urchin and avian myeloblastosis virus where the weak apparent activation by Ni(II) could well be an artifact. Some other metalloenzymes for which Ni(II) activation is either absent, uncertain, or still not well quantified are summarized below. Genuine Ni(II) activation will undoubtedly be found for many more enzymes if it is sought.

Some implications of results with alternative metal ions in enzymes are discussed in Sec. 8.3.

TABLE 6

Enzymatic Activity of Some Metal(II)-Substituted Enzymes

EC no.[a]	Enzyme and source	Native metal ion	Percent activity[b] of the substituted enzyme									Ref.
			Mg	Mn	Fe	Co	Ni	Cu	Zn	Cd	None	
1.1.1.1	Alcohol dehydrogenase[c] horse liver	Zn				149	99		100	3	1	216
1.1.1.42	Isocitrate dehydrogenase (NADP⁺) pig heart	?	100	135		67	10		236	86	0	217,218
2.1.3.2	Aspartate transcarbamylase Escherichia coli	Zn				101	95		100			219
2.7.1.37	Protein kinase bovine lung	?	100	7		24	30		<0.2	<0.2	<0.6	220
2.7.1.40	Pyruvate kinase[d] rabbit muscle (kinase) (hydrolase)	?	100 8	50 <0.1		60 83	10 100					221,222
2.7.7.7	DNA-directed DNA polymerase Escherichia coli[e] sea urchin[f] avian myeloblastosis virus[f]	Mg Mg Mg	100 100 100	153 21 42		57 58 27	<1 29 6	<0.1 <0.1 <0.1	4 9 2			223

EC No. / Enzyme	Metal										Ref.
3.4.11.10 Aminopeptidase, Aeromonas proteolytica[g]	Zn		<6	<6	500	980	1000	100	22	<6	224
3.4.17.1 Carboxypeptidase A, bovine pancreas	Zn					19	500	92	100	3	225,226
3.5.3.4 Allantoicase, Pseudomonas aeruginosa	Mn	37	100	20	51	28	85	33	95	2	227
4.2.1.6 Galactonate dehydratase, Pseudomonas sp.	?			100	84	0	28	7	0	0	228
4.4.1.5 Glyoxalase I, rat erythrocyte	?					100	46	85	40	<1	229
5.1.2.2 Mandelate racemase[h], Pseudomonas putida	?			100	48	?	60	59	0	0	230
5.4.2.2 Phosphoglucomutase, rabbit muscle	?			100	5	15	60	0.3	0.8	<0.001	231

[a] From Ref. 62
[b] Relative activity toward a particular substrate. Divalent metal ions are intended throughout.
[c] Reportedly substituted at the active site (see also Table 4, No. 3).
[d] Phosphoenolpyruvate is enzymatically either phosphorylated by ATP or slowly hydrolyzed in its absence.
[e] DNA polymerase I.
[f] Low enzymatic activity. Ni(II) could be releasing Mg(II) from contaminants in the enzyme preparation.
[g] This enzyme contains two binding sites for divalent metal ions, but maximal activity is restored when as little as 1 mol of Ni^{2+}, Cu^{2+}, or Zn^{2+} is added per mol of apoenzyme.
[h] The relative activity is buffer-dependent.

8.2.1. Malate Dehyrdogenase (Oxaloacetate Decarboxylating) (NADP⁺) (EC 1.1.1.40)

This enzyme catalyzes the oxidative conversion of (S)-malate
(= L-malate) to pyruvate and carbon dioxide [Eq. (8)]. The partially purified enzymes from a *Pseudomonas* sp. [232] and from a

$$^-OOC\text{-}CH(OH)\text{-}CH_2\text{-}COO^- + NADP^+ \rightleftharpoons {}^-OOC\text{-}C(O)\text{-}CH_3 + CO_2 + NADPH \qquad (8)$$

Mycobacterium sp. [233] are activated by Mg^{2+}, Mn^{2+}, Co^{2+}, Ni^{2+},
and Cd^{2+}. The relative activity depends on the concentration of
substrate [(S)-malate] and of KCl.

8.2.2. Amine Oxidase (Copper-Containint) (EC 1.4.3.6)

These enzymes catalyze the reaction of an alkylamine with oxygen
and water to form ammonia, hydrogen peroxide, and the corresponding
aldehyde [Eq. (9)]. The dimeric bovine serum enzyme contains 1 mol

$$R\text{-}CH_2\text{-}NH_2 + H_2O + O_2 \rightleftharpoons R\text{-}CHO + NH_3 + HOOH \qquad (9)$$

of Cu(II) per subunit whose replacement by Ni(II) leaves 2% residual
enzymatic activity which is almost certainly due to residual Cu(II).
For comparison, the Co(II) enzyme appears to have about 13% as much
activity as the Cu(II) enzyme [234]. It has been reported that
Cu(II) in the octameric pig liver enzyme may be replaced by Fe(II),
Co(II), or Ni(II) with retention of substantial enzymatic activity
[235]. This work should be interpreted cautiously because of
uncertain purity of the enzyme.

8.2.3. Acid Phosphatase (EC 3.1.3.2)

Phosphatases catalyze the hydrolysis of phosphate monoesters
[Eq. (10)]. Iron-containing acid phosphatases are a subset of

$$R\text{-}O\text{-}PO_3^{2-} + H_2O \rightleftharpoons R\text{-}OH + HO\text{-}PO_3^{2-} \qquad (10)$$

EC 3.1.3.2 noted for their violet color [236]. The active form of
the 40-kDa enzyme from pig allantoic fluid (uteroferrin) contains
one Fe(II) and one Fe(III) in a binuclear cluster [237]. The Fe(II)
may be removed rapidly and reversibly with complete loss of enzymatic

activity. Its replacement by Cu(II), Zn(II), and Hg(II) yields highly active enzyme. An iron-free apoenzyme has also been prepared. No metal ions other than Fe(II) and Fe(III) restore significant activity to it, but Zn(II) and Ni(II) bind tightly [238,239].

8.2.4. Phosphoprotein Phosphatase (EC 3.1.3.16)

These enzymes catalyze the reaction in Eq. (10) where R corresponds to an alcohol (serine, threonine) or phenol (tyrosine) side chain of a protein. Calcineurin is a Ca^{2+}- and calmodulin-dependent phosphoprotein phosphatase found in many mammalian tissues. It consists of two subunits: subunit A (~60 kDa) interacts with calmodulin, and subunit B (~19 kDa) binds up to four Ca^{2+} ions. Calcineurin from bovine brain has recently been found to contain nearly 1 mol of zinc and 1 mol of iron per mol of enzyme [240]. Calcineurin is activated by Ni(II) and, to a lesser degree, by Mn(II), Co(II), and Zn(II) [241,242], and definitive experiments in a well-defined system are needed.

8.2.5. C3 Convertase (EC 3.4.21.47)

The complement system of vertebrate blood is part of the immune defense against infection by microorganisms. Activation of the complement system involves two pathways, and the "alternative pathway" utilizes a serine proteinase known as C3b,Bb which consists of a 176-kDa subunit (C3b) and a 63-kDa subunit (Bb). This enzyme catalyzes the hydrolysis of a peptide bond in component C3 of the complement system and is therefore called a "C3 convertase" [243]. The formation of C3b,Bb from its precursors involves a proteolytic reaction and requires a divalent metal ion (traditionally Mg^{2+}). When Ni^{2+} is used in place of Mg^{2+}, the resulting 1:1 C3,Bb(Ni^{2+}) complex has enzymatic activity similar to that of the usual C3,Bb(Mg^{2+}) complex and is much more stable with respect to decomposition [244].

8.2.6. *Arginase (EC 3.5.3.1)*

Arginase catalyzes the hydrolysis of the guanidino group of arginine to form ornithine and urea [Eq. (11)]. The metal ion content of

$$(H_2N)_2C={}^+NH-R + H_2O \rightleftharpoons H_2N-C(O)-NH_2 + H_3N^+-R \qquad (11)$$
$$R = CH_2-CH_2-CH_2-CH({}^+NH_3)COO^-$$

arginase has not been determined, although the presence of Mn^{2+} during purification reduces inactivation of the enzyme [245]. However, bovine liver arginase is activated by both Ni^{2+} and Mn^{2+} at pH 7.5, and the inclusion of Ni^{2+} throughout the purification results in substantially higher arginase activity at each stage [246,247]. The metal ion in arginase appears to be much more loosely bound than the nickel ion in urease [74]. Since the guanidinium group of arginine is analogous to urea and since both arginase and urease are efficient nickel metalloenzymes, the metal ion may have a similar role in these two enzymes.

8.2.7. *dCMP Deaminase (EC 3.5.4.12)*

This enzyme catalyzes the deamination of deoxycytidine monophosphate to deoxyuridine monophosphate [Eq. (12)]. The enzyme from *Bacillus*

$$dCMP + H_2O \rightleftharpoons dUMP + NH_3 \qquad (12)$$

subtilis requires Zn^{2+} for maximal activity. Although the partially purified enzyme appears to be slightly activated by Ni^{2+}, there is apparently enough Zn^{2+} in the crude system to account for the results [248].

8.2.8. *Fructose-Bisphosphate Aldolase (EC 4.1.2.13)*

This enzyme catalyzes the aldol condensation between dihydroxyacetone phosphate and D-glyceraldehyde-3-phosphate to form D-fructose-1,6-bisphosphate [Eq. (13)]. Class II aldolases (from yeast and bacteria)

$$R-C(O)-CH_2-OH + HC(O)-R' \rightleftharpoons R-C(O)-CH(OH)-CH(OH)-R' \qquad (13)$$
$$R = CH_2-O-PO_3^{2-}$$
$$R' = CH(OH)-CH_2-O-PO_3^{2-}$$

contain a divalent metal ion. The enzyme from baker's yeast normally contains Zn^{2+}, but the apoenzyme is reactivated by several divalent metal ions ($Zn^{2+} > Co^{2+} > Mn^{2+} > Fe^{2+}$) but not by Ni^{2+} [249]. An earlier report of ~9% activation by Ni^{2+} [250] is apparently in error.

8.2.9. *Carbonic Anhydrase (EC 4.2.1.1)*

This enzyme catalyzes the reversible hydration of carbon dioxide to form carbonic acid [Eq. (14)]. Carbonic anhydrase normally contains

$$CO_2 + H_2O \rightleftharpoons HO\text{-}C(O)\text{-}OH \rightleftharpoons HO\text{-}COO^- + H^+ \tag{14}$$

one Zn^{2+} per molecule. The apoenzyme of bovine and human carbonic anhydrases forms a 1:1 complex at the active site with several divalent transition metal ions including Ni^{2+}, but only the Co^{2+} enzymes have catalytic properties similar to those of the native enzymes. The activity of the Ni^{2+} enzymes is at best very low and uncertain [128,251].

8.2.10. *Histidine Ammonia-Lyase (EC 4.3.1.3)*

Histidine ammonia-lyase catalyzes the elimination of ammonia from histidine to form urocanic acid [Eq. (15)]. The enzyme from *Pseudo-*

$$R\text{-}CH\text{-}CH(^+NH_3)\text{-}COO^- \rightleftharpoons R\text{-}CH=CH\text{-}COO^- + {}^+NH_4 \tag{15}$$

monas putida isolated in the absence of added divalent metal ions is active and is stable to exhaustive dialysis against EDTA [252,253]. Maximum specific activity is obtained when the enzyme has been treated with a sulfhydryl compound and then activated with Mn^{2+} or Cd^{2+}. However, the divalent cations Mg^{2+}, Ca^{2+}, Fe^{2+}, Co^{2+}, Ni^{2+}, Cu^{2+}, and Zn^{2+} also provide substantial activation. It is possible that the enzymatic reaction involves two different divalent metal ions, one of which is not removed by EDTA (Sec. 4.5) [74].

8.2.11. *Pyruvate Carboxylase (EC 6.4.1.1)*

This enzyme catalyzes the reaction of pyruvate with bicarbonate and ATP to form oxaloacetate, ADP, and orthophosphate [Eq. (16)].

$$^-OOC-C(O)-CH_3 + HO-CO_2^- + ATP \rightleftharpoons$$
$$^-OOC-C(O)-CH_2-COO^- + ADP + HO-PO_3^{2-} \qquad (16)$$

Pyruvate carboxylase contains one covalently bound biotin residue per subunit. The enzyme from chicken liver normally contains one Mn^{2+} per subunit, and evidence for an enzymatically active Ni^{2+} form of the enzyme was summarized in Sec. 4.5 [74,84].

8.2.12. *Ni(II) and Other Enzymes*

The rate of the ATP-dependent synthesis of phosphatidylserine by crude rat brain microsomes in vitro is enhanced by Ni^{2+} but not by several other divalent metal ions [254]. Definitive experiments will be required before any conclusion can be drawn.

The presence of Ni^{2+} during the extraction of nitrate reductase (EC 1.7.99.4) from sorghum significantly increases the yield of active enzyme. The function of the nickel may be to prevent inhibition by cyanide ion that is released from plant glycosides during homogenization of tissue [255].

Ni^{2+} appears to bind to crude guanylate cyclase (EC 4.6.1.2) from sea urchin sperm, but it affords no detectable enzymatic activity [256].

8.3. Some Implications of Metal Ion Substitution in Enzymes

A metal-substituted enzyme (Sec. 8.2) often has catalytic efficiency within a factor of 10 of that of the enzyme containing its "native" metal ion. Even if the normal metal ion is involved in a non-rate-limiting catalytic step, such a result requires that the substitute metal ion perform with respectable efficiency. Since most reasonable estimates of the efficiency of enzymatic catalysis yield factors of 10^8-10^{12} compared to appropriate model systems, in terms of efficient chemistry, the variations in activity of many metal-substituted enzymes may therefore be regarded as trivial. It is useful to discuss this phenomenon in terms of the ionic radii of metal ions (Table 7).

TABLE 7

Ionic Radii of Metal Ions (Å)[a]

Charge	11 Na	12 Mg	13 Al	14 Si
1+	0.97	0.82		
2+		0.66		0.65
3+			0.51	
4+				0.42

Charge	19 K	20 Ca	21 Sc	22 Ti	23 V	24 Cr	25 Mn	26 Fe	27 Co	28 Ni	29 Cu	30 Zn	31 Ga	32 Ge
1+	1.33	1.18		0.96		0.81					0.96	0.88	0.81	
2+		0.99		0.94	0.88	0.89	0.80	0.74	0.72	0.69	0.72	0.74		0.73
3+			0.73	0.76	0.74	0.63	0.66	0.64	0.63				0.62	
4+				0.68	0.63		0.60							0.53

Charge	37 Rb	38 Sr	39 Y	41 Nb	42 Mo	44 Ru	45 Rh	46 Pd	47 Ag	48 Cd	49 In	50 Sn
1+	1.47								1.26	1.14		
2+		1.12		1.00	0.93			0.80	0.89	0.97		0.93
3+			0.89				0.68				0.81	
4+				0.74	0.70	0.67		0.65				0.71
6+					0.62							

Charge	55 Cs	56 Ba	57 La	76 Os	77 Ir	78 Pt	79 Au	80 Hg	82 Pb
1+	1.67	1.53					1.37	1.27	
2+		1.34	1.39	0.88		0.80		1.10	1.20
3+			1.02				0.85		
4+				0.69	0.68	0.65			0.84

[a]Adapted from Ref. 257, p. F-179. Values are either experimental or theoretical, and may depend on the particular salt and on the methods of measurement and calculation.

The Zn^{2+} in the regulatory subunit of aspartate transcarbamylase
(EC 2.1.3.2 in Table 6; No. 2 in Table 4) has a *binding* rather than a
catalytic role, as defined in Sec. 2.3. The enzyme has virtually
identical specific activity and similar cooperativity with Co^{2+}, Ni^{2+},
and Zn^{2+} in this site. This presumably reflects the fact that these
divalent ions have a similar ionic radius (Table 7). Further, either
they have a common tetrahedral geometry or the binding site must be
flexible enough to accept other geometries.

The replaceable divalent metal ion binds to the apoenzyme and
is known or highly likely to have a catalytic role as a Lewis acid
in several of the enzymes listed in Table 6 and Sec. 8.2: alcohol
dehydrogenase, isocitrate dehydrogenase, carboxypeptidase A, galac-
tonate dehydratase, glyoxalase I, mandelate racemase, phosphogluco-
mutase, the iron-containing acid phosphatases, fructose bisphosphate
aldolase, carbonic anhydrase, histidine ammonia-lyase, and pyruvate
carboxylase.

The divalent cations of Mg, Mn, Fe, Co, Ni, Cu, and Zn all have
ionic radii between 0.66 and 0.80 Å (Table 7). The high catalytic
efficiency of various substituted enzymes listed above suggests that
for a Lewis acid role, the size and "preferred" geometry of the ion
may often not be a critical factor. There is clearly no systematic
discrimination against divalent transition metal ions as compared
with spherically symmetrical ions (Mg^{2+} and Zn^{2+}) in serving as Lewis
acids in enzymes [74].

Ca^{2+}, Cd^{2+}, Hg^{2+}, and Pb^{2+} all have ionic radii between 0.97
and 1.20 Å. Of the Cd^{2+}-substituted enzymes in Table 6, some have
very low activity while others have very high activity. Further,
the Hg-Fe form of pig allantoic acid phosphatase (Sec. 8.2.4) has
about 30% of the specific activity of the normal Fe-Fe form. However, conventional "heavy-metal" enzyme poisons have found little
attention in enzyme activation studies. It therefore appears likely
that large divalent metal ions (e.g., Ca^{2+}, Cd^{2+}, Hg^{2+}, and Pb^{2+})
may be found to activate more enzymes than previously suspected. In
this connection, the distorted tetrahedral coordination of Cu^{2+} in

the electron transfer protein, plastocyanin from poplar leaves is almost unchanged in the Hg^{2+}-substituted protein, which indicates flexibility [74] in the 10.5-kDa polypeptide [258]. The ability of several metal ions to activate an apometalloenzyme means (1) that it may sometimes be difficult to specify what the "native" metal ion is and (2) that the actual metal ion(s) utilized will be a function of the nutritional status of the organism (Sec. 2.3) [74]. The latter was strikingly demonstrated by the replacement of Mn^{2+} by Mg^{2+} in pyruvate carboxylase (Sec. 8.2.11). Further, bovine pancreatic procarboxypeptidase A was originally reported to contain essentially 1 mol of metal ion per mol of protein, with the following distribution of metals: Zn (~79%), Fe (~14%), Ni (~5%) [259].

Substitution of metal ions in enzymes in which the metal ion undergoes redox changes during catalysis must obviously have more stringent requirements than in those discussed thus far in Sec. 8. One such enzyme is superoxide dismutase (SOD) which catalyzes the dismutation of superoxide radical anion $(O_2^{\cdot})$ to oxygen (O_2) and hydrogen peroxide (HOOH). On the basis of the metal cofactor and polypeptide chain, these enzymes have been grouped into three families: those containing both copper and zinc (CuZnSOD), those containing Fe(FeSOD), and those containing Mn (MnSOD). Prokaryotes typically synthesize MnSOD or FeSOD or both, and there is considerable amino acid sequence homology among both enzymes from all sources. The molecular differences between MnSOD and FeSOD are as yet poorly defined, and recent evidence suggests that the same polypeptide chain may be activated by either metal under appropriate conditions [260,261].

9. BINDING SITES FOR METAL IONS

9.1. Metal Ion Ligands in Proteins

The potential ligands of metal ions in proteins are (1) side chains of amino acid residues, (2) the carbonyl oxygen or amide nitrogen of the peptide bond linking the amino acid residues, (3) the N-terminal amine, and (4) the C-terminal carboxylate group. There may be additional binding to the metal ion by (5) nonprotein nitrogen ligands of a macrocycle as in heme proteins, and (6) solvent, substrate, or other small molecules such as molecular oxygen, H_2S, or inorganic phosphate. The type of ligand will depend on the function of the metal ion in the protein (e.g., structural, catalytic, electron transfer, and so on); presumably proteins have evolved so that the ligands are appropriate to the metal ion and its function.

The metal ion ligands of some metalloproteins are listed in Table 8. This table elaborates earlier versions by Liljas and Rossmann [308] and Ibers and Holm [309]. Generally, the ligands have been determined by the x-ray crystallography of metalloproteins of known amino acid sequence.

In other metalloproteins where the coordination of protein groups to the constituent metal ion has been studied, the particular amino acid residues involved have not been so identified. Some examples of the latter systems are discussed below.

9.1.1. Iron, Molybdenum, and Manganese

Nitrogenase (EC 1.18.6.1) catalyzes the transfer of electrons from a reducing agent such as ferredoxin or flavodoxin to molecular nitrogen to form ammonia. The nitrogen half-reaction is given by Eq. (17).

$$N_2 + 6H^+ + 6e^- \rightleftharpoons 2NH_3 \qquad\qquad (17)$$

Nitrogenase is a complex consisting of an iron protein (55-60 kDa) and a molybdenum-iron protein (220-250 kDa) [310]. The iron protein is an α_2 dimer whose subunits each contain an Fe_4S_4 cluster (16) coordinated to the protein by Fe-cysteine bonds, and the iron-molybdenum protein is an $\alpha_2\beta_2$ tetramer with two molybdenum atoms and about 30 iron atoms and acid-labile sulfides. The molybdenum in nitrogenase of *Clostridium pasteurianum* exists as a complex of molybdenum ion with cysteine and inorganic sulfide (but no oxygen) ligands coupled to iron-sulfur clusters [311,312]. By way of

TABLE 8

Some Metal Ion Ligands in Proteins

Protein and source	Proposed metal ion ligands[a]	Geometry[b]	Ref.
Carboxypeptidase A, B bovine	Zn N[His-69,196] O[Glu-72; H_2O]	bip?	127,262, 263
Carbonic anhydrase B human erythrocytes	Zn N[His-94,96,119] O[H_2O]	tet	264
Carbonic anhydrase C human erythrocytes	Zn N[His-93,95,117] O[H_2O]	tet	265
Alcohol dehydrogenase horse liver	Zn_c N[His-67] O[H_2O] S[Cys-46,174] Zn_d S[Cys-97,100,103,111]	tet tet	266
Aspartate carbamoyltransferase *Escherichia coli*	Zn S[Cys-109,114,137,140]	tet	267
Alkaline phosphatase *Escherichia coli*	Zn N[His-331,372,412] X[H_2O, NH_3, Cl^-, or SO_4^{2-}?] Zn N[His-370] O[Asp-51?,369; Ser-102?] Zn or Mg O[Asp-51?,153; Glu-322; Thr-155]	tet tet tet	268
Superoxide dismutase bovine erythrocytes[e]	Cu N[His-44,46,61,118] Zn N[His-61,69,78] O[Asp-81]	pla tet	269,270
Thermus thermophilus	Mn N[His-28,83,169] O[Asp-165; H_2O]	bip	271
Staphylococcal nuclease *Staphylococcus aureus*	Ca O[Asp-19,21,40; Glu-43; Thr-41[f]; H_2O?]	oct	272
Concanavalin A jack bean	Mn N[His-24] O[Asp-10,19; Glu-8; $(H_2O)_2$] Ca O[Asn-14; Asp-10,19; Tyr-12[f]; $(H_2O)_2$]	oct oct	273,274
Azurin *Pseudomonas aeruginosa* *Alcaligenes denitrificans*	Cu N[His-46,117] S[Cys-112; Met-121] Cu N[His-46,117] S[Cys-112; Met-121]	tet tet	275 276
Plastocyanin poplar leaf	Cu N[His-37,87] S[Cys-84; Met-92]	tet	277
S100b protein bovine brain	Zn N[His-15; His-45,85 and/or 90]		278

TABLE 8 (continued)

Protein and source	Proposed metal ion ligands[a]	Geometry[b]	Ref.
Metallothionein			
rat	Zn S[Cys-13,24,26,29]	tet	279
	Zn S[Cys-5,7,21,24]	tet	
	Cd S[Cys-41,57,59,60]	tet	
	Cd S[Cys-33,36,37,41]	tet	
	Cd S[Cys-33,44,48,59]	tet	
	Cd S[Cys-34,37,48,50]	tet	
	Cd S[Cys-7,13,15,19]	tet	
Thermolysin			
Bacillus thermoproteolyticus	Zn N[His-142,146] O[Glu-166; H_2O]	tet	280,281
	Ca O[Asp-138,185; Glu-177,187[f],190; H_2O]	oct	281,282
	Ca O[Asn-183[f]; Asp-185; Glu-177,190; $(H_2O)_2$]	oct	
	Ca O[Asp-57,59; Gln-61[f]; $(H_2O)_3$]	oct	
	Ca O[Asp-200; Ile-197[f]; Thr^2-194,194[f]; Tyr-193[f]; $(H_2O)_2$]	oct?	
Parvalbumin			
carp muscle	Ca O[Asp-51,53; Glu-59,62; Phe-57[f]; Ser-55]	oct	283
	Ca O[Asp-90,92,94; Glu-101; Lys-96[f]; H_2O]	oct	
Troponin C			
rabbit muscle	Ca O[Asp-27,29,33[f]; Glu-38; Ser-35; H_2O]	oct	284
	Ca O[Asp-63,65,71; Glu-74; Ser-67; Thr-69[f]]	oct	
	Ca O[Asn-105; Asp-103,107,111; Glu-114; Tyr-109[f]]	oct	
	Ca O[Arg-145[f]; Asn-141; Asp-139,143,147; Glu-150]	oct	
Calcium-binding protein			
bovine intestine	Ca O[Asn-56; Asp-54,58; Glu-60[f],65]	oct	285
	Ca O[Ala-15[f]; Asp-19[f]; Glu-17[f],22[f],27]	oct	
Hemerythrin			
Themiste dyscritum	Fe N[His-25,54] O[Asp-106; Glu-58; $1/2\ O_2^{2-}$]	sqp	286,287
	Fe N[His-73,77,101] O[Asp-106; Glu-58; $1/2\ O_2^{2-}$]	oct	
Rubredoxin			
Clostridium pasteurianum	Fe S[Cys-6,9,39,42]	tet	288
Desulfovibrio vulgaris	Fe S[Cys-6,9,39,42]	tet	289
Micrococcus aerogeneses	Fe S[Cys-6,9,38,41]	tet	290
Amidophosphoribosyltransferase			
Bacillus subtilis	Fe_4S_4 S[Cys-393,445,448,451]	tet	291

Protein	Metal center and ligands	Geometry	Ref.
Ferredoxin			
Peptococcus aerogenes	Fe₄S₄ S[Cys-8,11,14,45]	tet	292
	Fe₄S₄ S[Cys-18,35,38,41]		
Azotobacter vinelandii	Fe₄S₄ S[Cys-24,39,42,45]	tet	293
	Fe₃S₃ S[Cys-8,11,16,20,49] O[H₂O?]g	tet	
"High-potential" iron protein			
Chromatium vinosum	Fe₄S₄ S[Cys-43,46,63,77]	tet	294
Metmyoglobin			
sperm whale	Fe N[porphyrin; His-F8] O[H₂O]	oct	295,296
Myoglobin			
sperm whale	Fe N[porphyrin; His-F8]	sqp	296,297
Methemoglobin			
horse	Fe N[porphyrin; His-F8] O[H₂O]	oct	298
Cytochromes			
calf liver (b₅)	Fe N[porphyrin; His-39,63]	oct	299
tuna heart (c)	Fe N[porphyrin; His-18] S[Met-80]	oct	300
Rhodospirillum rubrum (c₂)	Fe N[porphyrin; His-18] S[Met-91]	oct	301
Desulfovibrio desulfuricans	Fe N[porphyrin; His-36,48]	oct	302
(c₃)	Fe N[porphyrin; His-39,96]	oct	
	Fe N[porphyrin; His-49,67]	oct	
	Fe N[porphyrin; His-89,115]	oct	
Paracoccus denitrificans (c₅₅₀)	Fe N[porphyrin; His-19] S[Met-99]	oct	303
Cytochrome P-450			
Pseudomonas putida	Fe N[porphyrin] O[H₂O?] S[Cys-357]	oct	304

a Abbreviations: Ala, alanine; Arg, arginine; Asn, asparagine; Asp, aspartic acid; Cys, cysteine; Glu, glutamic acid; Gln, glutamine; His, histidine; Ile, isoleucine; Lys, lysine; Met, methionine; Phe, phenylalanine; Ser, serine; Thr, threonine; Tyr, tyrosine. The amino acid is followed by its residue number.
b bip, distorted trigonal bipyramidal; oct, distorted octahedral; pla, distorted square-planar; tet, distorted tetrahedral; sqp, distorted square-pyramidal.
c Catalytic zinc ion.
d Noncatalytic zinc ion.
e The same metal ion ligands are reported in enzyme from human erythrocytes, horse liver, swordfish liver, yeast, and *Neurospora crassa* [305].
f The ligand is the carbonyl oxygen of the peptide bond.
g An Fe₃S₄ structure with either 3 or 4 cysteine ligands has been proposed [306,307].

DMF = O-bonded *N,N*-dimethylformamide

16   17

parenthesis, a related vanadium-iron-sulfur cubane cluster (17) has recently been reported by Kovacs and Holm [313].

The mononuclear molybdenum in oxo transfer molybdoenzymes (e.g., sulfite oxidase, nitrate reductase, and xanthine dehydrogenase) is associated with a pterin and has both thiolate and oxygen (Mo=O) ligands [314].

Phosvitin is an iron(III)-binding protein where tetrahedral coordination to four oxygen ligands is suggested on the basis of the electronic absorption spectrum. These ligands are most likely phosphorylated serine residues which constitute approximately one-half of the total amino acid content of the protein and roughly twice the number of bound Fe(III) ions [315].

There is substantial evidence from resonance Raman and EPR studies for Fe(III)-tyrosine coordination in a number of nonheme iron proteins including 4-hydroxyphenylpyruvate dioxygenase (EC 1.13.11.27) from *Pseudomonas* sp. [316], protocatechuate 3,4-dioxygenase (EC 1.13.11.3) from *Pseudomonas aeruginosa* [317,318], pig allantoic fluid acid phosphatase (EC 3.1.3.2) [319], and the serum iron transport protein, transferrin. In conalbumin (ovotransferrin), Fe(III) is six-coordinated (with distorted octahedral geometry) presumably to two tyrosine residues, two histidine residues, an anion such as bicarbonate or oxalate, and water [58,320]. Mn(III)-tyrosine coordination has been reported in sweet potato acid phosphatase in which cysteine is also suggested to be a ligand [321]. The ligation of histidine to Mg(II) in pyruvate kinase

(EC 2.7.1.40) of rabbit muscle has also been suggested on the basis of NMR studies [322]. This site has been taken to be the same site as that which binds Mn(II). However, recent x-ray data of Hilary Muirhead do not show any histidine side chains close to the metal-binding site of the cat enzyme [323].

The heme iron proteins have axial protein ligands of either one histidine ligand (hemoglobin and myoglobin), two histidine ligands (cytochromes b_5 and c_3), or histidine and methionine (cytochromes c, c_2 and c_{550}) in addition to the four nonprotein nitrogen ligands of the porphyrin ring (Table 8). The heme Fe of cytochrome P-450 (camphor 5-monooxygenase, EC 1.14.15.1) from *Pseudomonas putida* is linked to the protein through a cysteine thiolate ligand [304]. The cytochrome a component of cytochrome c oxidase (EC 1.9.3.1) from *Saccharomyces cerevisiae* has two histidine ligands to the heme iron [324], whereas the copper ion in cytochrome c oxidase has at least one histidine and one cysteine ligand, and probably two of each [325].

9.1.2. Copper

Histidine and cysteine ligands have been implicated in the coordination of the copper ions of laccase (EC 1.10.3.2) from the lacquer tree (*Rhus vernicifera*) [326], whereas histidine (and not cysteine) ligands have been proposed along with unidentified nitrogen or oxygen ligands for galactose oxidase (EC 1.1.3.9) from *Polyporus circinatus* [327] (cf. Zn^{2+} in glyoxalase I (EC 4.4.1.5) from human erythrocytes [328]). The dioxygen-binding protein, hemocyanin from arthropods and molluscs, has a two-copper center with histidine ligands to the copper ions and a bridging oxygen ligand of perhaps tyrosine. Whereas oxyhemocyanin is four-coordinate, deoxyhemocyanin is two-coordinate having possibly lost one of the two histidine ligands per copper ion and the bridging oxygen ligand [329]. The EPR spectrum of the active site of tyrosinase from *Neurospora crassa* closely resembles that of the two-copper center found in hemocyanin [330]. From EXAFS studies of the *Neurospora crassa* enzyme [331], two nitrogen (histidine) and two oxygen ligands have been implicated in the two-copper center.

Histidine (and possibly tyrosine) residues have been suggested
to coordinate to the copper ions in ceruloplasmin, a metalloglycopro-
tein from human serum, since the modification of histidine residues
by iodination or photooxidation greatly reduces the binding of copper
ions and the oxidase activity of the protein [332]. Both copper and
zinc ions also bind to nonheme sites in sperm whale myoglobin, where
histidine ligands have been identified and the only other possible
ligands are lysine and asparagine, although they have not been
securely implicated in binding [333].

9.1.3. Zinc

Zinc ion-cysteine coordination has been tentatively proposed for the
metalloenzyme dihydroorotase (EC 3.5.2.3) from *Escherichia coli* from
the effect of bound zinc ion on the reactivity of certain cysteine
residues toward the thiol reagent DTNB [334]. Zinc ion is also
associated with crystals of hexameric insulin. In the "two-zinc"
form, there are three histidine (His-B10) residues (one from each of
three subunits) and three water ligands per zinc ion. The "four-
zinc" form has one zinc ion site equivalent to the two-zinc complex,
and three sites each composed of His-B5 and His-B10 of adjacent sub-
units and two water molecules [335]. Mn^{2+}, Fe^{2+}, Co^{2+}, Ni^{2+}, Cu^{2+},
or Cd^{2+} can replace Zn^{2+} as the divalent cation in the crystalliza-
tion of insulin [336]. A specific Ca^{2+}-binding site in the bovine
two-zinc hexamer of insulin has also been reported [337]. This site
consists of six glutamyl carboxylate ligands and is distinct from
the Zn^{2+}-binding sites.

9.1.4. Calcium

Calcium ion in proteins typically has oxygen ligands from oxygen-
containing amino acid side chains, from the peptide carbonyl oxygen
of the protein backbone, or from water (Table 8). As pointed out
by Kretsinger and Nelson [284], virtually all of the calcium ions
in proteins have at least one coordination site occupied by a carbonyl oxygen of the peptide chain. An exception is the calcium ion

site of the insulin hexamer. While Ca^{2+}-binding sites in the
"alkali" light chain of myosin are indicated by the amino acid
sequence homology of this protein compared with other calcium-binding
proteins (carp parvalbumin, rabbit muscle troponin C, and staphylococ-
cal nuclease) [284], the molecule apparently doesn't bind calcium
ions under physiologic conditions [338].

9.1.5. Other Metals

X-ray crystallographic studies of bacteriochlorophyll a protein from
Prosthecochloris aestuarii show that the magnesium ions in five of
the seven porphine rings (four nonprotein nitrogen ligands to a magnesium ion) are coordinated to histidine residues, while the remaining two magnesium centers each have one oxygen ligand: the carbonyl
oxygen of the peptide bond at Leu-242 has been suggested for one of
these, and water for the other [339].

   Metallothionein, a cysteine-rich metal-binding protein found
in plants, animals, and microorganisms, binds divalent ions of iron,
cobalt, nickel, copper, zinc, and cadmium (Sec. 3.4.3). Each metal
ion in the mammalian system is bound through four cysteine thiol
ligands as part of a polynuclear cluster (see Table 8 for cadmium,
zinc-metallothionein-II) [279]. However, in addition to cysteine,
histidine and glutamate residues may participate in metal ion binding by a metallothionein from Pseudomonas putida [340]. The cysteine
ligands to the metal ions of rat metallothionein form two domains,
each involving adjacent or closely spaced residues [53]. The binding
of Ni^{2+} to cysteine residues in metallothionein may have relevance
for the Ni^{2+}-binding site in urease which has a high cysteine
content (Sec. 11).

9.2. Ligands of Ni(II) in Proteins

The Ni^{2+} ligands in several Ni^{2+}-substituted proteins have been
described in Secs. 3 and 6.5. The protein side chains involved

include histidine (azurin, serum albumin, lectins, alcohol dehydrogenase), aspartic or glutamic acids (serum albumin, lectins), cysteine
(metallothionein, azurin, aspartate transcarbamylase, alcohol dehydrogenase), methionine (azurin), and tyrosine (conalbumin). Ni(II)
also coordinates to the α-amino group of the N-terminal aspartic
acid residue and the deprotonated amide nitrogen of Ala-2 and His-3
in serum albumin (Sec. 3.2.1).

X-ray diffraction data for Ni^{2+}- and Co^{2+}-carboxypeptidase A
reveal that these enzymes have the same ligands as the native Zn^{2+}
ion (side-chain nitrogens of His-69 and His-196, *both* side chain
oxygens of Glu-72) and a water molecule [127,341]. For the Co^{2+} and
Zn^{2+} enzymes, the two oxygens of the carboxylate ligand are reported
to be virtually equidistant (2.25 Å) from the metal ion which is
pentacoordinate [127]. Previously, the carboxylate ligand in Zn^{2+}-
carboxypeptidase was reported to be coordinated by a single oxygen
with the other oxygen located well away from the four-coordinate
metal ion [342]. The geometries of Co^{2+} and Zn^{2+} in carboxypeptidase
A are clearly identical within experimental error. The only significant difference reported in any M^{2+}-ligand bond length among the Co^{2+},
Ni^{2+}, and Zn^{2+} enzymes is that of Glu-72 with bond lengths of 2.49
and 2.11 Å in the Ni^{2+} enzyme: "Only the Ni^{2+} and its bound water
have moved about 0.5 Å, to make an octahedral-minus-one geometry"
[127]. However, these data scarcely constitute an argument for the
reevaluation of Gray's spectroscopic data, as suggested by the
authors. Despite popular notions, the static geometry deduced for
an active site metal ion and its complexes with inhibitors and substrates by any current physical technique, such as x-ray crystallography, does not necessarily have any simple relationship to the
mechanism of action of the enzyme. It may be recalled that the
mechanism of a reaction is a description of the detailed sequence of
changes in covalent and noncovalent interactions that occur during
a reaction. Catalysis by an enzyme commonly involves a series of
reactions which constitute a pathway, e.g., the acyl-enzyme pathway
in the hydrolysis of peptide substrates by α-chymotrypsin [343].

While physical techniques can provide vital information about starting materials, intermediates, and products on a pathway, a mechanism
is a picture of a transient event whose relationship to a static
picture may be established only through the techniques of chemical
dynamics. X-ray crystallography may successfully identify groups
close to the active sites of enzymes, but great caution is required
even in relating, let alone translating, these "structures" to
mechanism.

Where Ni^{2+} has been substituted for the catalytic Zn^{2+} in
horse liver alcohol dehydrogenase, and the regulatory Zn^{2+} in *Escherichia coli* aspartate carbamoyltransferase (Sec. 6.5), the Ni^{2+} ion
has two cysteine and a histidine, and four cysteine ligands, respectively. The native Ni^{3+} ion in hydrogenases from *Methanobacterium*
(Mb.) thermoautotrophicum and from *Desulfovibrio gigas* appears to
be coordinated to cysteine thiol ligands (Sec. 10.3).

9.3. Ni(II) Ligands in Urease

The electronic absorption spectrum of urease in relation to model
systems was discussed in Sec. 6. The difference spectrum of the
urease-2-mercaptoethanol complex shows major new absorptions which
are undoubtedly due to $RS^- \rightarrow Ni(II)$ charge transfer transitions
(Fig. 4, Table 5). Because comparable absorption peaks are not seen
in the absence of 2-mercaptoethanol [124], the clear implication is
that the active site Ni^{2+} in urease is not coordinated to cysteine.
However, the data do not rule out the possibility of sulfur-nickel
interaction of a different kind at the second nickel site [75].
Jack bean urease contains 15 thiol groups per subunit (Sec. 11), of
which nine are buried in the native protein [344]. The high content
of cysteine in urease suggests that the possibility of sulfur-Ni^{2+}
ion coordination should not be prematurely discarded in native urease
(cf. metallothionein, Sec. 9.1).

Jack bean urease has also been examined to see if inorganic
sulfide is coordinated to one or both of the nickel ions. The

method of King and Morris [345], which involves the spectrophotometric measurement of methylene blue produced by coupling of zinc
sulfide and N,N'-dimethyl-p-phenylenediamine, shows less than 0.23
mol inorganic sulfide per mol of subunits in urease [346].

On the basis of EXAFS and XANES spectra of urease and of
simpler complexes of benzimidazoles with Ni^{2+}, Piggott postulated
that the nickel ion(s) in urease may have some histidine ligands as
part of a distorted octahedral complex [347,348]. Mamiya [349] has
identified the unique active site cysteine as CysH-592 in the amino
acid sequence of jack bean urease (see Sec. 11). Several histidine
residues (His-585, 593, 594, 607) are located in this region of the
polypeptide chain and several other near-neighbor histidine residues
are potential Ni^{2+} ligands.

9.4. Observations on Metal Ion Ligands in Proteins

A common feature of protein-metal ion interactions is that at least
two of the metal-binding ligands are situated close together in the
polypeptide chain [308,350], and this is evident in the majority of
the examples listed in Table 8. The smallest distance between two
metal-binding amino acid side chains is six residues or less in all
of the nonheme proteins shown, with the exception of the catalytic
Zn^{2+} in horse liver alcohol dehydrogenase where the minimum distance
is 21 residues.

Perhaps the best example of the regularity of metal ion binding
sequences comes from the protein transcription factor IIIA of *Xenopus*
oocytes [351,352]. The 7S particle of *Xenopus laevis* oocytes contains 5S-RNA and a 40-kDa protein (factor A, transcription factor
IIIA or TFIIIA) required for in vitro transcription of the 5S-RNA.
TFIIIA has in tandem nine similar units of about 30 residues each
carrying two invariant pairs of cysteines and histidines, the
commonest ligands for zinc. The native 7S particle binds 7-11 zinc
ions. The data therefore clearly support the existence of nine
repetitive Zn^{2+} binding domains and these make up the major part

of the protein. The authors point out that such a structure explains how TFIIIA can bind to the long internal control region of the 5S RNA gene and remain bound during transcription [351]. A systematic search for potential metal ion binding domains by Berg [353] has revealed five classes of proteins involved in nucleic acid binding or gene regulation which contain sequences similar to those in TFIIIA. Moreover, other sequences homologous to the repeats in TFIIIA have also been reported [354,355].

In their 1974 review, Liljas and Rossmann [308] noted the clear preference of a particular metal for certain protein ligands which was apparent in the available data at that time. The protein side-chain ligands of native metal ions found or implicated in metalloproteins discussed in this review are summarized in Table 9. None of the earlier observations would appear to be refuted by the additional data now available.

10. NICKEL(III) ENZYMES

10.1. Introduction

Nickel ion was reported to be a constituent of the *hydrolytic* enzyme jack bean urease in 1975 [158,159], and nickel was subsequently found, or strongly indicated by inhibition and nutrition studies, in urease from a host of sources (Sec. 11). Spectral studies of native jack bean urease and of several of its complexes clearly indicate that the enzyme contains Ni(II) (Sec. 6.3) [124]. Since then nickel has been discovered in three other enzymes where its oxidation state may be variable. These enzymes will be referred to as "nickel(III) enzymes". While they are not the burden of this review, a brief discussion of them is included to give some balance to the position of nickel in enzymology.

A specific requirement for nickel in the growth of *Alcaligenes eutrophus* (*Hydrogenomonas*) on carbon dioxide and hydrogen as the sole source of carbon and energy was shown in 1965 [356]. Ni^{2+}

TABLE 9

Side Chain Ligands for Metal Ions in Proteins[a]

Ligand	Mg^{2+}	Mn^{2+}	Mn^{3+}	Fe^{2+}	Fe^{3+}	Ni^{2+}	Ni^{3+}	Cu^{2+}	Zn^{2+}	Ca^{2+}	Mo^{2+}/Mo^{3+}
N[His]	+	+	+	+	+	(+)		+	+		
O[Asp,Glu]	(+)	+	+	+	+			+	+	+	
O[Asn,Gln]										+	
S[Cys]			(+)		+		+	+	+		+
S[Met]				+				+			
O[Tyr]			+		+			(+)			
O[Ser,Thr]	(+)								(+)	+	
O[$^-$O-PO$_3^{2-}$]					+						

[a] "+", positive identification of ligand; "(+)", probable ligand.

could not be replaced by Mn^{2+}, Co^{2+}, Cu^{2+}, or Zn^{2+}. However, nickel ion was not required when an organic carbon source (succinate or fumarate) was available to the bacterium. In 1981, work from Friedrich's laboratory correlated the nickel requirement for *Alcaligenes eutrophus* (grown on hydrogen and carbon dioxide) with the formation of a hydrogenase [357], and in the following year the nickel content of the purified enzyme was determined [358]. Graf and Thauer found in 1981 that purified hydrogenase from *Methanobacterium (Mb.) thermoautotrophicum* was a nickel enzyme [359].

In 1979, Diekert et al. [360] demonstrated that the formation of carbon monoxide dehydrogenase by *Clostridium pasteurianum* was dependent on nutrient nickel ion and proposed that it might be a nickel enzyme. In the following year, Drake et al. reported that nickel ion was a constituent of the partially purified enzyme from *Clostridium thermoaceticum* [361].

At about this same time, Wolfe and coworkers [362,363] showed that a nickel-containing factor isolated several years earlier was a coenzyme for S-methyl coenzyme M methylreductase from *Mb. thermoautotrophicum*.

The properties of some nickel(III) enzymes from various sources are summarized in Table 10. In the discussion of each enzyme that follows, methanogenic bacteria are those of the genera *Methanobacterium (Mb.)*, *Methanobrevibacterium (Mbr.)*, *Methanococcus (Mc.)*, *Methanospirillum (Msp.)*, or *Methanosarcina (Ms.)* [410]. The methanogens have been recently grouped with the extreme halophiles (genus *Halobacterium*) and the thermoacidophiles (genera *Thermoplasma* and *Sulfolobus*) in the kingdom Archaebacteria [410,411]. The Archaebacteria are distinct from members of the Eubacteria kingdom which includes all remaining bacteria, cyanobacteria, and mycoplasmas.

10.2. Carbon Monoxide Dehydrogenase

Carbon monoxide dehydrogenase (EC 1.2.99.2) catalyzes the oxidation of carbon monoxide to carbon dioxide as in Eq. (18) [412,413].

TABLE 10

Examples of Nickel(III) Enzymes

Enzyme and source	Subunit or molecular weight (kDa)	Ni[a]	Iron and inorganic sulfur content	Ref.
Carbon Monoxide Dehydrogenase				
Clostridium thermoaceticum	$(78,71)_{2\ \text{or}\ 3}$[f]	F[b]	FeS clusters[c]	361,364-366
Clostridium pasteurianum		F		360,365
Clostridium formicoaceticum		Ni[d]		367
Acetobacterium woodii	$(80,68)_{1,3}$	Ni[e]	FeS clusters[c]	366,368,369
Methanobrevibacter arboriphilicus		Ni[d]		370
Hydrogenase				
Rhizobium japonicum	60,30	Ni		371,372
Alcaligenes eutrophus	$(63,56,30,26)_1$	Ni	$(Fe_4S_4)_2\ (Fe_2S_2)_2$	358,373,374
Desulfovibrio gigas	$(63,26)_1$[g]	Ni[h]	$(Fe_4S_4)_2\ Fe_3S_x$[g,i]	375,376
Desulfovibrio desulfuricans	75.5[j]	Ni	$(Fe_4S_4)_2\ Fe_3$[j]	377
	58[k]	Ni	6 Fe 6 S[c,k]	378
Vibrio succinogenes	$(60,30)_1$	Ni	FeS clusters[c]	379
Methanobacterium thermoautotrophicum	$(40,31,26)_1$	Ni[m]	FeS clusters[c]	380-383
	$(52,40)$[n]	Ni[m]	FeS clusters[c]	383
Rhodopseudomonas capsulata	60[o]	Ni		359, 384,385
Chromatium vinosum	62	Ni[p]	Fe_4S_4[q]	386-388

Escherichia coli	$64^{r,s}$	Ni		389
Xanthobacter autotrophicus		Ni^d		390
Azotobacter chroococcum		Ni^d		391
Methanosarcina barkeri	60^t	Ni	$(Fe_4S_4)_2$	392
Nocardia opaca 1b	$(64,56,31,27)_1$	Ni^u	Fe_4S_4, Fe_2S_2	373,393
Anabaena cylindrica		Ni^d		394
S-Methyl Coenzyme M Methylreductase				
Methanobacterium thermoautotrophicum	$(68,45,38.5)_2^v$	F_{430}^w		362,363,395,396
Methanobacterium bryantii		F_{430}^x		397
Methanosarcina barkeri	$(64,50,42)_2^y$	F_{430}^w		398,399
Methanospirillum hungatii		F_{430}^z		398
Methanobrevibacter ruminantium		F_{430}^z		400
Methanobrevibacter smithii		F_{430}^z		398
Methanococcus vanielii		F_{430}^z		398

aNi: The enzyme contains nickel ion. F430: The Ni is part of factor 430 (see text). F: Ni in the enzyme is apparently part of a dissociable factor which is not F430 (see text).
bThere are two Ni ions and an additional 1-3 zinc ions per dimer [364].
cThe arrangement of iron and inorganic sulfur in the cluster(s) has not been determined.
dSynthesis of the enzyme is dependent on nutrient nickel ion.
eThere are 1.4 Ni, a variable amount of Zn or Mg ion, and 1-4 Ca ions per dimer [369].
fStrain H16. A hydrogenase from strain Z1 is reported as a mixed dimer, $(65,30 kDa)_2$ [401].
gSubunits of 62 and 26 kDa, and three Fe_4S_4 clusters were originally proposed [402].
hThere is one Ni ion per 89-kDa molecule, and x-ray absorption studies indicate that the Ni ion has sulfur ligands to the protein [403].
iThe Fe_3S_x center is similar to that of beef heart aconitase, ferredoxin I of Azotobacter vinelandii, and ferredoxin II of Desulfovibrio gigas [307,376].

TABLE 10 (continued)

[j]Strain ATCC No. 27774. The Fe_3 cluster in this enzyme may have oxygen or nitrogen ligands [377]. A lower specific activity hydrogenase (77.6 kDa) which also contained nickel ion and FeS clusters was isolated from the same source, and both hydrogenases were composed of two polypeptide chains of unreported molecular weight [377].

[k]Strain Norway. Another hydrogenase (~52 kDa) from strain N.R.C. 49001 has 12 Fe and 12 S per mol, but the nickel content was not analyzed.

[l]Strain ΔH. The hydrogenase is the F_{420}-reducing form, and contains approximately one nickel ion and one FAD per ~100-kDa trimer.

[m]Sulfur and nitrogen-containing ligands to the nickel ion in the F_{420}-reducing hydrogenase have been proposed [380,404], whereas no nitrogen-containing ligands are indicated in the methyl viologen reducing form [404].

[n]Strain ΔH. The hydrogenase is the methyl viologen reducing form, and the 40-kDa subunit is different from the large subunit of the F_{420}-reducing form [405].

[o]Strain Marburg, DSM 2133. EPR spectral studies gave no evidence for FeS centers in this hydrogenase [406].

[p]Oxidation states for the nickel of 3+, 2+, 1+, and possibly zero have been proposed [407].

[q]The Fe_4S_4 cluster is irreversibly degraded to Fe_3S_x in the inactive enzyme [387]. A hydrogenase has been isolated from Chromatium sp. with two 50-kDa subunits and an Fe_4S_4 cluster [408], but the nickel content is not known.

[r]Strain K12. Two different hydrogenases both containing Ni were reported [389].

[s]A hydrogenase from strain MRE 600 is a dimer with 56.5-kDa subunits and contains 12 Fe and 12 S [409]. The nickel content has not been measured.

[t]The molecular weight is approximately 800 kDa [392].

[u]The NAD^+-reducing hydrogenase is activated in vitro by Ni^{2+}, Co^{2+}, Mg^{2+}, and Mn^{2+}, or by high salt concentration [373].

[v]This hexamer is component C of the active enzyme complex [362]. Subunit weights of 68, 47, and 38 kDa [396] and of 61, 50, and 36 kDa [399] have also been reported for component C.

[w]Two bound F_{430} complexes per mole of enzyme [363,399].

[x]Two species containing nickel were reported [397] but were probably a result of heat treatment during preparation of F_{430} [398,400].

[y]An F_{430}-containing protein which was probably component C of S-methyl coenzyme M methylreductase was isolated [399].

[z]F_{430} was isolated and is probably associated with S-methyl coenzyme M methylreductase, but the enzyme has not been securely identified [398].

$$CO + H_2O \rightleftharpoons CO_2 + 2H^+ + 2e^- \qquad (18)$$

This enzyme is found in several aerobic and anaerobic bacteria which use carbon monoxide as a carbon and energy source [405]. Not all carbon monoxide dehydrogenases contain nickel ion. For example, the purified enzyme from the aerobe *Pseudomonas carboxydovorans* [414,415] does not use nickel but is reported to contain molybdenum, zinc, copper, iron, and inorganic sulfide, and a flavin-adenine dinucleotide (FAD) cofactor [415].

A nickel-containing carbon monoxide dehydrogenase from *Clostridium thermoaceticum*, *Clostridium pasteurianum*, and *Acetobacterium woodii* has been characterized and is strongly implicated through nutrition studies to be present in *Clostridium formicoaceticum* (Table 10). In the *Clostridium thermoaceticum* and *Clostridium pasteurianum* enzymes [360,365], the nickel was postulated to be part of a dissociable cofactor which is spectrally different [416] from the nickel-containing factor F_{430} found in some methanogenic bacteria (see Sec. 10.4). The dissociated nickel factor was detected by electrophoresis after the enzyme was oxidized by molecular oxygen [365] or acid-treated [364]. However, this factor may be an artifact of the separation procedure caused by complexation of released nickel ion [365]. The nickel-free enzyme was inactive in both studies.

The nickel ion in *Clostridium thermoaceticum* [366,416] and *Acetobacterium woodii* [366,369] carbon monoxide dehydrogenases has variable oxidation states as shown by EPR studies. The enzyme catalytically reduces low-potential viologens and ferredoxins [364], whereas the carbon monoxide dehydrogenases from *Pseudomonas carboxydovorans* [414,415] or *Pseudomonas carboxydohydrogena* [417], which do not contain nickel ion, reduce only acceptors such as methylene blue or toluylene blue with much higher potentials (by greater than 0.4 V). The *Pseudomonas carboxydovorans*, *Clostridium thermoaceticum*, and *Acetobacterium woodii* enzymes all contain FeS clusters, although the extent and effect of any interaction of the FeS cluster with the nickel ion in the latter two enzymes has not been defined. Evidence from EPR studies points to the formation of a paramagnetic iron-nickel-carbon complex when carbon monoxide reacts with the carbon monoxide dehydrogenase from *Clostridium thermoaceticum* and strongly indicates an active role for the nickel ion in enzymatic catalysis [412,413,418]. Carbon monoxide is a nickel ligand in a number of simple compounds such as $Ni(CO)_4$ [112]. Finally, carbon monoxide dehydrogenase in *Clostridium thermoaceticum* is involved in the synthesis of acetate from carbon monoxide and a methyl donor [412, 413,418].

Carbon monoxide is oxidized to carbon dioxide in many methanogenic, sulfate-reducing (*Desulfovibrio* sp.), and other bacteria (e.g., *Eubacterium limosum, Butyribacterium methylotrophicum, Clostridium welchii, Nocardia* sp., and *Rhodopseudomonas* sp.) [361,364, 405 and references therein]. However the involvement of nickel in most systems has not been investigated.

10.3. Hydrogenase

Hydrogenases (EC 1.12) are a class of enzymes which use molecular hydrogen as an electron donor [Eq. (19)] to reduce one or more of

$$H_2 \rightleftharpoons 2H^+ + 2e^- \tag{19}$$

a variety of naturally occurring or artificial electron acceptors [419-421]. These electron acceptors include the bacterial coenzyme 8-hydroxy-5-deazaflavin (factor 420, F_{420}, 18), so named because of

18

an absorption maximum at 420 nm (ε_{420} = 45,500 M^{-1} cm^{-1} at pH 8.85)
in the oxidized form [422]. F_{420} is found in various methanogenic
bacteria and in the nonmethanogenic *Streptomyces aureofaciens* [423],
Streptomyces griseus [424], and other *Streptomyces* sp. [405]. It
has a redox potential [425] of about -0.340 V vs. NHE, and accepts
electrons derived from hydrogen or formate and transfers them
enzymatically to molecules such as $NADP^+$ [382]. In the methanogenic
bacteria, F_{420} provides electrons for the reduction of carbon dioxide
to methane [383]. F_{420} is a substrate for one of two hydrogenases
from *Mb. thermoautotrophicum* [383]. Other possible electron acceptors
include ferredoxin, flavodoxins, NAD^+, and the artificial viologen
dyes [405]. Hydrogenases can also catalyze the reverse reaction:
the formation of H_2 from $2H^+$ and $2e^-$ during the fermentative growth
of *Escherichia coli* removes unneeded reducing equivalents [421,426,
427].

Not all hydrogenases are nickeloenzymes. For example, the
hydrogenase from *Desulfovibrio vulgaris* has no nickel ion but does
contain Fe_4S_4 centers [428], and two forms of the enzyme purified
from *Clostridium pasteurianum* were also free of nickel [429]. The
hydrogenase of *Acetobacterium woodii* is apparently not a nickel
enzyme because the measured specific activity was independent of
the availability of nutrient nickel ion [368]. A hydrogenase from
Mc. vannielii is reported to contain selenium as part of a selenocysteine complex [430] but has not been analyzed for nickel.

Some nickel-containing hydrogenases are listed in Table 10.
These include enzymes from anaerobic sulfate-reducing bacteria
(*Desulfovibrio* sp.), methanogenic bacteria, and *Escherichia coli,*
the aerobe *Alcaligenes eutrophus,* and the photosynthetic bacterium
Chromatium vinosum. Further, the nitrogen-fixing bacteria *Azospiril-
lum brasilense, A. lipoferum,* and *Derxia gummosa* show hydrogenase
activity dependent on the availability of nutrient Ni^{2+} [431].

The nickel-containing hydrogenases (Table 10) are a diverse
group of enzymes with respect to their physical and biological
properties:

(1) The number and size of subunits of the hydrogenases range from monomers or oligomers of one type of subunit (e.g., *Ms. barkeri*) to aggregates of several types of subunit (e.g., *Mb. thermoautotrophicum* and *Alcaligenes eutrophus*).

(2) The presence of iron and inorganic sulfur has been detected in virtually all of the nickel-containing hydrogenases in which it has been sought. An exception is the hydrogenase from *Mb. thermoautotrophicum* strain Marburg, where EPR spectra failed to detect characteristic FeS signals [359,406]. In the other enzymes where the arrangement of FeS has been determined, there are one (*Chromatium vinosum*) or two (*Ms. barkeri*) Fe_4S_4 centers, and in some cases an additional Fe_3 cluster (*Desulfovibrio gigas, Desulfovibrio desulfuricans*). Two Fe_2S_2 and two Fe_4S_4 clusters have been proposed for the *Alcaligenes eutrophus* enzyme [374].

(3) There are many different electron acceptors associated with different nickelohydrogenases (F_{420}, ferredoxins, FAD, NAD^+).

(4) The location of the enzyme in the cell, either bound to the cell membrane or internal membranes, or in solution, is variable. Electron microscopy of whole cells labeled with fluorescent redox dyes indicates an association of a hydrogenase from *Mb. thermoautotrophicum* with internal membranes [432]. In many species, there appears to exist both membrane-bound and soluble forms of the enzyme [373,433].

The variation seen in hydrogenases outlined in points 1-4 above may occur between different bacterial species, between different strains of the same species, and between different forms of the enzyme in the same strain.

The ligands coordinated to nickel in hydrogenases are for the most part unknown. Multiple sulfur ligands (about three) and a nitrogen ligand (at ≥ 3.5 Å) to the nickel ion with distorted octahedral or square-pyramidal geometry have been proposed for the F_{420}^- reducing hydrogenase from *Mb. thermoautotrophicum* [380,404], whereas no nitrogen-containing ligands are indicated in the methyl viologen-reducing enzyme from the same source [404]. The nitrogen-containing ligand in the former enzyme might be FAD [404], which is stoichio-

metrically bound to this enzyme but absent from the methyl viologen-
reducing form [383]. X-ray studies of *Desulfovibrio gigas* hydrogenase indicate that the nickel ion is coordinated to the protein
through four sulfur ligands [403]. Values of the oxidation potentials of nickel-thiolate model complexes measured in aprotic media
are consistent with four cysteine sulfur ligands (or possibly three,
if other anionic residues are involved in the binding) to the Ni(III)
in some hydrogenases [434]. From EPR studies, it has tentatively
been postulated that the Ni(III) in hydrogenase from *Chromatium*
vinosum has distorted octahedral coordination with five protein
ligands, and perhaps water as the sixth ligand [387].

Work on the redox potential of the Ni(III)/Ni(II) couple in
the hydrogenase from *Desulfovibrio gigas* has resulted in conflicting
interpretations of the data [376,435-438].

The redox potentials of Ni(III)/Ni(II) model systems (including
nickel-peptide complexes) in aqueous solution at 4°C or at 20°C range
from about +0.6 to +1.2 V [175,439] and are substantially higher than
any values proposed for nickel in hydrogenases (about -0.2 to -0.4 V).
The lowest known value of a redox potential for a nonenzymic Ni(III)/
Ni(II) couple is +0.48 V vs. NHE [182,440] (see Sec. 6.8).

The role of the nickel ion in hydrogenase is not understood,
although there is evidence for its involvement as a redox center in
catalysis. In hydrogenase from *Mb. thermoautotrophicum* strain
Marburg [406] and from *Desulfovibrio gigas* [403], Ni(III) is apparently reduced to Ni(II) by hydrogen. In the tentative hypothesis
presented for *Desulfovibrio gigas* hydrogenase activity [376], up to
four forms of enzyme with different oxidation states of the nickel
and three FeS clusters may be present. A Ni(III)-hydride complex
may be an intermediate in the mechanism of action [376].

Nickel in hydrogenase from *Chromatium vinosum* can exist in
the 3+, 2+, 1+, and possibly zero oxidation state [407]. The Ni(I)
enzyme is unstable and is oxidized to Ni(II) unless hydrogen is
present. It is also light-sensitive, and differences in the rate
of the low-temperature photoreaction in D_2O and H_2O have been interpreted in terms of the breaking of a Ni-hydrogen bond [407].

The nickel(III) ion and an Fe_4S_4 cluster have a weak spin-spin interaction in the active form of hydrogenase from *Chromatium vinosum* [441]. The magnitude of the EPR signal of both moieties decreases in tandem when the enzyme is treated with 2-mercaptoethanol [387].

10.4. S-Methyl Coenzyme M Methylreductase

A yellow compound was isolated from *Mb. thermoautotrophicum* in 1978 and the factor was named F_{430} because of an absorption maximum at 430 nm [442]. In 1980, stoichiometric nickel was discovered in F_{430} (from *Mb. thermoautotrophicum* [395,443] and *Mb. bryantii* [397]). This substance was the first example of a naturally occurring low molecular weight nickel cofactor. The complete structure (19) of

19

F_{430} from *Mb. thermoautotrophicum* was determined shortly after, and the nickel ion was found to be part of a tetrapyrrole ring with nickel coordinated to four nitrogen ligands [396,444,445].

Reported values for the molar absorption coefficient of F_{430} at 430 nm are about 23,000 M^{-1} cm^{-1} [396,446]. The spectrum is independent of pH between pH 1 and 12 [446]. The factor is heat-sensitive [398,400], and chromatography on Bio-Gel P6 yielded an apparent molecular weight of greater than 3,000 [400,446]. This has led to some confusion over the molecular weight of F_{430}, which has now been confirmed as 905 [396,400,444].

In 1982, work from Wolfe's laboratory [363] showed that F_{430} was the prosthetic group of the enzyme S-methyl coenzyme M methylreductase from *Mb. thermoautotrophicum*, and in the following year Moura [399] isolated F_{430}-containing proteins from *Mb. thermoautotrophicum* and *Ms. barkeri,* which were probably this enzyme. S-Methyl coenzyme M methylreductase catalyzes the two-electron reduction of S-methyl coenzyme M ($H_3C-S-CH_2CH_2-SO_3^-$) to methane and coenzyme M ($HS-CH_2CH_2-SO_3^-$) as in Eq. (20). The enzyme is active only when

$$H_3C-S-CH_2CH_2-SO_3^- + 2H^+ + 2e^- \rightleftharpoons CH_4 + HS-CH_2CH_2-SO_3^- \qquad (20)$$

component A (composed of FAD and three proteins including an F_{420}^- reducing hydrogenase [447]) and component B (an oxygen-sensitive cofactor of unknown function [447,448]) are present. F_{430} is found in several species of methanogenic bacteria (Table 10), but the factor has not always been specifically associated with an enzymatic activity [398].

The enzymes from *Mb. thermoautotrophicum* [449] and *Ms. barkeri* [450] also require Mg^{2+} and ATP. F_{430} from *Mb. thermoautotrophicum* has been reported to be very tightly and stoichiometrically bound to coenzyme M [451-453]. Later reports clearly demonstrate that F_{430} and coenzyme M are not covalently linked [396,400].

There is approximately 2 mol of bound F_{430} per mol of S-methyl coenzyme M methylreductase from *Mb. thermoautotrophicum* [363] and from *Ms. barkeri* [399], and F_{430} in an additional nonprotein pool in *Mb. thermoautotrophicum* is indistinguishable from that extracted from the enzyme [396]. Coenzyme M is associated in stoichiometric amounts with the purified reductase [396,452].

Recent EXAFS studies [454] on protein-bound and free F_{430} from *Mb. thermoautotrophicum* and on a number of characterized Ni^{2+} complexes provide evidence about the association of F_{430} with the holoenzyme. The coordination of F_{430} and related models to other species has also been studied [451,455,456].

Enzymatically active molecules of purified S-methyl coenzyme M methylreductase from *Mb. thermoautotrophicum* display an EPR spectrum consistent with either Ni(I) or Ni(III) in F_{430} [453].

Few model systems are available, but the redox potential measured for Ni(III)/Ni(II) and for Ni(II)/Ni(I) couples for a variety of macrocyclic complexes in acetonitrile at 25°C depend on the ring size of the compound; the number, position, and charge of ring substituents; the coordination geometry of the nickel ion; and on the degree of saturation of the ligand [172,456,457]. The size of the macrocyclic ring apparently affects the stability of the nickel complex: a larger ring system forms a more stable complex with Ni(I) where the ionic radius is larger [456].

A recent review of S-methyl coenzyme M methylreductase has been given by Walsh's group [458].

11. NICKEL(II) METALLOENZYMES: UREASE

11.1. Introduction

In 1926, Sumner isolated crystalline urease (EC 3.5.1.5, urea amido-hydrolase) from the ground meal of jack beans and demonstrated it to be a protein devoid of organic coenzymes and, incidentally, metal ions [459]. In this landmark achievement [75], Sumner considered the possibility that the urease protein might contain metal ions, but his analyses for Mn and Fe were negative. It was not until 1975 [158,159] that careful metal analysis and evaluation of the equivalent weight of the enzyme clearly demonstrated the presence of stoichiometric nickel ion in jack bean urease (2.00 ± 0.12 g-atoms of nickel per 96.6-kDa subunit) [460]. Urease was the first example of a nickel-containing enzyme, and thus provided the first specific biological role for nickel.

Ureases from a variety of sources (mycoplasma, bacteria, fungi, algae, higher plants, and animals) may be safely suggested to be nickel-containing enzymes. In these examples, nickel has been found

by analysis, a nutritional requirement for nickel has been demonstrated, or the presence of nickel has been strongly implicated
through inhibition of enzymatic activity by hydroxamic acids or
phosphorodiamidates. It is likely that all hydrolytic ureases are
nickeloenzymes [75].

Urease catalyzes the *hydrolysis* of urea to carbamate ion and
ammonium ion as shown in Eq. (21). A detailed review of the nonenzymatic synthesis and degradation of urea, of the urease-catalyzed

$$H_2N-\overset{\overset{\text{O}}{\|}}{C}-NH_2 + H_2O \rightleftharpoons H_2N-CO_2^- + {}^+NH_4 \qquad (21)$$

reaction, of the physicochemical properties of the enzyme, and of
a proposed mechanism of action will very soon be published [75],
and these topics will not be considered here.

The present discussion will concentrate on the latest developments in urease enzymology, namely, the determination of the amino
acid sequence of the enzyme and its reactions with phosphorus substrates.

11.2. Amino Acid Sequence of Jack Bean Urease

Mamiya's group [349] has determined the complete amino acid sequence
of jack bean urease as shown in Fig. 5. The polypeptide of 840
residues has a molecular weight of 90,790, which is equivalent to
a subunit weight of 90,910 for the subunit carrying two nickel ions
[75]. Other values reported for the subunit weight of the jack bean
enzyme [160] are 95,000 ± 5,000 (equilibrium ultracentrifugation),
~93,000 (SDS-PAG electrophoresis), and ~96,000 (SDS-PAG electrophoresis). The equivalent weight of jack bean urease is 96,600
based on spectrophotometric titration with the chromophoric inhibitor
trans-cinnamoylhydroxamic acid, or on the binding of [^{14}C]aceto-
hydroxamic acid and [^{32}P]phosphoramidate.

The amino acid composition of urease measured in our laboratory
is compared with that determined by Mamiya's group in Table 11.

258   ANDREWS, BLAKELEY, AND ZERNER

```
  1  MKLSPREVEK  LGLHNAGYLA  QKRLARGVRL  NYTEAVALIA  SQIMEYARDG
 51  EKTVAQLMCL  GQHLLGRRQV  LPAVPHLLNA  VQVEATFPDG  TKLVTVHDPI
101  SRENGELQEA  LFGSLLPVPS  LDKFAETKED  NRIPGEILCE  DECLTLNIGR
151  KAVILKVTSK  GDRPIQVGSH  YHFIEVNPYL  TFDRRKAYGM  RLNIAAGTAV
201  RFEPGDCKSV  TLVSIEGNKV  IRGGNAIADG  PVNETNLEAA  MHAVRSRGFG
251  HEEEKDASEG  FTKEDPNCPF  NTFIHRKEYA  NKYGPTTGDK  IRLGDTNLLA
301  EIEKDYALYG  DECVFGGGKV  IRDGMGQSCG  HPPAISLDTV  ITNAVIIDYT
351  GIIKADIGIK  DGLIASIGKA  GNPDIMNGVF  SNMIIGANTE  VIAGEGLIVT
401  AGAIDCHVHY  ICPQLVYEAI  SSGITTLVGG  GTGPAAGTRA  TTCTPSPTQM
451  RLMLQSTDDL  PLNFGFTGKG  SSSKPDELHE  IIKAGAMGLK  LHEDWGSTPA
501  AIDNCLTIAE  HHDIQINIHT  DTLNEAGFVE  HSIAAFKGRT  IHTYHSEGAG
551  GGHAPDIIKV  CGIKNVLPSS  TNPTRPLTSN  TIDEHLDMLM  VCHHLDREIP
601  EDLAFAHSRI  RKKTIAAEDV  LNDIGAISII  SSDSQAMGRV  GEVISRTWQT
651  ADKMKAQTGP  LKCDSSDNDN  FRIRRYIAKY  TINPAIANGF  SQYVGSVEVG
701  KLADLVMWKP  SFFGTKPEMV  IKGGMVAWAD  IGDPNASIPT  PEPVKMRPMY
751  GTLGKAGGAL  SIAFVSKAAL  DQRVNVLYGL  NKRVEAVSNV  RKLTKLDMKL
801  NDALPEITVD  PESYTVKADG  KLLCVSEATT  VPLSRNYFLF
```

*CysH-592

CHARGED RESIDUES	HYDROPHOBIC RESIDUES		AMBIVALENT RESIDUES	
D Asp	F Phe	V Val	A Ala	S Ser
E Glu	I Ile	W Trp	C Cys	T Thr
H His	L Leu		G Gly	Y Tyr
K Lys	M Met		N Asn	
R Arg	P Pro		Q Gln	

FIG. 5. Amino acid sequence of jack bean urease. The 25 histidine residues are in bold face and the 15 cysteine residues are underlined. Urease is inactive when CysH-592* is covalently blocked. [Adapted from Ref. 349 with permission and correction of residue 258 to serine and 269 to proline (personal communication from Professor G. Mamiya to B. Z., December, 1986).]

Generally, the results from the two laboratories are in excellent agreement. Asp + Asn, Pro, CysH, Ile, and Trp are the amino acids where the concordance is poorest. Significantly, Mamiya finds no cystine disulfide bond in the native subunit, whereas we previously reported one such bond on the basis of titration studies [344].

Two implications of Mamiya's data are (1) that the best enzyme so far obtained [specific activity of 93 (mkat/liter)/A_{280}] is about 94% pure; and (2) that the isoelectric point of jack bean urease should be ~6 (cf. the measured value of 5.0-5.1 [75]). While the first result is in the right direction, the latter implies

TABLE 11

Amino Acid Composition of Jack Bean Urease

Amino acid	Residues per 90.91-kDa subunit[a] Mamiya et al.	Residues per 96.6-kDa subunit	
		Mamiya et al.[b]	Dixon et al.[c]
Asp	50	53.13 ⎫ 94.57	91.08
Asn	39	41.44 ⎭	
Thr	55	58.44	58.29
Ser	46	48.88	48.72
Glu	50	53.13 ⎫ 72.26	73.18
Gln	18	19.13 ⎭	
Pro	42	44.63	49.37
Gly	79	83.94	83.55
Ala	74	78.63	80.56
CysH	15	15.94	17.05[d]
Val	55	58.44	59.12
Met	21	22.31	22.79
Ile	66	70.13	67.75
Leu	69	73.32	72.58
Tyr	21	22.31	22.21
Phe	24	25.50	25.40
Lys	49	52.07	51.51
His	25	26.56	26.66
Arg	38	40.38	39.53
Trp	4	4.25	5.01

[a]From Ref. 349. Allowing for two nickel ions per subunit.
[b]Calculated from Ref. 349, allowing for two nickel ions per subunit.
[c]Calculated from Ref. 160.
[d]Cysteine + half-cystine.

some uncertainty in the charged side-chain content of the sequenced protein.

Kobashi's group published a partial amino acid sequence about what is reported to be the essential active site cysteine residue of jack bean urease [461]. The peptide is completely different

from that found about the essential cysteine (CysH-592) in Mamiya's
sequence, and most resembles that in the vicinity of CysH-207:

Kobashi: Phe-Glu-Pro-Gly-Asp-CysH-Asn-Ser-Thr-Phe-Lys

Mamiya(CysH-207): Phe-Glu-Pro-Gly-Asp-CysH-Lys-Ser-Val-Thr-Leu

Mamiya(CysH-592): Asp-Met-Leu-Met-Val-CysH-His-His-Leu-Asp-Arg

The discrepancy in the analyses is probably a result of the thiol
reagent (diazonium-1H-tetrazole) used by Kobashi and coworkers to
modify the active site cysteine residue. The reagent is known to
be nonspecific in its reactions [346,462].

The detailed sequence (Fig. 5) shows that 13 of the 25 histidine residues in the urease subunit occur between residue 479 and
607, and Mamiya reasonably suggests that this region may contain
the metal-binding site of the enzyme (see Sec. 9.3).

11.3. Substrates for Urease

For a long time, urease was considered to be absolutely specific
for urea [75], but the eventual discovery of a range of alternative
substrates (beginning with N-hydroxyurea in 1965 [162-164]) was of
major importance in indicating a role for nickel ion(s) in urease
catalysis [29,463]. Evidence for the coordination of urea to an
active site nickel was later derived from spectral studies [124,161].

Compounds known to be substrates for jack bean urease are
shown in Table 12. Recent additions to this list are the amides and
esters of phosphoric acid, with the general structure shown below
(20). Phosphoramidate (20a) inhibits urease by virtue of stoichio-

$$X - \overset{\overset{\displaystyle O}{\|}}{\underset{\underset{\displaystyle Y}{|}}{P}} - NH_2$$

a	X = HO Y = HO	d	X = H₂N Y = ArO
b	X = HO Y = H₂N	e	X = H₂N Y = AlkC(O)NH
c	X = H₂N Y = H₂N	f	X = H₂N Y = ArC(O)NH

20

TABLE 12

Substrates for Urease

Substrate	k_{cat}[a] (sec^{-1})	K_m[a] (sec^{-1})	$\dfrac{k_{cat} \text{ pH } 5.2}{k_{cat} \text{ pH } 7.0}$	$10^4 \, k_{react.}$[b] (sec^{-1})
Urea	5870	0.0029	0.6	
Semicarbazide	30	0.060	1.7[c]	
Formamide	92	1.06	2.4	
Acetamide	0.55	0.75		
N-Methylurea	0.075	0.22		
N-Hydroxyurea[d]				
Phenyl phosphorodiamidate				0.36 ± 0.03[e]
N-(3-Methyl-2-butenyl) phosphoric triamide[f]				0.34 ± 0.04[e]
Phosphoric triamide				0.36 ± 0.03[e]
N-Benzoyl phosphoric triamide[f]				
Diamidophosphate[g]				7 ± 1[h]
Phosphoramidate[g,i]				8.2 ± 0.5[j]

[a]pH 7.00, 38°C [463].
[b]First order rate constant for the reactivation of the enzyme-inhibitor complex which is rapidly formed in the presence of excess enzyme [165].
[c]The lower pH was 5.0.
[d]Refs. 162-164.
[e]In oxygen-free N-ethylmorpholinium chloride buffer (pH 7.0; 1 mM in EDTA).
[f]N-Acylamide was reported as a product [464].
[g]1:1 complex.
[h]In oxygen-free N-ethylmorpholinium chloride buffer, pH 7.0.
[i]$k_{cat}/K_m = 8 \pm 2 \, M^{-1} \, sec^{-1}$ ($[S]_0 = 8.4$ mM; 38.0°C).
[j]In oxygen-free N-ethylmorpholinium chloride buffer (pH 7.11; 1mM in EDTA, 5 mM in 2-mercaptoethanol, 0.1 M in KCl).

metric coordination to active site nickel [29,75,158,159,161], and
a large fraction of 20a is released intact upon reactivation of the
enzyme under appropriate conditions [154]. We have recently established that 20a is a very poor substrate for the enzyme under conditions where the nonenzymatic decomposition is not catalyzed by buffer
salts or other additives. Further, abolition of the urease-catalyzed
hydrolysis of 20a by preequilibration of the enzyme with acetohydroxamic acid confirms that 20a is a genuine substrate. At 38°C, k_{cat}/K_m =
8 ± 2 M^{-1} sec^{-1} ($[20a]_0$ = 8.4 mM), and this value may be compared with
those for other substrates which range from 2.0 x 10^6 M^{-1} sec^{-1} for
urea to 0.34 M^{-1} sec^{-1} for N-methylurea [463].

Phenyl phosphorodiamidate (20d, Ar = Ph, PPD), N-(3-methyl-2-
butenyl)phosphoric triamide [20e, Alk = $(H_3C)_2$C=CH, MBPT], and phosphoric triamide (20c) each rapidly and stoichiometrically inactivates
urease, but the inactive enzyme slowly regains full activity at pH 7
and 38°C. The consonance of the rate constants for the reactivation
of the enzyme inhibited by these three compounds identifies a
diamidophosphate-nickel complex ($20b \cdots Ni^{2+}$) as the species responsible for inhibition. A different, significantly more labile complex
is formed when the inhibitor is not derived from substrate. The
identity of the rate constants for reactivation of urease inhibited
by 20a and 20b (Table 12) implies that 20b is itself a substrate.
The isomeric diamidophosphate-nickel complexes probably differ in
strength by virtue of N vs. O coordination to the metal ion, and
constitute the first such enzymatic data [165, cf. 164].

The definition of substrates for this enzyme has proved to be
an exceedingly difficult task because of the frequently observed
concomitant inhibition and hydrolysis. However, the securely identified substrates for urease have provided a mine of information
implicating nickel in its mechanism of action.

12. CONCLUSION

An understanding of the role of nickel in enzymology, and consequently in biology, has increased enormously in the decade since the discovery of stoichiometric nickel in urease in 1975. But the bridling of the rogue has just begun. We hope that this chapter will make a modest contribution to its continuation and to the applications of the principles of chemistry to the enzymology of this and other metal ions.

ACKNOWLEDGMENTS

The authors thank Dr. Gunji Mamiya for supplying his data on the sequence of the urease protein.

We thank our many colleagues who have contributed to the work in this area, and two of us would like to acknowledge our incalculable debt to two great teachers, Myron L. Bender and Frank Westheimer.

The Wellcome Trust (London), the National Institutes of Health (Bethesda), the National Health and Medical Research Council, the Australian Research Grants Committee, and the University of Queensland have supported this work.

REFERENCES

1. "Gmelins Handbuch der Anorganischen Chemie", 8th ed., System 57, Part AI (A. Kotowski, ed.), Verlag Chemie, Weinheim, 1967, pp. 4-5.
2. F. B. Howard-White, "Nickel, An Historical Review", Methuen, London, 1963, pp. 9-12.
3. J. Needham, "Science and Civilisation in China", Vol. 5, Part II, Cambridge Univ. Press, London, 1974, p. 231.

4. J. W. Mellor, "A Comprehensive Treatise on Inorganic Chemistry", Vol. XV, Longmans, Green, London, 1936, Chap 68, pp. 1-32.

5. M. E. Weeks (revised by H. M. Leicester), "Discovery of the Elements", 7th ed., Journal of Chemical Education, Easton, PA, 1968, pp. 147-175.

6. W. Mertz, *Fed. Proc., 29,* 1482 (1970).

7. J. P. Riley and R. Chester, "Introduction to Marine Chemistry", Academic, London, 1971, Chap. 4.

8. H. J. M. Bowen, "Environmental Chemistry of the Elements", Academic, London, 1979.

9. J. Buffle, in "Metal Ions in Biological Systems", Vol. 18 (H. Sigel, ed.), Marcel Dekker, New York, 1984, Chap. 6.

10. J. Versieck and R. Cornelis, *Anal. Chim. Acta, 116,* 217 (1980).

11. "Biology Data Book", 2nd ed., Vol. III (P. L. Altman and D. S. Dittmer, eds.), Federation of American Societies for Experimental Biology, Bethesda, 1974, pp. 1751-1754.

12. C. N. Leach, Jr., J. V. Linden, S. M. Hopfer, M. C. Crisostomo, and F. W. Sunderman, Jr., *Clin. Chem., 31,* 556 (1985).

13. A. Kabata-Pendias and H. Pendias, "Trace Elements in Soil and Plants", CRC, Boca Raton, FL, 1984, p. 246 et seq.

14. M. Anke, M. Grün, B. Groppel, and H. Kronemann, in "Biological Aspects of Metals and Metal-Related Diseases" (B. Sarkar, ed.), Raven, New York, 1983, p. 89 ff.

15. R. R. Brooks, in "Nickel in the Environment" (J. O. Nriagu, ed.), John Wiley, New York, 1980, Chap. 15.

16. M. Stoeppler, in "Nickel in the Environment" (J. O. Nriagu, ed.), John Wiley, New York, 1980, Chap. 29.

17. A. Banin and J. Navrot, *Science, 189,* 550 (1975).

18. D. W. Jenkins, in "Nickel in the Environment" (J. O. Nriagu, ed.), John Wiley, New York, 1980, Chap. 11.

19. D. H. S. Richardson, P. J. Beckett, and E. Nieboer, in "Nickel in the Environment" (J. O. Nriagu, ed.), John Wiley, New York, 1980, Chap. 14.

20. F. W. Sunderman, Jr., M. C. Crisostomo, M. C. Reid, S. M. Hopfer, and S. Nomoto, *Ann. Clin. Lab. Sci., 14,* 232 (1984).

21. M. R. Wills, C. S. Brown, R. L. Bertholf, R. Ross, and J. Savory, *Clin. Chim. Acta, 145,* 193 (1985).

22. F. W. Sunderman, Jr., F. Coulston, G. L. Eichhorn, J. A. Fellows, E. Mastromatteo, H. T. Reno, and M. H. Samitz, "Nickel", National Academy of Sciences, Washington, D.C., 1975, Chap. 3.

23. G. N. Schrauzer, in "Biochemistry of the Essential Ultratrace Elements" (E. Frieden, ed.), Plenum, New York, 1984, Chap. 2.

24. D. Ankel-Fuchs and R. K. Thauer, in "Bioinorganic Chemistry of Nickel" (J. R. Lancaster, Jr., ed.), VCH, New York, 1988, Chap. 5, in press.

25. J. W. Spears, *J. Anim. Sci.*, *59*, 823 (1984).

26. F. H. Nielsen, *J. Nutr.*, *115*, 1239 (1985).

27. R. K. Andrews, G. J. King, R. L. Blakeley, and B. Zerner, *Proc. Aust. Biochem. Soc.*, *19*, 11 (1987).

28. (a) J. L. Hall, J. A. Swisher, D. G. Brannon, and T. M. Liden, *Inorg. Chem.*, *1*, 409 (1962). (b) B. E. Fischer, U. K. Häring, and H. Sigel, *Eur. J. Biochem.*, *94*, 523 (1979).

29. R. K. Andrews, R. L. Blakeley, and B. Zerner, in "Advances in Inorganic Biochemistry", Vol. 6 (G. L. Eichhorn and L. G. Marzilli, eds.), Elsevier, New York, 1984, Chap. 7.

30. T. R. Peters, Jr., in "The Plasma Proteins", 2nd ed., Vol. 1 (F. W. Putnam, ed.), Academic, New York, 1975, Chap. 3.

31. J. D. Glennon and B. Sarkar, *Biochem. J.*, *203*, 15 (1982).

32. J.-P. Laussac and B. Sarkar, *Biochemistry*, *23*, 2832 (1984).

33. S. Nomoto, M. D. McNeely, and F. W. Sunderman, Jr., *Biochemistry*, *10*, 1647 (1971).

34. F. W. Sunderman, Jr., M. I. Decsy, and M. D. McNeely, *Ann. N.Y. Acad. Sci.*, *199*, 300 (1972).

35. M. I. Decsy and F. W. Sunderman, Jr., *Bioinorg. Chem.*, *3*, 95 (1974).

36. J. Travis and G. S. Salvesen, *Ann. Rev. Biochem.*, *52*, 655 (1983).

37. C. W. Pratt and S. V. Pizzo, *Biochim. Biophys. Acta*, *791*, 123 (1984).

38. A. F. Parisi and B. L. Vallee, *Biochemistry*, *9*, 2421 (1970).

39. M. K. Burch and W. T. Morgan, *Biochemistry*, *24*, 5919 (1985).

40. W. T. Morgan, *Biochemistry*, *20*, 1054 (1981).

41. S. L. Guthans and W. T. Morgan, *Arch. Biochem. Biophys.*, *218*, 320 (1982).

42. Cf. Ref. 22, Chap. 4.

43. F. W. Sunderman, Jr., B. L. K. Mangold, S. H. Y. Wong, S. K. Shen, M. C. Reid, and I. Jansson, *Res. Commun. Chem. Pathol. Pharmacol.*, *39*, 477 (1983).

44. D. M. Templeton and B. Sarkar, *Biochem. J.*, *230*, 35 (1985).

45. A. W. Abdulwajid and B. Sarkar, *Proc. Natl. Acad. Sci. USA*, *80*, 4509 (1983).

46. M.-C. Herlant-Peers, H. F. Hildebrand, and J.-P. Kerckaert, *Carcinogenesis*, *4*, 387 (1983).

47. L. Bhattacharyya, C. F. Brewer, R. D. Brown, III, and S. H. Koenig, *Biochemistry*, *24*, 4974 (1985).

48. C. F. Brewer, R. D. Brown, III, and S. H. Koenig, *Biochemistry*, *22*, 3691 (1983).

49. J. W. Becker, G. N. Reeke, Jr., J. L. Wang, B. A. Cunningham, and G. M. Edelman, *J. Biol. Chem.*, *250*, 1513 (1975).

50. E. T. Adman, G. W. Canters, H. A. O. Hill, and N. A. Kitchen, *FEBS Lett.*, *143*, 287 (1982).

51. J. A. Blaszak, E. L. Ulrich, J. L. Markley, and D. R. McMillin, *Biochemistry*, *21*, 6253 (1982).

52. D. L. Tennent and D. R. McMillin, *J. Am. Chem. Soc.*, *101*, 2307 (1979).

53. D. H. Hamer, *Ann. Rev. Biochem.*, *55*, 913 (1986).

54. M. Vašák, J. H. R. Kagi, B. Holmquist, and B. L. Vallee, *Biochemistry*, *20*, 6659 (1981).

55. B. P. Monia, T. R. Butt, D. J. Ecker, C. K. Mirabelli, and S. T. Crooke, *J. Biol. Chem.*, *261*, 10957 (1986).

56. J. A. Shelnutt, K. Alston, J.-Y. Ho, N.-T. Yu, T. Yamamoto, and J. M. Rifkind, *Biochemistry*, *25*, 620 (1986).

57. A. T. Tan and R. C. Woodworth, *Biochemistry*, *8*, 3711 (1969).

58. P. Aisen and I. Listowsky, *Ann. Rev. Biochem.*, *49*, 357 (1980).

59. J. J. West, *Biochemistry*, *10*, 3547 (1971).

60. M. Miki and P. Wahl, *Biochim. Biophys. Acta*, *828*, 188 (1985).

61. L. M. Kozloff, *J. Biol. Chem.*, *253*, 1059 (1978).

62. "Enzyme Nomenclature, 1984" (E. C. Webb, ed.), Academic, New York, 1984.

63. M. D. Lane and F. Lynen, *Proc. Natl. Acad. Sci. USA*, *49*, 379 (1963).

64. M. Waite and S. J. Wakil, *J. Biol. Chem.*, *238*, 77 (1963).

65. J. Salach, W. H. Walker, T. P. Singer, A. Ehrenberg, P. Hemmerich, S. Ghisla, and U. Hartmann, *Eur. J. Biochem.*, *26*, 267 (1972).

66. H. Möhler, M. Brühmüller, and K. Decker, *Eur. J. Biochem.*, *29*, 152 (1972).

67. N. P. B. Dudman and B. Zerner, *Biochim. Biophys. Acta*, *310*, 248 (1973).

68. F. Wold, *Ann. Rev. Biochem.*, *50*, 783 (1981).

69. D. E. Metzler, "Biochemistry: The Chemical Reactions of Living Cells", Academic, New York, 1977, Chap. 8.

70. B. L. Vallee and W. E. C. Wacker, in "The Proteins", 2nd ed., Vol. 5 (H. Neurath, ed.), Academic, New York, 1970, Chap. 3.

71. M. Cohn and G. H. Reed, *Ann. Rev. Biochem.*, *51*, 365 (1982).

72. B. D. Sykes, in "Magnetic Resonance in Biology", Vol. 1 (J. S. Cohen, ed.), John Wiley, New York, 1980, pp. 171-196.

73. R. J. Magee and C. R. Dawson, *Arch. Biochem. Biophys.*, *99*, 338 (1962).

74. N. E. Dixon, C. Gazzola, R. L. Blakeley, and B. Zerner, *Science*, *191*, 1144 (1976).

75. R. K. Andrews, R. L. Blakeley, and B. Zerner, in "Bioinorganic Chemistry of Nickel" (J. R. Lancaster, Jr., ed.), VCH, New York, 1988, Chap. 7, in press.

76. C. B. Anfinsen, *Science, 181,* 223 (1973).

77. C. T. Duda and A. Light, *J. Biol. Chem.*, *257*, 9866 (1982) and references therein.

78. B. P. N. Ko, A. Yazgan, P. L. Yeagle, S. Chace Lottich, and R. W. Henkins, *Biochemistry, 16,* 1720 (1977).

79. P. S. Kim and R. L. Baldwin, *Ann. Rev. Biochem.*, *51*, 459 (1982).

80. R. Jaenicke and R. Rudolph, in "Protein Folding, Proc. 28th Conf. German Biochem. Soc." (R. Jaenicke, ed.), Elsevier, Amsterdam, 1980, pp. 525-548.

81. N. E. Dixon, Ph.D. thesis, Univ. of Queensland, Brisbane, Australia, 1976.

82. P. W. Riddles, Ph.D. thesis, Univ. of Queensland, Brisbane, Australia, 1980.

83. N. E. Dixon, C. Gazzola, C. J. Asher, D. S. W. Lee, R. L. Blakeley, and B. Zerner, *Can. J. Biochem.*, *58*, 474 (1980).

84. M. C. Scrutton, P. Griminger, and J. C. Wallace, *J. Biol. Chem.*, *247*, 3305 (1972).

85. A. Payen and J. F. Persoz, *Ann. Chim. (Phys.)*, *53*, 73 (1833) through Ref. 87.

86. J. J. Berzelius, "Lehrbuch der Chemie", 3rd ed., Vol. 6, Arnoldischen Buchhandlung, Dresden, 1837, pp. 19-25, through Ref. 87.

87. M. Dixon, in "The Chemistry of Life" (J. Needham, ed.), Cambridge Univ. Press, Cambridge, 1970, Chap. 2.

88. J. R. Partington, "A History of Chemistry", Vol. 4, Macmillan, London, 1964, pp. 595-600.

89. F. H. Westheimer, *Trans. N.Y. Acad. Sci.*, *18*, 15 (1955).

90. W. P. Jencks, "Catalysis in Chemistry and Enzymology", McGraw-Hill, New York, 1969.

91. D. P. N. Satchell and R. S. Satchell, *Ann. Rep. Prog. Chem., Sect. A: Phys. Inorg. Chem., 75,* 25 (1978).

92. J. Suh and K. H. Chun, *J. Am. Chem. Soc., 108,* 3057 (1986).

93. L. M. Sayre, *J. Am. Chem. Soc., 108,* 1632 (1986).

94. D. A. Buckingham, J. MacB. Harrowfield, and A. M. Sargeson, *J. Am. Chem. Soc., 96,* 1726 (1974).

95. D. P. N. Satchell and I. I. Secemski, *J. Chem. Soc. (B),* 1306 (1970).

96. D. Hopgood and D. L. Leussing, *J. Am. Chem. Soc., 91,* 3740 (1969).

97. R. S. McQuate and D. L. Leussing, *J. Am. Chem. Soc., 97,* 5117 (1975).

98. J. MacB. Harrowfield, V. Norris, and A. M. Sargeson, *J. Am. Chem. Soc., 98,* 7282 (1976).

99. M. F. Dunn, *Struct. Bonding, 23,* 61 (1975).

100. R. B. Martin, *J. Inorg. Nucl. Chem., 38,* 511 (1976).

101. B. S. Cooperman, *Met. Ions Biol. Syst., 5,* 79-125 (1976).

102. D. A. Buckingham, in "Biological Aspects of Inorganic Chemistry" (D. Dolphin, ed.), John Wiley, New York, 1977, pp. 141-196.

103. N. E. Dixon and A. M. Sargeson, in "Zinc Enzymes" (T. G. Spiro, ed.), John Wiley, New York, 1983, Chap. 7.

104. D. D. Roberts, S. D. Lewis, D. P. Ballou, S. T. Olson, and J. A. Shafer, *Biochemistry, 25,* 5595 (1986).

105. B. Zerner and M. L. Bender, *J. Am. Chem. Soc., 83,* 2267 (1961).

106. T. C. Bruice and S. J. Benkovic, "Bioorganic Mechanisms", Vol. 1, W. A. Benjamin, New York, 1966.

107. M. L. Bender, "Mechanisms of Homogeneous Catalysis from Protons to Proteins", Wiley-Interscience, New York, 1971.

108. C. J. Boreham, D. A. Buckingham, and F. R. Keene, *J. Am. Chem. Soc., 101,* 1409 (1979).

109. A. R. Fersht and A. J. Kirby, *Prog. Bioorg. Chem., 1,* 1 (1971).

110. D. Wester and G. J. Palenik, *J. Am. Chem. Soc., 96,* 7565 (1974).

111. H. J. Callot, T. Tschamber, B. Chevrier, and R. Weiss, *Angew. Chem., Int. Ed. Engl., 14,* 567 (1975).

112. F. A. Cotton and G. Wilkinson, "Advanced Inorganic Chemistry", 4th ed., John Wiley, New York, 1980, Chap. 21.

113. R. J. Butcher, C. J. O'Connor, and E. Sinn, *Inorg. Chem., 18,* 492 (1979).

114. K. O. Joung, C. J. O'Connor, E. Sinn, and R. L. Carlin, *Inorg. Chem., 18,* 804 (1979).

115. J. A. Bertrand, A. P. Ginsberg, R. I. Kaplan, C. E. Kirkwood, R. L. Martin, and R. C. Sherwood, *Inorg. Chem.*, *10*, 240 (1971).

116. R. G. Wilkins, *Accounts Chem. Res.*, *3*, 408 (1970).

117. L. G. Sillén and A. E. Martell, "Stability Constants of Metal-Ion Complexes", Supplement No. 1 (Special Publication No. 25), The Chemical Society, London, 1971.

118. Cf. Ref. 112, Chap. 3.

119. D. W. Margerum, G. R. Cayley, D. C. Weatherburn, and G. K. Pagenkopf, in "Coordination Chemistry", Vol. 2 (A. E. Martell, ed.), ACS Monograph 174, Am. Chem. Soc., Washington, D.C., 1978, Chap. 1.

120. M. Eigen, *Pure Appl. Chem.*, *6*, 105 (1963).

121. Cf. Ref. 112, Chap. 28.

122. H. Sigel, B. E. Fischer, and B. Prijs, *J. Am. Chem. Soc.*, *99*, 4489 (1977).

123. R. S. Drago, "Physical Methods in Chemistry", Saunders, Philadelphia, 1977, Chap. 10.

124. R. L. Blakeley, N. E. Dixon, and B. Zerner, *Biochim. Biophys. Acta*, *744*, 219 (1983).

125. W. J. Ray, Jr. and J. S. Multani, *Biochemistry*, *11*, 2805 (1972).

126. R. C. Rosenberg, C. A. Root, and H. B. Gray, *J. Am. Chem. Soc.*, *97*, 21 (1975).

127. K. D. Hardman and W. N. Lipscomb, *J. Am. Chem. Soc.*, *106*, 463 (1984).

128. J. E. Coleman, in "Inorganic Biochemistry", Vol. 1 (G. L. Eichhorn, ed.), Elsevier, New York, 1973, Chap. 16.

129. C. K. Jørgensen, *Adv. Chem. Phys.*, *5*, 33 (1963).

130. P. L. Meredith and R. A. Palmer, *Inorg. Chem.*, *10*, 1049 (1971).

131. B. J. McCormick and B. P. Stormer, *Inorg. Chem.*, *11*, 729 (1972).

132. D. P. Graddon and I. A. Siddiqi, *Aust. J. Chem.*, *30*, 2133 (1977).

133. V. V. Savant and C. C. Patel, *Ind. J. Chem.*, *9*, 261 (1971).

134. G. A. Melson, N. P. Crawford, and B. J. Geddes, *Inorg. Chem.*, *9*, 1123 (1970).

135. R. E. Wagner and J. C. Bailar, Jr., *J. Am. Chem. Soc.*, *97*, 533 (1975).

136. C. K. Jørgensen, *Acta Chem. Scand.*, *9*, 1362 (1955).

137. R. S. Drago, D. W. Meek, M. D. Joesten, and L. LaRoche, *Inorg. Chem.*, *2*, 124 (1963).

138. H. C. Freeman, in "Inorganic Biochemistry", Vol. 1 (G. L. Eichhorn, ed.), Elsevier, Amsterdam, 1973, Chap. 4.

139. J. M. T. Raycheba and D. W. Margerum, *Inorg. Chem.*, *19*, 497 (1980).

140. R. B. Martin, M. Chamberlin, and J. T. Edsall, *J. Am. Chem. Soc.*, *82*, 495 (1960).

141. J. M. T. Raycheba and D. W. Margerum, *Inorg. Chem.*, *19*, 837 (1980).

142. R. S. Johnson and H. K. Schachman, *Proc. Natl. Acad. Sci. USA*, *77*, 1995 (1980).

143. H. Dietrich, W. Maret, H. Kozlowski, and M. Zeppezauer, *J. Inorg. Biochem.*, *14*, 297 (1981).

144. Y. Sugiura and Y. Hirayama, *J. Am. Chem. Soc.*, *99*, 1581 (1977).

145. S. H. Laurie, D. H. Prime, and B. Sarkar, *Can. J. Chem.*, *57*, 1411 (1979).

146. C. A. Root and D. H. Busch, *Inorg. Chem.*, *7*, 789 (1968).

147. D. L. Leussing, *J. Am. Chem. Soc.*, *81*, 4208 (1959).

148. R. A. Libby and D. W. Margerum, *Biochemistry*, *4*, 619 (1965).

149. D. L. Leussing, *J. Am. Chem. Soc.*, *80*, 4180 (1958).

150. D. D. Perrin and I. G. Sayce, *J. Chem. Soc. (A)*, 82 (1967).

151. D. L. Leussing, R. E. Laramy, and G. S. Alberts, *J. Am. Chem. Soc.*, *82*, 4826 (1960).

152. D. L. Leussing and G. S. Alberts, *J. Am. Chem. Soc.*, *82*, 4458 (1960).

153. M. F. Fox and E. Hayon, *J. Chem. Soc. Faraday Trans.*, *1*, 75, 1380 (1979).

154. N. E. Dixon, R. L. Blakeley, and B. Zerner, *Can. J. Biochem.*, *58*, 481 (1980).

155. G. Anderegg, F. L'Eplattenier, and G. Schwarzenbach, *Helv. Chim. Acta*, *46*, 1400 (1963).

156. S. Mizukami and K. Nagata, *Coord. Chem. Rev.*, *3*, 267 (1968).

157. R. Berger and H. P. Fritz, *Z. Naturforsch. B*, *27*, 608 (1972).

158. N. E. Dixon, C. Gazzola, J. J. Watters, R. L. Blakeley, and B. Zerner, *J. Am. Chem. Soc.*, *97*, 4130 (1975).

159. N. E. Dixon, C. Gazzola, R. L. Blakeley, and B. Zerner, *J. Am. Chem. Soc.*, *97*, 4131 (1975).

160. N. E. Dixon, J. A. Hinds, A. K. Fihelly, C. Gazzola, D. J. Winzor, R. L. Blakeley, and B. Zerner, *Can. J. Biochem.*, *58*, 1323 (1980).

161. R. L. Blakeley and B. Zerner, *J. Mol. Catal.*, *23*, 263 (1984).

162. W. N. Fishbein, T. S. Winter, and J. D. Davidson, *J. Biol. Chem.*, *240*, 2402 (1965).

163. W. N. Fishbein and P. P. Carbone, *J. Biol. Chem.*, *240*, 2407 (1965).

164. R. L. Blakeley, H. E. Kunze, E. C. Webb, and B. Zerner, *Biochemistry*, *8*, 1991 (1969).

165. R. K. Andrews, A. Dexter, R. L. Blakeley, and B. Zerner, *J. Am. Chem. Soc.*, *108*, 7124 (1986).

166. L. A. Dominey and K. Kustin, *Inorg. Chem.*, *23*, 103 (1984).

167. P. Dasgupta and R. B. Jordan, *Inorg. Chem.*, *24*, 2717 (1985).

168. R. I. Haines and A. McAuley, *Coord. Chem. Rev.*, *39*, 77 (1981).

169. A. H. Maki, N. Edelstein, A. Davison, and R. H. Holm, *J. Am. Chem. Soc.*, *86*, 4580 (1964).

170. E. I. Stiefel, J. H. Waters, E. Billig, and H. B. Gray, *J. Am. Chem. Soc.*, *87*, 3016 (1965).

171. E. S. Gore and D. H. Busch, *Inorg. Chem.*, *12*, 1 (1973).

172. D. H. Busch, *Accounts Chem. Res.*, *11*, 392 (1978).

173. A. McAuley, P. R. Norman, and O. Olubuyide, *J. Chem. Soc. Dalton Trans.*, 1501 (1984).

174. F. P. Bossu and D. W. Margerum, *J. Am. Chem. Soc.*, *98*, 4003 (1976).

175. F. P. Bossu and D. W. Margerum, *Inorg. Chem.*, *16*, 1210 (1977).

176. A. G. Lappin, C. K. Murray, and D. W. Margerum, *Inorg. Chem.*, *17*, 1630 (1978).

177. Y. Sugiura and Y. Mino, *Inorg. Chem.*, *18*, 1336 (1979).

178. F. P. Bossu, E. B. Paniago, D. W. Margerum, S. T. Kirksey, Jr., and J. L. Kurtz, *Inorg. Chem.*, *17*, 1034 (1978).

179. G. L. Burce, E. B. Paniago, and D. W. Margerum, *J. Chem. Soc. Chem. Commun.*, 261 (1975).

180. S. T. Kirksey, Jr., T. A. Neubecker, and D. W. Margerum, *J. Am. Chem. Soc.*, *101*, 1631 (1979).

181. E. Kimura, A. Sakonaka, R. Machida, and M. Kodama, *J. Am. Chem. Soc.*, *104*, 4255 (1982).

182. E. Kimura, R. Machida, and M. Kodama, *J. Am. Chem. Soc.*, *106*, 5497 (1984).

183. E. Kimura, A. Sakonaka, and M. Nakamoto, *Biochim. Biophys. Acta*, *678*, 172 (1981).

184. A. L. Lehninger, *Physiol. Rev.*, *30*, 393 (1950).

185. E. L. Smith, *Adv. Enzymol. Relat. Areas Mol. Biol.*, *12*, 191 (1951).

186. H. Kroll, *J. Am. Chem. Soc.*, *74*, 2036 (1952).

187. J. M. White, R. A. Manning, and N. C. Li, *J. Am. Chem. Soc.*, *78*, 2367 (1956).

188. M. L. Bender, *Adv. Chem. Ser.*, *37*, 19 (1962).

189. H. Kroll, *J. Am. Chem. Soc.*, *74*, 2034 (1952).

190. W. D. McElroy and B. Glass (eds.), "A Symposium on the Mechanism of Enzyme Action", Part III, Johns Hopkins Press, Baltimore, 1954, pp. 221-318.

191. J. F. Speck, *J. Biol. Chem.*, *149*, 315 (1949).

192. J. E. Prue, *J. Chem. Soc.*, 2331 (1952).

193. L. Meriwether and F. H. Westheimer, *J. Am. Chem. Soc.*, *78*, 5119 (1956).

194. B. L. Vallee and A. Galdes, *Adv. Enzymol. Relat. Areas Mol. Biol.*, *56*, 283 (1984).

195. B. L. Vallee and R. J. P. Williams, *Proc. Natl. Acad. Sci. USA*, *59*, 498 (1968).

196. R. Rawls, *Chem. Eng. News*, *59* (May 18), 46 (1981).

197. B. L. Vallee in "Zinc Enzymes" (T. G. Spiro, ed.), John Wiley, New York, 1983, Chaps. 1 & 2.

198. R. J. Angelici and D. B. Leslie, *Inorg. Chem.*, *12*, 431 (1973).

199. T. J. Przystas and T. H. Fife, *J. Am. Chem. Soc.*, *102*, 4391 (1980).

200. T. H. Fife and T. J. Przystas, *J. Am. Chem. Soc.*, *104*, 2251 (1982).

201. R. Breslow, R. Fairweather, and J. Keana, *J. Am. Chem. Soc.*, *89*, 2135 (1967).

202. R. Breslow and M. Schmir, *J. Am. Chem. Soc.*, *93*, 4960 (1971).

203. M. A. Schwartz, *Bioorg. Chem.*, *11*, 4 (1982).

204. N. P. Gensmantel, P. Proctor, and M. I. Page, *J. Chem. Soc. Perkin Trans. 2*, 1725 (1980).

205. M. A. Wells and T. C. Bruice, *J. Am. Chem. Soc.*, *99*, 5341 (1977).

206. J. T. Groves and R. R. Chambers, Jr., *J. Am. Chem. Soc.*, *106*, 630 (1984).

207. C. L. Perrin and O. Nuñez, *J. Am. Chem. Soc.*, *108*, 5997 (1986) and references therein.

208. R. L. Blakeley, A. Treston, R. K. Andrews, and B. Zerner, *J. Am. Chem. Soc.*, *104*, 612 (1982).

209. R. Breslow and L. E. Overman, *J. Am. Chem. Soc.*, *92*, 1075 (1970).

210. F. Cramer and G. Mackensen, *Angew. Chem., Int. Ed. Engl.*, *5*, 601 (1966).

211. R. Breslow and M. P. Mehta, *J. Am. Chem. Soc.*, *108*, 2485 (1986).

212. R. Breslow and M. P. Mehta, *J. Am. Chem. Soc.*, *108*, 6417 (1986).

213. R. Breslow and M. P. Mehta, *J. Am. Chem. Soc.*, *108*, 6418 (1986).

214. *Chem. Eng. News*, *64* (December 8), 2, 6 (1986).

215. I. Bertini and C. Luchinat, in "Advances in Inorganic Biochemistry", Vol. 6 (G. L. Eichhorn and L. G. Marzilli, eds.), Elsevier, New York, 1984, Chap. 2.

216. M. F. Dunn, H. Dietrich, A. K. H. MacGibbon, S. C. Koerber, and M. Zeppezauer, *Biochemistry*, *21*, 354 (1982).

217. D. B. Northrop and W. W. Cleland, *Fed. Proc., Fed. Am. Soc. Exp. Biol.*, *29*, 408 (1970).

218. M. H. O'Leary and J. A. Limburg, *Biochemistry*, *16*, 1129 (1977).

219. R. S. Johnson and H. K. Schachman, *J. Biol. Chem.*, *258*, 3528 (1983).

220. D. Bhatnagar, D. B. Glass, R. Roskoski, Jr., R. A. Lessor, and N. J. Leonard, *Biochemistry*, *24*, 1122 (1985).

221. C.-Y. Kwan, K. Erhard, and R. C. Davis, *J. Biol. Chem.*, *250*, 5951 (1975).

222. K. Erhard and R. C. Davis, *J. Biol. Chem.*, *250*, 5945 (1975).

223. M. A. Sirover and L. A. Loeb, *Biochem. Biophys. Res. Commun.*, *70*, 812 (1976).

224. J. M. Prescott, F. W. Wagner, B. Holmquist, and B. L. Vallee, *Biochemistry*, *24*, 5350 (1985).

225. M. L. Ludwig and W. N. Lipscomb, in "Inorganic Biochemistry", Vol. 1 (G. L. Eichhorn, ed.), Elsevier, Amsterdam, 1973, Chap. 15.

226. D. S. Auld and B. Holmquist, *Biochemistry*, *13*, 4355 (1974).

227. C. van der Drift and G. D. Vogels, *Biochim. Biophys. Acta*, *198*, 339 (1970).

228. A. Donald, D. Sibley, D. E. Lyons, and A. S. Dahms, *J. Biol. Chem.*, *254*, 2132 (1979).

229. L.-P. B. Han, C. M. Schimandle, L. M. Davison, and D. L. Vander Jagt, *Biochemistry*, *16*, 5478 (1977).

230. J. A. Fee, G. D. Hegeman, and G. L. Kenyon, *Biochemistry*, *13*, 2528 (1974).

231. W. J. Ray, Jr., *J. Biol. Chem.*, *244*, 3740 (1969).

232. E. Massarini and J. J. Cazzulo, *Experientia*, *31*, 1126 (1975).

233. R. Parvin, S. V. Pande, and T. A. Venkitasubramanian, *Biochim. Biophys. Acta*, *92*, 260 (1964).

234. S. Suzuki, T. Sakurai, A. Nakahara, T. Manabe, and T. Okuyama, *Biochemistry*, *22*, 1630 (1983).

235. L.-H. Chew, W. R. Carper, and H. J. Issaq, *Biochem. Biophys. Res. Commun.*, *66*, 217 (1975).

236. H. D. Campbell and B. Zerner, *Biochem. Biophys. Res. Commun.*, *54*, 1498 (1973).

237. P. G. Debrunner, M. P. Hendrich, J. de Jersey, D. T. Keough, J. T. Sage, and B. Zerner, *Biochim. Biophys. Acta, 745*, 103 (1983).

238. D. T. Keough, D. A. Dionysius, J. de Jersey, and B. Zerner, *Biochem. Biophys. Res. Commun.*, *94*, 600 (1980).

239. J. L. Beck, D. T. Keough, J. de Jersey, and B. Zerner, *Biochim. Biophys. Acta, 791*, 357 (1984).

240. M. M. King and C. Y. Huang, *J. Biol. Chem.*, *259*, 8847 (1984).

241. M. M. King and C. Y. Huang, *Biochem. Biophys. Res. Commun.*, *114*, 955 (1983).

242. C. J. Pallen and J. H. Wang, *J. Biol. Chem.*, *259*, 6134 (1984).

243. K. B. M. Reid and R. R. Porter, *Ann. Rev. Biochem.*, *50*, 433 (1981).

244. Z. Fishelson, M. K. Pangburn, and H. J. Müller-Eberhard, *J. Biol. Chem.*, *258*, 7411 (1983).

245. W. Grassmann, H. Hörmann, and O. Janowsky, *Hoppe-Seyler's Z. Physiol. Chem.*, *312*, 273 (1958).

246. D. T. Keough and B. Zerner, unpublished results.

247. R. K. Andrews, Ph.D. thesis, Univ. of Queensland, Brisbane, Australia, 1986.

248. H. Møllgaard and J. Neuhard, *J. Biol. Chem.*, *253*, 3536 (1978).

249. J. T. Kadonaga and J. R. Knowles, *Biochemistry, 22*, 130 (1983).

250. R. D. Kobes, R. T. Simpson, B. L. Vallee, and W. J. Rutter, *Biochemistry, 8*, 585 (1969).

251. S. Lindskog, in "Advances in Inorganic Biochemistry", Vol. 4 (G. L. Eichhorn and L. G. Marzilli, eds.), Elsevier, New York, 1982, Chap. 4.

252. C. B. Klee, *J. Biol. Chem.*, *247*, 1398 (1972).

253. I. L. Givot, T. A. Smith, and R. H. Abeles, *J. Biol. Chem.*, *244*, 6341 (1969).

254. R. K. Pullarkat, M. Sbaschnig-Agler, and H. Reha, *Biochim. Biophys. Acta, 663*, 117 (1981).

255. J. W. Maranville, *Plant Physiol.*, *45*, 591 (1970).

256. D. L. Garbers and J. G. Hardman, *J. Biol. Chem.*, *250*, 2482 (1975).

257. R. C. Weast (ed.), "CRC Handbook of Chemistry and Physics", 63rd ed., CRC, Boca Raton, FL, 1982, p. F-179.

258. W. B. Church, J. M. Guss, J. J. Potter, and H. C. Freeman, *J. Biol. Chem.*, *261*, 234 (1986).

259. R. Piras and B. L. Vallee, *Biochemistry*, *6*, 348 (1967).

260. D. A. Clare, J. Blum, and I. Fridovich, *J. Biol. Chem.*, *259*, 5932 (1984).

261. M. E. Martin, B. R. Byers, M. O. J. Olson, M. L. Salin, J. E. L. Arceneaux, and C. Tolbert, *J. Biol. Chem.*, *261*, 9361 (1986).

262. W. N. Lipscomb, *Proc. Natl. Acad. Sci. USA*, *70*, 3797 (1973).

263. M. F. Schmid and J. R. Herriott, *J. Mol. Biol.*, *103*, 175 (1976).

264. K. K. Kannan, B. Notstrand, K. Fridborg, S. Lövgren, A. Ohlsson, and M. Petef, *Proc. Natl. Acad. Sci. USA*, *72*, 51 (1975).

265. K. K. Kannan, A. Liljas, I. Waara, P.-C. Bergstén, S. Lövgren, B. Strandberg, U. Bengtsson, U. Carlbom, K. Fridborg, L. Järup, and M. Petef, *Cold Spring Harbor Symp. Quant. Biol.*, *36*, 221 (1971).

266. H. Eklund, B. Nordström, E. Zeppezauer, G. Söderlund, I. Ohlsson, T. Boiwe, B.-O. Söderberg, O. Tapia, C.-I. Brändén, and Å. Åkeson, *J. Mol. Biol.*, *102*, 27 (1976).

267. H. L. Monaco, J. L. Crawford, and W. N. Lipscomb, *Proc. Natl. Acad. Sci. USA*, *75*, 5276 (1978).

268. J. M. Sowadski, M. D. Handschumacher, H. M. K. Murthy, B. A. Foster, and H. W. Wyckoff, *J. Mol. Biol.*, *186*, 417 (1985).

269. J. S. Richardson, K. A. Thomas, B. H. Rubin, and D. C. Richardson, *Proc. Natl. Acad. Sci. USA*, *72*, 1349 (1975).

270. K. M. Beem, D. C. Richardson, and K. V. Rajagopalan, *Biochemistry*, *16*, 1930 (1977).

271. W. C. Stallings, K. A. Pattridge, R. K. Strong, and M. L. Ludwig, *J. Biol. Chem.*, *260*, 16424 (1985).

272. F. A. Cotton, C. J. Bier, V. W. Day, E. E. Hazen, Jr., and S. Larsen, *Cold Spring Harbor Symp. Quant. Biol.*, *36*, 243 (1971).

273. G. N. Reeke, Jr., J. W. Becker, B. A. Cunningham, G. R. Gunther, J. L. Wang, and G. M. Edelman, *Ann. N.Y. Acad. Sci.*, *234*, 369 (1974).

274. G. N. Reeke, Jr., J. W. Becker, and G. M. Edelman, *J. Biol. Chem.*, *250*, 1525 (1975).

275. E. T. Adman, R. E. Stenkamp, L. C. Sieker, and L. H. Jensen, *J. Mol. Biol.*, *123*, 35 (1978).

276. G. E. Norris, B. F. Anderson, and E. N. Baker, *J. Mol. Biol.*, *165*, 501 (1983).

277. P. M. Colman, H. C. Freeman, J. M. Guss, M. Murata, V. A. Norris, J. A. M. Ramshaw, and M. P. Venkatappa, *Nature*, *272*, 319 (1978).

278. J. Baudier, N. Glasser, and D. Gerard, *J. Biol. Chem.*, *261*, 8192 (1986).

279. W. F. Furey, A. H. Robbins, L. L. Clancy, D. R. Winge, B. C. Wang, and C. D. Stout, *Science*, *231*, 704 (1986).

280. P. M. Colman, J. N. Jansonius, and B. W. Matthews, *J. Mol. Biol.*, *70*, 701 (1972).

281. B. W. Matthews, L. H. Weaver, and W. R. Kester, *J. Biol. Chem.*, *249*, 8030 (1974).

282. B. W. Matthews and L. H. Weaver, *Biochemistry*, *13*, 1719 (1974).

283. P. C. Moews and R. H. Kretsinger, *J. Mol. Biol.*, *91*, 201 (1975).

284. R. H. Kretsinger and D. J. Nelson, *Coord. Chem. Rev.*, *18*, 29 (1976).

285. D. M. E. Szebenyi, S. K. Obendorf, and K. Moffat, *Nature*, *294*, 327 (1981).

286. R. E. Stenkamp, L. C. Sieker, L. H. Jensen, J. D. McCallum, and J. Sanders-Loehr, *Proc. Natl. Acad. Sci. USA*, *82*, 713 (1985).

287. I. M. Klotz and D. M. Kurtz, Jr., *Accounts Chem. Res.*, *17*, 16 (1984).

288. K. D. Watenpaugh, L. C. Sieker, and L. H. Jensen, *J. Mol. Biol.*, *131*, 509 (1979).

289. E. T. Adman, L. C. Sieker, L. H. Jensen, M. Bruschi, and J. LeGall, *J. Mol. Biol.*, *112*, 113 (1977).

290. H. Bachmayer, L. H. Piette, K. T. Yasunobu, and H. R. Whiteley, *Proc. Natl. Acad. Sci. USA*, *57*, 122 (1967).

291. C. A. Makaroff, J. L. Paluh, and H. Zalkin, *J. Biol. Chem.*, *261*, 11416 (1986).

292. E. T. Adman, L. C. Sieker, and L. H. Jensen, *J. Biol. Chem.*, *248*, 3987 (1973).

293. D. Ghosh, S. O'Donnell, W. Furey, Jr., A. H. Robbins, and C. D. Stout, *J. Mol. Biol.*, *158*, 73 (1982).

294. C. W. Carter, Jr., J. Kraut, S. T. Freer, N.-H. Xuong, R. A. Alden, and R. G. Bartsch, *J. Biol. Chem.*, *249*, 4212 (1974).

295. T. Takano, *J. Mol. Biol.*, *110*, 537 (1977).

296. T. Takano, *J. Mol. Biol.*, *110*, 569 (1977).

297. J. C. Norvell, A. C. Nunes, and B. P. Schoenborn, *Science*, *190*, 568 (1975).

298. R. C. Ladner, E. J. Heidner, and M. F. Perutz, *J. Mol. Biol.*, *114*, 385 (1977).

299. F. S. Mathews, M. Levine, and P. Argos, *J. Mol. Biol.*, *64*, 449 (1972).

300. R. Swanson, B. L. Trus, N. Mandel, G. Mandel, O. B. Kallai, and R. E. Dickerson, *J. Biol. Chem., 252,* 759 (1977).

301. F. R. Salemme, *Ann. Rev. Biochem., 46,* 299 (1977).

302. M. Pierrot, R. Haser, M. Frey, F. Payan, and J.-P. Astier, *J. Biol. Chem., 257,* 14341 (1982).

303. R. Timkovich and R. E. Dickerson, *J. Biol. Chem., 251,* 4033 (1976).

304. T. L. Poulos, B. C. Finzel, I. C. Gunsalus, G. C. Wagner, and J. Kraut, *J. Biol. Chem., 260,* 16122 (1985).

305. K. Lerch and E. Schenk, *J. Biol. Chem., 260,* 9559 (1985).

306. H. Beinert, M. H. Emptage, J.-L. Dryer, R. A. Scott, J. E. Hahn, K. O. Hodgson, and A. J. Thomson, *Proc. Natl. Acad. Sci. USA, 80,* 393 (1983).

307. M. K. Johnson, R. S. Czernuszewicz, T. G. Spiro, J. A. Fee, and W. V. Sweeney, *J. Am. Chem. Soc., 105,* 6671 (1983).

308. A. Liljas and M. G. Rossmann, *Ann. Rev. Biochem., 43,* 475 (1974).

309. J. A. Ibers and R. H. Holm, *Science, 209,* 223 (1980).

310. V. K. Shah, R. A. Ugalde, J. Imperial, and W. J. Brill, *Ann. Rev. Biochem., 53,* 231 (1984).

311. W. G. Zumft, *Eur. J. Biochem., 91,* 345 (1978).

312. S. P. Cramer, K. O. Hodgson, W. O. Gillum, and L. E. Mortenson, *J. Am. Chem. Soc., 100,* 3398 (1978).

313. J. A. Kovacs and R. H. Holm, *J. Am. Chem. Soc., 108,* 340 (1986).

314. J. M. Berg and R. H. Holm, *J. Am. Chem. Soc., 107,* 917 (1985).

315. J. Webb, J. S. Multani, P. Saltman, N. A. Beach, and H. B. Gray, *Biochemistry, 12,* 1797 (1973).

316. F. C. Bradley, S. Lindstedt, J. D. Lipscomb, L. Que, Jr., A. L. Roe, and M. Rundgren, *J. Biol. Chem., 261,* 11693 (1986).

317. W. E. Keyes, T. M. Loehr, and M. L. Taylor, *Biochem. Biophys. Res. Commun., 83,* 941 (1978).

318. Y. Tatsuno, Y. Saeki, M. Iwaki, T. Yagi, M. Nozaki, T. Kitagawa, and S. Otsuka, *J. Am. Chem. Soc., 100,* 4614 (1978).

319. B. C. Antanaitis, T. Strekas, and P. Aisen, *J. Biol. Chem., 257,* 3766 (1982).

320. I. Bertini, C. Luchinat, L. Messori, R. Monnanni, and A. Scozzafava, *J. Biol. Chem., 261,* 1139 (1986).

321. Y. Sugiura, H. Kawabe, and H. Tanaka, *J. Am. Chem. Soc., 102,* 6581 (1980).

322. S. Meshitsuka, G. M. Smith, and A. S. Mildvan, *J. Biol. Chem., 256,* 4460 (1981).

323. H. Muirhead, cited in T. M. Dougherty and W. W. Cleland, *Biochemistry, 24,* 5870 (1985).

324. C. T. Martin, C. P. Scholes, and S. I. Chan, *J. Biol. Chem., 260,* 2857 (1985).

325. T. H. Stevens, C. T. Martin, H. Wang, G. W. Brudvig, C. P. Scholes, and S. I. Chan, *J. Biol. Chem., 257,* 12106 (1982).

326. L. Nestor, J. A. Larrabee, G. Woolery, B. Reinhammar, and T. G. Spiro, *Biochemistry, 23,* 1084 (1984).

327. D. J. Kosman, J. Peisach, and W. B. Mims, *Biochemistry, 19,* 1304 (1980).

328. L. Garcia-Iniguez, L. Powers, B. Chance, S. Sellin, B. Mannervik, and A. S. Mildvan, *Biochemistry, 23,* 685 (1984).

329. G. L. Woolery, L. Powers, M. Winkler, E. I. Solomon, and T. G. Spiro, *J. Am. Chem. Soc., 106,* 86 (1984).

330. R. S. Himmelwright, N. C. Eickman, C. D. LuBien, K. Lerch, and E. I. Solomon, *J. Am. Chem. Soc., 102,* 7339 (1980).

331. G. L. Woolery, L. Powers, M. Winkler, E. I. Solomon, K. Lerch, and T. G. Spiro, *Biochim. Biophys. Acta, 788,* 155 (1984).

332. I. M. Vasiletz, M. M. Shavlovsky, and S. A. Neifakh, *Eur. J. Biochem., 25,* 498 (1972).

333. L. J. Banaszak, H. C. Watson, and J. C. Kendrew, *J. Mol. Biol., 12,* 130 (1965).

334. M. W. Washabaugh and K. D. Collins, *J. Biol. Chem., 261,* 5920 (1986).

335. G. Bentley, E. Dodson, G. Dodson, D. Hodgkin, and D. Mercola, *Nature, 261,* 166 (1976).

336. T. Blundell, G. Dodson, D. Hodgkin, and D. Mercola, in "Advances in Protein Chemistry", Vol. 26 (C. B. Anfinsen, Jr., J. T. Edsall, and F. M. Richards, eds.), Academic, New York, 1972, pp. 279 ff.

337. J. L. Sudmeier, S. J. Bell, M. C. Storm, and M. F. Dunn, *Science, 212,* 560 (1981).

338. A. G. Weeds and A. D. McLachlan, *Nature, 252,* 646 (1974).

339. S. T. Daurat-Larroque, K. Brew, and R. E. Fenna, *J. Biol. Chem., 261,* 3607 (1986).

340. D. P. Higham, P. J. Sadler, and M. D. Scawen, *Science, 225,* 1043 (1984).

341. D. C. Rees, M. Lewis and W. N. Lipscomb, *J. Mol. Biol., 168,* 367 (1983).

342. J. A. Hartsuck and W. N. Lipscomb, in "The Enzymes", 3rd ed., Vol. 3 (P. D. Boyer, ed.), Academic, New York, 1971, Chap. 1.

343. B. Zerner and M. L. Bender, *J. Am. Chem. Soc., 86,* 3669 (1964).

344. P. W. Riddles, R. K. Andrews, R. L. Blakeley, and B. Zerner, *Biochim. Biophys. Acta*, *743*, 115 (1983).

345. T. E. King and R. O. Morris, *Methods Enzymol.*, *10*, 634 (1967).

346. R. K. Andrews, R. L. Blakeley, and B. Zerner, unpublished results.

347. S. S. Hasnain and B. Piggott, *Biochem. Biophys. Res. Commun.*, *112*, 279 (1983).

348. L. Alagna, S. S. Hasnain, B. Piggott, and D. J. Williams, *Biochem. J.*, *220*, 591 (1984).

349. G. Mamiya, K. Takishima, M. Masakuni, T. Kayumi, K. Ogawa, and T. Sekita, *Proc. Jpn. Acad., Ser. B*, *61*, 395 (1985).

350. I. Waara, S. Lövgren, A. Liljas, K. K. Kannan, and P.-C. Bergstén, in "Advances in Experimental Medicine and Biology, Vol. 28, Hemoglobin and Red Cell Structure and Function" (G. J. Brewer, ed.), Plenum, New York, 1972, pp. 169-187.

351. J. Miller, A. D. McLachlan, and A. Klug, *EMBO J.*, *4*, 1609 (1985).

352. R. S. Brown, C. Sander, and P. Argos, *FEBS Lett.*, *186*, 271 (1985).

353. J. M. Berg, *Science*, *232*, 485 (1986).

354. A. Vincent, H. V. Colot, and M. Rosbash, *J. Mol. Biol.*, *186*, 149 (1985).

355. U. B. Rosenberg, C. Schröder, A. Preiss, A. Kienlin, S. Côté, I. Riede, and H. Jäckle, *Nature*, *319*, 336 (1986).

356. R. Bartha and E. J. Ordal, *J. Bacteriol.*, *89*, 1015 (1965).

357. B. Friedrich, E. Heine, A. Finck, and C. G. Friedrich, *J. Bacteriol.*, *145*, 1144 (1981).

358. C. G. Friedrich, K. Schneider, and B. Friedrich, *J. Bacteriol.*, *152*, 42 (1982).

359. E. G. Graf and R. K. Thauer, *FEBS Lett.*, *136*, 165 (1981).

360. G. B. Diekert, E. G. Graf, and R. K. Thauer, *Arch. Microbiol.*, *122*, 117 (1979).

361. H. L. Drake, S.-I. Hu, and H. G. Wood, *J. Biol. Chem.*, *255*, 7174 (1980).

362. W. L. Ellefson and R. S. Wolfe, *J. Biol. Chem.*, *256*, 4259 (1981).

363. W. L. Ellefson, W. B. Whitman, and R. S. Wolfe, *Proc. Natl. Acad. Sci. USA*, *79*, 3707 (1982).

364. S. W. Ragsdale, J. E. Clark, L. G. Ljungdahl, L. L. Lundie, and H. L. Drake, *J. Biol. Chem.*, *258*, 2364 (1983).

365. H. L. Drake, *J. Bacteriol.*, *149*, 561 (1982).

366. S. W. Ragsdale, L. G. Ljungdahl, and D. V. DerVartanian, *Biochem. Biophys. Res. Commun.*, *115*, 658 (1983).

367. G. Diekert and R. K. Thauer, *FEMS Microbiol. Lett.*, *7*, 187 (1980).

368. G. Diekert and M. Ritter, *J. Bacteriol.*, *151*, 1043 (1982).

369. S. W. Ragsdale, L. G. Ljungdahl, and D. V. DerVartanian, *J. Bacteriol.*, *155*, 1224 (1983).

370. K. E. Hammel, K. L. Cornwell, G. B. Diekert, and R. K. Thauer, *J. Bacteriol.*, *157*, 975 (1984).

371. A. R. Harker, L.-S. Xu, F. J. Hanus, and H. J. Evans, *J. Bacteriol.*, *159*, 850 (1984).

372. L. W. Stults, E. B. O'Hara, and R. J. Maier, *J. Bacteriol.*, *159*, 153 (1984).

373. K. Schneider, R. Cammack, and H. G. Schlegel, *Eur. J. Biochem.*, *142*, 75 (1984) and references therein.

374. K. Schneider, R. Cammack, H. G. Schlegel, and D. O. Hall, *Biochim. Biophys. Acta*, *578*, 445 (1979).

375. R. Cammack, D. Patil, R. Aguirre, and E. C. Hatchikian, *FEBS Lett.*, *142*, 289 (1982).

376. M. Teixeira, I. Moura, A. V. Xavier, B. H. Huynh, D. V. DerVartanian, H. D. Peck, Jr., J. LeGall, and J. J. G. Moura, *J. Biol. Chem.*, *260*, 8942 (1985).

377. H.-J. Krüger, B. H. Huynh, P. O. Ljungdahl, A. V. Xavier, D. V. DerVartanian, I. Moura, H. D. Peck, Jr., M. Teixeira, J. J. G. Moura, and J. LeGall, *J. Biol. Chem.*, *257*, 14620 (1982).

378. W. V. Lalla-Maharajh, D. O. Hall, R. Cammack, K. K. Rao, and J. LeGall, *Biochem. J.*, *209*, 445 (1983).

379. G. Unden, R. Böcher, J. Knecht, and A. Kröger, *FEBS Lett.*, *145*, 230 (1982).

380. P. A. Lindahl, N. Kojima, R. P. Hausinger, J. A. Fox, B. K. Teo, C. T. Walsh, and W. H. Orme-Johnson, *J. Am. Chem. Soc.*, *106*, 3062 (1984).

381. R. W. Spencer, L. Daniels, G. Fulton, and W. H. Orme-Johnson, *Biochemistry*, *19*, 3678 (1980).

382. F. S. Jacobson, L. Daniels, J. A. Fox, C. T. Walsh, and W. H. Orme-Johnson, *J. Biol. Chem.*, *257*, 3385 (1982).

383. N. Kojima, J. A. Fox, R. P. Hausinger, L. Daniels, W. H. Orme-Johnson, and C. T. Walsh, *Proc. Natl. Acad. Sci. USA*, *80*, 378 (1983).

384. S. Takakuwa and J. D. Wall, *FEMS Microbiol Lett.*, *12*, 359 (1981).

385. A. Colbeau and P. M. Vignais, *Biochim. Biophys. Acta, 748,* 128 (1983).

386. H. Van Heerikhuizen, S. P. J. Albracht, E. C. Slater, and P. S. Van Rheenen, *Biochim. Biophys. Acta, 657,* 26 (1981).

387. S. P. J. Albracht, J. W. van der Zwaan, and R. D. Fontijn, *Biochim. Biophys. Acta, 766,* 245 (1984).

388. S. P. J. Albracht, K. J. Albrecht-Ellmer, D. J. M. Schmedding, and E. C. Slater, *Biochim. Biophys. Acta, 681,* 330 (1982).

389. S. P. Ballantine and D. H. Boxer, *J. Bacteriol., 163,* 454 (1985).

390. Y. Nakamura, J. Someya, and T. Suzuki, *Agric. Biol. Chem., 49,* 1711 (1985).

391. C. D. P. Partridge and M. G. Yates, *Biochem. J., 204,* 339 (1982).

392. G. Fauque, M. Teixeira, I. Moura, P. A. Lespinat, A. V. Xavier, D. V. DerVartanian, H. D. Peck, Jr., J. LeGall, and J. J. G. Moura, *Eur. J. Biochem., 142,* 21 (1984).

393. M. Aggag and H. G. Schlegel, *Arch. Microbiol., 100,* 25 (1974).

394. A. Daday and G. D. Smith, *FEMS Microbiol. Lett., 20,* 327 (1983).

395. G. Diekert, B. Klee, and R. K. Thauer, *Arch. Microbiol., 124,* 103 (1980).

396. R. P. Hausinger, W. H. Orme-Johnson, and C. Walsh, *Biochemistry, 23,* 801 (1984).

397. W. B. Whitman and R. S. Wolfe, *Biochem. Biophys. Res. Commun., 92,* 1196 (1980).

398. G. Diekert, U. Konheiser, K. Piechulla, and R. K. Thauer, *J. Bacteriol., 148,* 459 (1981).

399. I. Moura, J. J. G. Moura, H. Santos, A. V. Xavier, G. Burch, H. D. Peck, Jr., and J. LeGall, *Biochim. Biophys. Acta, 742,* 84 (1983).

400. R. Hüster, H.-H. Gilles, and R. K. Thauer, *Eur. J. Biochem., 148,* 107 (1985).

401. E. V. Pinchukova, S. D. Varfolomeev, and E. N. Kondratjeva, *Biochimia (USSR), 44,* 605 (1979), through V. O. Popov, I. V. Berezin, A. M. Zaks, I. G. Gazaryan, I. B. Utkin, and A. M. Egorov, *Biochim. Biophys. Acta, 744,* 298 (1983).

402. E. C. Hatchikian, M. Bruschi, and J. LeGall, *Biochem. Biophys. Res. Commun., 82,* 451 (1978).

403. R. A. Scott, S. A. Wallin, M. Czechowski, D. V. DerVartanian, J. LeGall, H. D. Peck, Jr., and I. Moura, *J. Am. Chem. Soc., 106,* 6864 (1984).

404. S. L. Tan, J. A. Fox, N. Kojima, C. T. Walsh, and W. H. Orme-Johnson, *J. Am. Chem. Soc., 106,* 3064 (1982).

405. L. Daniels, R. Sparling, and G. D. Sprott, *Biochim. Biophys. Acta, 768,* 113 (1984).

406. S. P. J. Albracht, E. G. Graf, and R. K. Thauer, *FEBS Lett., 140,* 311 (1982).

407. J. W. van der Zwaan, S. P. J. Albracht, R. D. Fontijn, and E. C. Slater, *FEBS Lett., 179,* 271 (1985).

408. P. H. Gitlitz and A. I. Krasna, *Biochemistry, 14,* 2561 (1975).

409. M. W. W. Adams and D. O. Hall, *Biochem. J., 183,* 11 (1979).

410. P. J. Large, "Aspects of Microbiology 8: Methylotrophy and Methanogenesis", Van Nostrand Reinhold, Wokingham, 1983, pp. 14-15.

411. C. R. Woese, L. J. Magrum, and G. E. Fox, *J. Mol. Evol., 11,* 245 (1978).

412. G. Diekert, in "Bioinorganic Chemistry of Nickel" (J. R. Lancaster, Jr., ed.), VCH, New York, 1988, Chap. 13, in press.

413. S. W. Ragsdale, H. G. Wood, T. A. Morton, L. G. Ljungdahl, and D. V. DerVartanian, in "Bioinorganic Chemistry of Nickel" (J. R. Lancaster, Jr., ed.), VCH, New York, 1988, Chap. 14, in press.

414. O. Meyer and H.-G. Schlegel, *J. Bacteriol., 141,* 74 (1980).

415. O. Meyer, *J. Biol. Chem., 257,* 1333 (1982).

416. S. W. Ragsdale, L. G. Ljungdahl, and D. V. DerVartanian, *Biochem. Biophys. Res. Commun., 108,* 658 (1982).

417. Y. M. Kim and G. D. Hegeman, *J. Bacteriol., 148,* 904 (1981).

418. S. W. Ragsdale, H. G. Wood, and W. E. Antholine, *Proc. Natl. Acad. Sci. USA, 82,* 6811 (1985).

419. R. Cammack, V. M. Fernandez and K. Schneider, in "Bioinorganic Chemistry of Nickel" (J. R. Lancaster, Jr., ed.), VCH, New York, 1988, Chap. 8, in press.

420. J. J. G. Moura, M. Teixeira, I. Moura, and J. LeGall, in "Bioinorganic Chemistry of Nickel" (J. R. Lancaster, Jr., ed.), VCH, New York, 1988, Chap. 9, in press.

421. N. R. Bastian, D. A. Wink, L. P. Wackett, D. J. Livingston, L. M. Jordan, J. Fox, W. H. Orme-Johnson, and C. T. Walsh, in "Bioinorganic Chemistry of Nickel" (J. R. Lancaster, Jr., ed.), VCH, New York, 1988, Chap. 10, in press.

422. L. D. Eirich, G. D. Vogels, and R. S. Wolfe, *Biochemistry, 17,* 4583 (1978).

423. J. R. D. McCormick and G. O. Morton, *J. Am. Chem. Soc., 104,* 4014 (1982).

424. A. P. M. Eker, A. Pol, P. van der Meyden, and G. D. Vogels, *FEMS Microbiol. Lett., 8,* 161 (1980).

425. F. S. Jacobson and C. T. Walsh, *Biochemistry*, *23*, 979 (1984).

426. R. G. Sawers, S. P. Ballantine, and D. H. Boxer, *J. Bacteriol.*, *164*, 1324. (1985).

427. J. M. Odom and H. D. Peck, Jr., *FEMS Microbiol. Lett.*, *12*, 47 (1981).

428. B. H. Huynh, M. H. Czechowski, H.-J. Krüger, D. V. DerVartanian, H. D. Peck, Jr., and J. LeGall, *Proc. Natl. Acad. Sci. USA*, *81*, 3728 (1984).

429. M. W. W. Adams and L. E. Mortenson, *J. Biol. Chem.*, *259*, 7045 (1984).

430. S. Yamazaki, *J. Biol. Chem.*, *257*, 7926 (1982).

431. F. O. Pedrosa and M. G. Yates, *FEMS Microbiol. Lett.*, *17*, 101 (1983).

432. H. J. Doddema, C. van der Drift, G. D. Vogels, and M. Veenhuis, *J. Bacteriol.*, *140*, 1081 (1979).

433. K. Schneider and H. G. Schlegel, *Arch. Microbiol.*, *112*, 229 (1977).

434. T. Yamamura, *Chem. Lett.*, 801 (1986).

435. R.-M. Mege and C. Bourdillon, *J. Biol. Chem.*, *260*, 14701 (1985).

436. M. Teixeira, I. Moura, A. V. Xavier, D. V. DerVartanian, J. LeGall, H. D. Peck, Jr., B. H. Huynh, and J. J. G. Moura, *Eur. J. Biochem.*, *130*, 481 (1983).

437. T. Lissolo, S. Pulvin, and D. Thomas, *J. Biol. Chem.*, *259*, 11725 (1984).

438. Y. M. Berlier, G. Fauque, P. A. Lespinat, and J. LeGall, *FEBS Lett.*, *140*, 185 (1982).

439. T. Sakurai, J.-I. Hongo, A. Nakahara, and Y. Nakao, *Inorg. Chim. Acta*, *46*, 205 (1980).

440. L. Fabbrizzi, *J. Chem. Soc. Chem. Commun.*, 1063 (1979).

441. S. P. J. Albracht, *Biochem. Soc. Trans.*, *13*, 582 (1985).

442. R. P. Gunsalus and R. S. Wolfe, *FEMS Microbiol. Lett.*, *3*, 191 (1978).

443. G. Diekert, B. Weber, and R. K. Thauer, *Arch. Microbiol.*, *127*, 273 (1980).

444. D. A. Livingston, A. Pfaltz, J. Schreiber, A. Eschenmoser, D. Ankel-Fuchs, J. Moll, R. Jaenchen, and R. K. Thauer, *Helv. Chim. Acta*, *67*, 334 (1984).

445. A. Pfaltz, in "Bioinorganic Chemistry of Nickel" (J. R. Lancaster, Jr., ed.), VCH, New York, 1988, Chap. 12, in press.

446. G. Diekert, H.-H. Gilles, R. Jaenchen, and R. K. Thauer, *Arch. Microbiol.*, *128*, 256 (1980).

447. D. P. Nagle and R. S. Wolfe, *Proc. Natl. Acad. Sci. USA*, *80*, 2152 (1983).

448. R. P. Gunsalus and R. S. Wolfe, *J. Biol. Chem.*, *255*, 1891 (1980).

449. R. P. Gunsalus and R. S. Wolfe, *J. Bacteriol.*, *135*, 851 (1978).

450. S. Shapiro and R. S. Wolfe, *J. Bacteriol.*, *141*, 728 (1980).

451. J. T. Keltjens, G. Caerteling, A. M. van Kooten, H. F. van Dijk, and G. D. Vogels, *Biochim. Biophys. Acta*, *743*, 351 (1983).

452. J. T. Keltjens, W. B. Whitman, C. G. Caerteling, A. M. van Kooten, R. S. Wolfe, and G. D. Vogels, *Biochem. Biophys. Res. Commun.*, *108*, 495 (1982).

453. S. P. J. Albracht, D. Ankel-Fuchs, J. W. Van der Zwaan, R. D. Fontijn, and R. K. Thauer, *Biochim. Biophys. Acta*, *870*, 50 (1986).

454. M. K. Eidsness, R. J. Sullivan, J. R. Schwartz, P. L. Hartzell, R. S. Wolfe, A.-M. Flank, S. P. Cramer, and R. A. Scott, *J. Am. Chem. Soc.*, *108*, 3120 (1986).

455. S. J. Cole, G. C. Curthoys, E. A. Magnusson, and J. N. Phillips, *Inorg. Chem.*, *11*, 1024 (1972).

456. F. V. Lovecchio, E. S. Gore, and D. H. Busch, *J. Am. Chem. Soc.*, *96*, 3109 (1974).

457. E. K. Barefield, F. V. Lovecchio, N. E. Tokel, E. Ochiai, and D. H. Busch, *Inorg. Chem.*, *11*, 283 (1972).

458. L. P. Wackett, J. F. Honek, T. P. Begley, S. L. Shames, E. C. Niederhoffer, R. P. Hausinger, W. H. Orme-Johnson, and C. T. Walsh, in "Bioinorganic Chemistry of Nickel" (J. R. Lancaster, Jr., ed.), VCH, New York, 1988, Chap. 11, in press.

459. J. B. Sumner, *J. Biol. Chem.*, *69*, 435 (1926).

460. N. E. Dixon, R. L. Blakeley, and B. Zerner, *Can. J. Biochem.*, *58*, 469 (1980).

461. K. Sakaguchi, K. Mitsui, N. Nakai, and K. Kobashi, *J. Biochem. (Tokyo)*, *96*, 73 (1984).

462. S. F. Andres and M. Z. Atassi, *Biochemistry*, *12*, 942 (1973).

463. N. E. Dixon, P. W. Riddles, C. Gazzola, R. L. Blakeley, and B. Zerner, *Can. J. Biochem.*, *58*, 1335 (1980).

464. K. Kobashi, K. Sakaguchi, S. Takebe, J. Hase, and M. Sato, *Abstr. Int. Congr. Biochem. Perth, 12th*, 305 (1982).

7

Nickel-Containing Hydrogenases

José J. G. Moura, Isabel Moura,
Miguel Teixeira, Antonio V. Xavier,
Centro de Quimica Estrutural
Complexo I and UNL
Av. Rovisco Pais
1096 Lisboa, Portugal

Guy D. Fauque,
A.R.B.S., Section Enzymologie et Biochimie Bacterienne
C.E.N. Cadarache, 13108 Saint-Paul-les-Durance Cedex, France

and

Jean LeGall
School of Chemical Sciences
Department of Biochemistry, University of Georgia
Athens, Georgia 30602

1. INTRODUCTION

Nickel is now well recognized as a transition metal required for a
wide range of biological functions [1,2]. The discovery of nickel
in purified proteins is relatively recent and many current studies
are centered on the characterization of the nickel-containing active
center as well as the search for structure-function relationship.
Nickel-containing hydrogenases constitute a choice enzyme group
since the metal, besides being a constitutive element, also plays
a diversified role in the redox process involved and in the regulation of the catalytic activity [3].

Over the past 10 years there has been a renewed interest in
the physiology and biochemistry of hydrogenases with emphasis on
the catalytic properties and the mechanisms involved [3-6], and the
applications of the enzyme to bioconversion (chloroplast-hydrogenase
interplay in biophotolysis) [7] as well as other biotechnologically
oriented processes [8]. Homogeneous preparations have been purified
from strict and facultative aerobic and anaerobic organisms and in
particular sulfate-reducing, methanogenic, and photosynthetic bacteria. It is generally believed that hydrogenases represent a
diverse group of proteins (in contrast with other enzyme groups,
e.g., nitrogenase) differing not only in respect to their metal
content and type of active centers but also in subunits composition,
stability, and reactivity.

2. NICKEL INVOLVEMENT IN HYDROGENASES

Hydrogenases are generally classified as iron-sulfur-containing
proteins with 4-12 iron atoms in different cluster arrangements.
Nickel was considered until recently an uncommon metal in biology.
The involvement of nickel in the hydrogenase system was established
through accumulated evidences of different origins: physiologic
studies (nickel requirement for autotrophic bacterial growth [9]
and nickel-dependent hydrogenase synthesis [10,11]), chemical determination, and mainly electron paramagnetic resonance (EPR) spectroscopic studies (see below). The detection of nickel EPR signals in
biological systems was first reported by Lancaster [12] in membrane
preparations of *Methanobacterium* (*Mb.*) *bryantii*; the assignment of
the unusual rhombic signal with g values at 2.3, 2.2, and 2.0 was
based on data available for nickel model compounds and on spectral
characteristic of the signal (g values and relaxation). The signal
tentatively attributed to Ni(III) could be detected up to 120°K.
The definitive assignment of the observed rhombic signal was achieved
when *Mb. bryantii* was grown in a medium supplemented with ^{61}Ni
(I = 3/2) [13]. Similar signals have been observed for bacterial
hydrogenases of different origins (see Sec. 6.1). The response and
alterations of the EPR features associated with nickel upon hydrogen
exposure are also indicative of direct participation of the nickel
center (see Sec. 6.2). Recently nickel was also suggested to have
a role on binding of subunits in the NAD-linked hydrogenase purified
from *Nocardia opaca* [14].

3. SPECTROSCOPIC TOOLS AND INSTRUMENTAL PROBES:
ISOTOPIC SUBSTITUTIONS

EPR is a reasonably sensitive method for detecting paramagnetic
nickel species [Ni(III) and Ni(I)] and for giving information about
the metal environment. The technique does not allow the unambiguous
assignment to one of these two redox states and some controversy is

still unresolved [3,15,16]. Midpoint redox properties can be readily obtained from EPR measurements [17,18].

A better understanding of the structural features and functional properties of a metalloenzyme containing this "new biologically relevant metal" was achieved by the knowledge of coordination chemistry and redox properties of nickel model compounds. A wealth of information has been built in this way and proved useful in order to predict reasonable geometries and oxidation states for the nickel site [3,19].

Mössbauer spectroscopy has also been an indispensable tool for the characterization of the iron-sulfur centers present in the enzyme and its interrelation with the nickel site. The conjunction of the two techniques is very powerful in order to characterize complex systems and to study spin-coupled structures [20]. Also, important information can be extracted about the iron-sulfur oxidation states present in EPR silent states.

The use of isotopic replacements has also been a valuable source of information (see Sec. 6). ^{61}Ni isotope (1.2% natural abundance, I = 3/2) enabled unambiguous assignment of EPR active species to the nickel site due to the appearance of resolved nuclear hyperfine structure. This step of assignment may be crucial when the observed EPR spectrum is the superimposition of signals originated from different nuclei. ^{57}Fe isotope (2.1% natural abundance, I = 1/2) has an obvious advantage for increasing the experimental sensitivity of the Mössbauer experiments as well as introducing specific broadening in the EPR-detectable iron species. ^{33}S isotope (0.76% natural abundance, I = 3/2) has recently been used in order to explore the nickel coordination sphere [21].

In order to probe the catalytic active site properties, the isotopic exchange D_2-H^+ has been used and the amount of HD plus H_2 formed measured by mass spectrometry [22].

Magnetic circular dichroism (MCD) and extended x-ray absorption fine structure (EXAFS) spectroscopies have also been used in order to probe the nickel site [23,24].

4. ACTIVE CENTER COMPOSITION: TYPES OF HYDROGENASES

The metabolism of hydrogen in sulfate-reducing bacteria is regulated by reversible hydrogenases and *Desulfovibrio* sp. enzymes are clearly representative of the complexity involved in this process. At least three different types are now recognized within this bacterial group (Table 1):

1. [Fe] hydrogenases--containing only iron-sulfur centers purified from *Desulfovibrio* (*D.*) *vulgaris* (Hildenborough) [25,26]. The metal cores are arranged as two [4Fe-4S] clusters and the third one considered atypical [25].

2. [NiFe] hydrogenases--containing one nickel and iron-sulfur centers generally arranged as one [3Fe-xS] and two [4Fe-4S] clusters, purified from *D. gigas* (NCIB 9332) [18,27-29], *D. desulfuricans* (ATCC 27774) [30], and *D. multispirans* n.sp. (NCIB 12078) [31].

3. [NiFeSe] hydrogenases--containing iron-sulfur centers and equimolecular amounts of nickel and selenium, have been purified from *D. desulfuricans* (Norway 4) [32,33], *D. baculatus* (DSM 1743)* [34], and *D. salexigens* (British Guiana) [35]. No definitive proof has been presented for the quantitative presence of a [3Fe-xS] core, but two distinct [4Fe-4S] cores have been observed.

The [NiFe] hydrogenase isolated from *Chromatium* (*Ch.*) *vinosum* has the minimal active center composition reported so far: one nickel and one [4Fe-4S] center [36]. The bidirectional [Fe] hydrogenase (hydrogenase I) isolated from *Clostridium* (*Cl.*) *pasteurianum* has an active center composition close to the *D. vulgaris* enzyme [37]. [Fe] hydrogenase II from *Cl. pasteurianum* (unidirectional) lacks one of the [4Fe-4S] cores [38].

---

*Desulfovibrio baculatus (DSM 1743) was formerly called *Desulfovibrio* strain 9974.

TABLE 1

Comparison of Physicochemical Properties of *Desulfovibrio* sp. [NiFe] and [NiFeSe] Hydrogenases

Property	D. gigas (NCIB 9332)	D. multispirans n.sp. (NCIB 12078)	D. desulfuricans (ATCC 27774)	D. baculatus (DSM 1743)	D. desulfuricans (Norway 4)	D. salexigens (British Guiana NCIB 8403)
Type	[NiFe]	[NiFe]	[NiFe]	[NiFeSe]	[NiFeSe]	[NiFeSe]
Localization	Periplasm	Cytoplasm	NR*	Periplasm[a]	Membrane[b]	Periplasm
Molecular weight (kDa)	89.50	82.50	77.60	100	58	98
Subunits	2	2	2	2	1	2
Nickel	1	1	1	1	+ (EPR)	1
Selenium	0	0	0	1	Present	1
Nonheme iron	11	11	12	12	6	12-15
[3Fe-xS]	1	1	1	+[c]	"g = 2.02"	+[c]
[4Fe-4S]	2	+(probably 2)	2	+(probably 2)	+	+(probably 2)
Specific activity (μmol H_2/min·mg)						
Evolution	440	790	152	527	70	1830
Consumption	1500	586	NR*	190	200	1300
Exchange H_2/HD	0.22-0.40	0.3-0.4	0.2-0.4	1.3-1.6	NR	> >1
References	[17,18,27]	[31]	[30]	[34,64]	[33]	[35,65,66]

*Not reported.
[a] Also a cytoplasmic and a membrane-bound form were purified [34].
[b] Also a soluble form was purified [32].
[c] Weak signal.

5. CELL LOCALIZATION OF HYDROGENASES: EXISTENCE OF MULTIPLE FORMS AND GENETIC INFORMATION

The hydrogenase activity isolated from sulfate-reducing bacteria of the genus *Desulfovibrio* is most often localized in the periplasmic space [6], although membrane-bound enzymes have been reported [33, 39-41]. A cell localization study was reported for the [NiFe] hydrogenase from *D. multispirans* n.sp. where the cytoplasmic localization of the enzyme was confirmed [31]. Furthermore, multiple forms of hydrogenase have been reported within a single organism. For example, two soluble forms of hydrogenase were found in *D. desulfuricans* (ATCC 27774) [30]. A soluble and a membrane-bound hydrogenase were reported in *D. desulfuricans* (Norway 4) [32-33], and there was evidence for the presence of a periplasmic enzyme [32]. Recently, a membrane-bound hydrogenase was identified in *D. vulgaris* (Hildenborough) [39], and in an independent study three new hydrogenases were isolated from the membranes of the same organism [41]. Two of them can react with antibodies to the [NiFe] periplasmic hydrogenase of *D. gigas* and the third one reacts with antibodies to the [NiFeSe] hydrogenase of *D. baculatus* (DSM 1743) [41]. In the last strain the hydrogenase activity was shown to be distributed in the periplasm, cytoplasm, and membranes [34].

The genes encoding for the large and small subunits of the periplasmic hydrogenases of *D. gigas* and *D. baculatus* have been cloned and partially sequenced (C. Li, N. Menon, J. LeGall, H. D. Peck, Jr., and A. Przybyla, unpublished data). As suggested by immunological studies, there appears to be little sequence homology between the two types of nickel-containing hydrogenases. The relationship among these multiple forms of hydrogenase within the same bacterium is not yet clear. Also, their presence contrasts with the more simplistic idea that in sulfate-reducing bacteria of the genus *Desulfovibrio* one enzyme is responsible for both the utilization and production of molecular hydrogen, and is coupled to low molecular weight electron carriers such as ferredoxin, flavodoxin, and rubredoxin through the tetraheme cytochrome c_3 [42]. From a physiologic point of view,

multiple forms of hydrogenase with different molecular properties
may be required to provide regulatory mechanisms for the various
metabolic pathways involving the production and utilization of
hydrogen.

6. EPR AND MÖSSBAUER STUDIES: NICKEL
AND [FeS] CENTERS

A reasonable understanding of D. gigas hydrogenase metal centers
constitution has already emerged from EPR and Mössbauer spectroscopic
studies, indicating the presence of four noninteracting redox centers
in the isolated state: one nickel, one [3Fe-xS] center (EPR active),
and two $[4Fe-4S]^{2+}$ centers (EPR silent). The enzyme has a molecular
weight of 89 kDa and two subunits (62 and 26 kDa), the maximal specific activity is 440 μmol H_2 evolved/min. mg protein at pH 7.6 and
32°C (using reduced methylviologen as an electron donor).

Comparison with other [NiFe] hydrogenases indicates that the
Ni(III) site may exist in different environments. The EPR-detectable
species are correlated with an inactive form of the enzyme [3,28].

6.1. Native State

In the native state, periplasmic D. gigas hydrogenase shows a slow-
relaxing rhombic EPR signal with g values at g_1 = 2.31, g_2 = 2.23,
and g_3 = 2.02 (Ni signal A) detected from low temperature up to
~100°K [18,27-29] (Fig. 1A). Similar signals are observed in D.
baculatus (membrane-bound form) [34], D. desulfuricans (Norway 4)
(membrane-bound form) [33], and D. multispirans n.sp. [31] (see
Table 2 and Fig. 2). Isotopic substitution by [61]Ni (I = 3/2)
induced line broadening in the feature at g_1 = 2.31 and resolved
hyperfine structures in the g_2 = 2.23 ($^{61}A_2$ = 1.5 mT) and g_3 = 2.02
($^{61}A_3$ = 2.7 mT) in the enzymes isolated from D. gigas [43,44],
D. desulfuricans (ATCC 27774) [30], and Mb. thermoautotrophicum

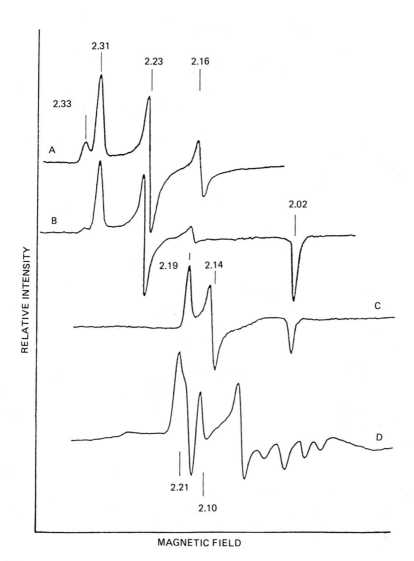

FIG. 1. X-band EPR spectra of *D. gigas* hydrogenase. A, B—Different
preparations of native hydrogenase, 100°K (Ni signals A and B).
C, D—Hydrogen-reduced state, 77°K (Ni signal C) (C) and 4.2°K
("g = 2.21" signal) (D). Modulation amplitude 1 mT; microwave
power 2 mW.

TABLE 2

EPR g-Values of Native (as Isolated) [NiFe]
and [NiFeSe] Hydrogenases

Organism	Ni center			[3Fe-xS] core (g = 2.02)	Ref.
	g_1	g_2	g_3		
SULFATE REDUCERS					
[NiFe] hydrogenases					
D. gigas (NCIB 9332)	2.31	2.23	2.02	+	[18,27]
	2.33	2.16	2.02		
D. desulfuricans (ATCC 27774)	2.32	2.16	2.01	+	[30]
D. multispirans n.sp. (NCIB 12078)	2.31	2.22	2.00	+	[31]
[NiFeSe] hydrogenases					
D. desulfuricans (Norway 4)					
soluble	2.22	2.07	2.016	+[a]	[32]
membrane	2.32	2.23	2.014	+[a]	[33]
D. salexigens (British Guiana)	EPR silent				[35]
D. baculatus (DSM 1743)					
membrane	2.34	2.15	2.0	+[a]	[34]
	2.33	2.24	2.0		
periplasmic	2.20	2.06	2.0	+[a]	[34]
cytoplasmic	EPR silent				[34]
METHANOGENS					
Mb. thermoautotrophicum (Marburg)	2.3	2.2	2.0		[45]
Mb. thermoautotrophicum ΔH	2.309	2.237	2.07		[46]
Ms. barkeri (DSM 800)	2.24	2.20	2.02		[47]

[+]present; [a]weak intensity.

strains Marburg and ΔH [45,46], clearly showing that the paramagnetic nickel is at the origin of those signals. This species was attributed to nickel(III) with a tetragonally distorted octahedral symmetry in the presence of a strong field, resulting in an $S = 1/2$ system with one unpaired electron in a d_{z^2} orbital.

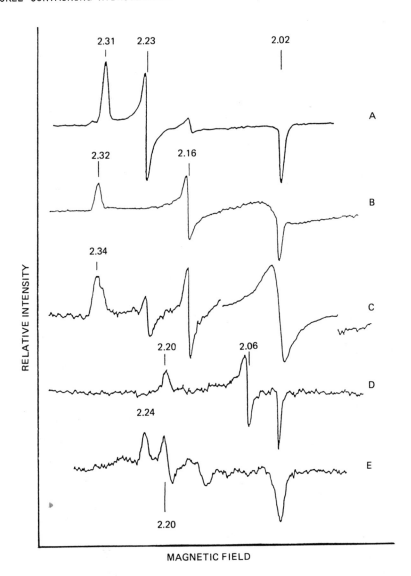

FIG. 2. Nickel EPR signals from native [NiFe] and [NiFeSe] hydrogenases from *Desulfovibrio species*. A: *D. gigas* 100°K. B: *D. desulfuricans* (ATCC 27774), 77°K. C: *D. baculatus* (DSM 1743) (membrane)
4.2°K. D: *D. baculatus* (DSM 1743) (periplasm), 43°K. E: *Ms. barkeri*
(DSM 800), 25°K. Modulation amplitude 1 mT; microwave power 2 mW.

In addition to Ni signal A another rhombic species with g
values at 2.33, 2.16, and 2.0 (Ni signal B) (Fig. 1A) is observed
for the periplasmic hydrogenase from *D. gigas* [28], *D. baculatus*
(DSM 1743) (membrane-bound form) [34], and *D. desulfuricans* (Norway
4) (membrane-bound form) [33] (Fig. 2). This signal is the main
component detected in the soluble hydrogenase from *D. desulfuricans*
(ATCC 27774) [30]. Ni signal A can be interconverted into Ni signal
B by anaerobic redox cycling of the hydrogenase [28].

Other nickel EPR signals were observed in the native hydrogenase from *D. desulfuricans* (Norway 4) (soluble form) and *D. baculatus*
(DSM 1743) (periplasmic) hydrogenase with g values at 2.20, 2.06, and
2.0 [32,34]. The hydrogenase isolated from *Methanosarcina* (*Ms.*)
barkeri (DSM 800) shows a nickel signal at g values 2.24, 2.12, and
2.0 [47]. These signals (Fig. 2) may represent variations of the
coordination sphere and/or geometry of the nickel(III) site as compared to those originating from Ni signals A and B.

The intensity of these native state nickel signals can vary
from 0.2 up to 0.9 spin/mol in *D. gigas, D. desulfuricans* (ATCC
27774), and *D. multispirans* n.sp., depending on the enzyme preparations. In native *D. baculatus* (DSM 1743) and *D. desulfuricans*
(Norway 4) membrane-bound hydrogenases the nickel EPR signals are
less than 10% of the chemically detectable nickel and *D. baculatus*
(DSM 1743) (cytoplasmic) and *D. salexigens* hydrogenases do not
reveal rhombic EPR signals being almost EPR silent as isolated.

The EPR spectrum at low temperatures (<30°K) is dominated by
an intense signal centered at g = 2.02 for the *D. gigas* [17,18,27],
D. desulfuricans (ATCC 27774) [30], and *D. multispirans* n.sp. hydrogenases [31]. By conjunction of detailed EPR and Mössbauer spectroscopic studies on the *D. gigas* [18,48] and *D. desulfuricans* (ATCC
27774) enzymes [30], the presence of an oxidized [3Fe-xS] center
with S = 1/2 was definitively established. Hyperfine broadening is
observed in the EPR spectra of [57]Fe-enriched samples [3,30].

The 4.2°K Mössbauer spectra of native natural abundance and
[57]Fe-enriched samples [18,48] indicate that besides a paramagnetic
component extending from -2 to +3 mm/sec due to the [3Fe-xS] cluster,

an intense quadrupole doublet at the center of the spectra accounts
for up to 70-80% of the total iron absorption. The observed quadrupole splitting (Eq = 1.16 mm/sec) and the isomer shift (δ = 0.46
mm/sec) indicate that D. gigas hydrogenase contains two oxidized
EPR silent (S = 0) $[4Fe-4S]^{2+}$ centers in the native state. The
conjunction of EPR and Mössbauer studies also show that those four
redox centers are magnetically noninteracting in this enzyme state.

D. baculatus (DSM 1743) membrane and periplasmic hydrogenases
and D. desulfuricans (Norway 4) (membrane-bound and soluble) hydrogenases show weak EPR signals at g = 2.02 (less than 0.05 spin/mol).
In the soluble hydrogenase isolated from D. desulfuricans (Norway 4)
Mössbauer spectroscopy was unable to detect the presence of 3Fe
centers either in the oxidized or in the reduced forms of the enzyme
[49].

Mössbauer studies performed on the periplasmic hydrogenase
from D. gigas [18,48], D. desulfuricans (ATCC 27774) (soluble) [30],
and D. desulfuricans (Norway 4) [49] (soluble) clearly show the
presence of two [4Fe-4S] clusters in the 2+ oxidation state.

6.2. Intermediate Redox Species Generated
Under Hydrogen

The occurring redox processes are related to the redox-linked activation steps required for full expression of activity. Special emphasis is given to the interpretation of the EPR and Mössbauer results
on the active state of the enzyme. Existence of a proton and a
hydride acceptor site are postulated in agreement with the heterolytic cleavage of the hydrogen molecule.

The EPR signals observed in native states of hydrogenases are
related to inactive species [16,18]. The first event taking place
upon interaction with hydrogen (reductive activation of D. gigas
hydrogenase in the presence of hydrogen) is the disappearance of the
g = 2.02 EPR signal associated with the [3Fe-xS] center [E_0' = -70 mV)
followed by the disappearance of the Ni signals A and B (for Ni

signal A $E_0' = -220$ mV) [17,18], attaining an EPR silent active state.
On the basis of EPR and Mössbauer studies of the enzyme in reduced
states, the redox process at -70 mV was assigned to the reduction of
the [3Fe-xS] center. Mössbauer spectra of dithionite and hydrogen-
reduced samples show that the [3Fe-xS] center remains reduced (integer
spin state, $S > 1$) at redox potentials below -80 mV, being *not* converted to a $[4Fe-4S]^{1+}$ cluster [48]. This is the first example of a
[3Fe-xS]-reduced center in a catalytically active state. The Mössbauer data obtained for fully reduced samples (-400 mV) indicate that
all the iron-sulfur centers are reduced: two $[4Fe-4S]^{1+}$ clusters and
one [3Fe-xS] reduced [48]. When the sample is poised at -270 mV,
~70% of a [4Fe-4S] cluster is reduced. The corresponding sample
shows a concomitant 70% reduction of the native nickel signal. No
other signals are observed. The fact that the [4Fe-4S] cluster is
in a 1+ state but does not exhibit a corresponding EPR signal requires
further investigation (see below).

Following the EPR silent state, further reduction of *D. gigas*
hydrogenase results in the appearance of a new, slow-relaxing transient rhombic EPR signal with g values at $g_1 = 2.19$, $g_2 = 2.14$, and
$g_3 = 2.02$ (termed Ni signal C) (Fig. 1C), which subsequently disappears upon longer exposure to hydrogen gas or in the presence of
excess sodium dithionite [18]. This EPR signal was also assigned to
nickel by isotopic substitution ($^{61}A_3 = 2.0$ mT) [43] and is readily
observable at intermediate redox states of the following hydrogenases
from the sulfate-reducing bacteria: *D. gigas* [43], *D. desulfuricans*
(ATCC 27774) [30], *D. multispirans* n.sp. [31], *D. salexigens* [35],
and *D. baculatus* (DSM 1743) [34] (Fig. 3). EPR studies conducted at
low temperature (generally below 10°K) at redox levels concomitant
and below the development of Ni signal C reveal another complex EPR
signal termed "g = 2.21" signal (Fig. 1D). The observation of the
"g = 2.21" signal requires a high microwave power (fast-relaxing
species) (Table 3).

Due to the heterolytic mechanism deduced from H^+-D_2 isotopic
exchange experiments [22] the Ni signal C was proposed to represent

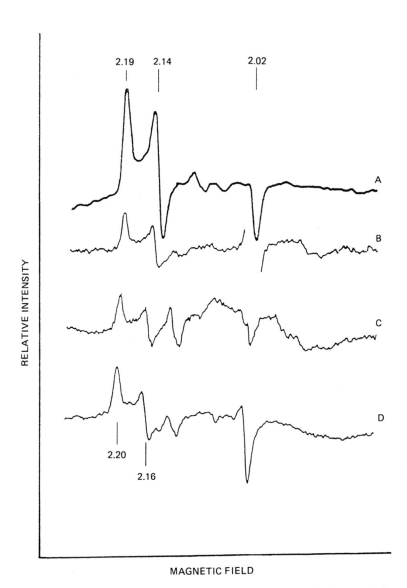

FIG. 3.   EPR spectra of hydrogen-reduced [niFe] and [NiFeSe] hydrogenases. A: *D. desulfuricans* (ATCC 27774) 85°K. B: *D. salexigens*,
25°K. C: *D. baculatus* (DSM 1743) (periplasm), 37°K. D: *D. baculatus*
(DSM 1743) (cytoplasm), 27°K. Modulation amplitude 1 mT; microwave
power 2 mW.

TABLE 3

EPR g Values of H_2-Reduced States of [NiFe]
and [NiFeSe] Hydrogenases

Organism	Ni center			[4Fe-4S] core	Ref.
	g_1	g_2	g_3		
SULFATE REDUCERS					
[NiFe] hydrogenases					
D. gigas (NCIB 9332)	2.19[b]	2.14	2.02	+(2)	[43]
D. desulfuricans (ATCC 27774)	2.19[b]	2.14	2.02	+(2)	[30]
D. multispirans n.sp.	2.19[b]	2.14	2.01	+	[31]
D. africanus[a]	2.21	2.171	2.01	n.r.[*]	[57]
[NiFeSe] hydrogenases					
D. desulfuricans (Norway 4)					
soluble	2.34	2.12	2.12	+	[32]
membrane	2.20	2.15	~2.0		[33]
D. salexigens (British Guiana)	2.22[b]	2.16	2.0	+(2)	[35]
D. baculatus (DSM 1743)					
membrane	2.20[b]	2.16	2.0	+(2)	[34]
periplasm	2.20[b]	2.16	2.0	+(2)	[34]
cytoplasm	2.20[b]	2.16	2.0	+(2)	[34]
METHANOGENS					
Mb. thermoautotrophicum ΔH	2.196	2.140	(2.0)	+	[46]
Ms. barkeri (DSM 800)	2.33	2.127	2.01		
	2.33	2.085	2.04	+	[47]

[*]Not reported.
[a]D. africanus enzyme is EPR silent in the native state [57].
[b]A fast relaxing component ("g = 2.21") type is observed below 10°K.

a nickel hydride [28,44]. The nature of the "g = 2.21" signal is still
unclear, but its relaxation behavior indicates that this signal represents a spin-spin interacting species and not a simple S = 1/2 para-
magnet. The appearance of this "g = 2.21" signal has also been interpreted as a splitting of Ni signal C ("g = 2.19") by spin-spin interaction with a [4Fe-4S][+] cluster [16]. However, the relative intensi-
ties of the Ni signal C and of the "g = 2.21" one vary with the redox

potential and may have different origins. Redox states have been
observed for [NiFe] hydrogenases where the "g = 2.21" signal is
observable at low temperature without showing the g = 2.19 counterpart [50]. The same applies for *D. baculatus* (DSM 1743) hydrogenases
[34].

EPR signals with g_{av} values around 1.94 have been observed in
reduced states of hydrogenases. Temperature and microwave power
dependence studies reveal the presence of two types of clusters in
D. gigas [28], *D. baculatus* (DSM 1743) [34], and *D. salexigens* [35]
enzymes. Mössbauer studies fully support the presence of two [4Fe-4S]
clusters in *D. gigas* [48] and *D. desulfuricans* (ATCC 27774) [30]
hydrogenases. The presence of two [4Fe-4S] clusters was also revealed
by Mössbauer studies of [57]Fe-enriched *D. desulfuricans* (Norway 4) soluble hydrogenase, but upon H_2 reduction a 50% reduced state was
attained showing a single fast-relaxing rhombic EPR signal assigned
to an iron-sulfur center with g values at 2.03 and 1.89 [32].

Although different in the oxidized (native) state, upon H_2
reduction most of the [NiFe] and [NiFeSe] hydrogenases share identical
intermediates, suggesting that a common mechanism is operative.

7. MIDPOINT REDOX POTENTIALS

In order to characterize the intermediate redox species and place it
in a catalytic framework, the redox potential value is a necessary
parameter. Redox titrations (performed under partial pressures of
hydrogen or using dithionite as chemical reductant) were conducted
with the hydrogenases isolated from *D. gigas* (periplasmic) [17,18],
D. salexigens (periplasmic) [35], and *D. baculatus* (DSM 1743) (cytoplasmic) [34]. The results are summarized as follows:

1. The isotropic signal at g = 2.02 has a pH-independent
midpoint redox potential of -70 mV [7,28].

2. Ni signal A disappears by a one-electron process at around
-220 mV at pH = 8.5 (60 mV/pH unit) [17,18,28,44]. The
interpretation of this redox process has been questioned

[28]. Although a similar value was reported for the
Ch. vinosum enzyme [15], the nickel center in the Mb.
formicicum enzyme is reduced at -400 mV [51].

3. Ni signal C shows a bell-shaped redox titration curve
   appearing at around -300 mV, attaining maximal intensity
   around -350 to -400 mV and disappearing below -450 mV, in
   a process which is also pH-dependent [3,28].

4. The "g = 2.21" signal as observed for D. gigas [28] and
   D. salexigens [35] hydrogenases appears at slightly more
   negative potentials than the Ni signal C and is still
   observable around -450 mV.

Lissolo et al. [52] showed that for D. gigas periplasmic
hydrogenase the activation step is a redox and pH-dependent process
(60 mV/pH unit, E_0' = -350 mV at pH 8.0) [52]. The enzyme is also
deactivated by another redox-linked step (E_0' = -220 mV) which is
also pH-dependent [53]. These values are closely related to the
redox transitions depicted for Ni signals A and C.

8. NICKEL SITE COORDINATION

The nickel EPR active species observed for nickel-containing hydrogenases (Ni signals A, B, and C) differ both in g values, line shape
(more or less rhombic), and nuclear hyperfine coupling constants
(Figs. 1 and 2). The alterations observed by EPR must reflect variations in the nickel environment (different coordination numbers
and/or ligations). Preliminary EXAFS data on D. gigas [24] and Mb.
thermoautotrophicum (ΔH) hydrogenases indicate that the Ni(III) coordination sphere is dominated by sulfur atoms (4-6) although N and O
atoms where not excluded at this point [24]. However, no superhyper-
fine structure due to nitrogen coordination has been observed by EPR.
Also, EPR studies on the hydrogenase isolated from cells of Wolinella
succinogenes grown on ^{33}S also support the existence of sulfur-nickel
coordination [21]. However, most quantitative conclusions must await
systematic studies of novel Ni-S and Ni-N model compounds. Additionally, a small narrowing in the g = 2.19 (0.2 mT) and g = 2.0 (0.3 mT)

regions of the Ni signal C was observed in D_2-reduced *D. gigas*
hydrogenase relative to the H_2-reduced enzyme, which may indicate
a weak proton coupling [54]. Nevertheless, the Ni site coordination
of hydrogenases is different from the one observed for other nickel-
containing enzymes. EPR spectroscopic studies on *Cl. thermoaceticum*
CO-dehydrogenase indicate that the enzyme contains a nickel-iron-
carbon complex since magnetic hyperfine broadening was induced in the
Ni EPR signal by ^{57}Fe nuclear spin and by ^{13}CO binding [55]. The
comparison of ^{57}Fe and ^{56}Fe *D. gigas* hydrogenase EPR spectra clearly
shows that there is no observable broadening due to the ^{57}Fe nucleus
in Ni signals A, B, and C, suggesting that the EPR active Ni species
are magnetically isolated from the iron-sulfur clusters [48].

9. DISCUSSION OF A MECHANISTIC FRAMEWORK FOR THE NICKEL-CONTAINING HYDROGENASES

Activation and catalytic schemes are proposed showing the involvement
of the metal centers. Since the spectral data obtained upon hydrogen
reduction are similar for [NiFe] and [NiFeSe] hydrogenases, a common
mechanism is suggested.

9.1. Activation Step

Nickel-containing hydrogenases are reversibly inactivated by oxygen.
For example, *D. gigas* hydrogenase is mainly isolated in an *inactive*
[28,56] or *unready* [16] state and the catalytic competent form of the
enzyme is only attained after a lag phase consisting of two steps:
a deoxygenation step demonstrated by the use of oxygen scavengers
such as either glucose plus glucose oxidase or *Desulfovibrio* tetraheme cytochrome c_3, and a reductive step occurring under H_2 or D_2
[56]. This was rationalized in terms of a hypothetical activation
mechanism [16,28]. Ni signal A is associated with an inactive or
unready form of the enzyme (oxygenated). Ni signal B represents a
ready state of the enzyme, in the sense that the active state of

SCHEME

ACTIVATION CYCLE

(*D. gigas* periplasmic hydrogenase)

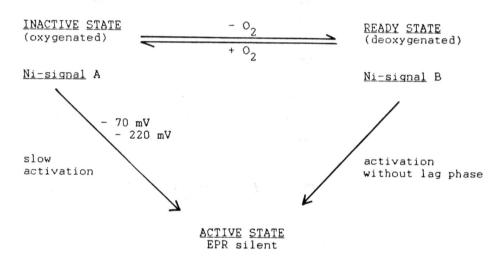

INACTIVE STATE $- O_2$ READY STATE
(oxygenated) $+ O_2$ (deoxygenated)

Ni-signal A   Ni-signal B

     $- 70$ mV
     $- 220$ mV

slow   activation
activation   without lag phase

ACTIVE STATE
EPR silent

the enzyme can rapidly be attained starting from this form [28]
(see Scheme).

The [4Fe-4S] clusters are in the 2+ state and are EPR silent
for both the unready and ready states. The $[3Fe-xS]_{oxid}$ cluster is
EPR active and exhibits an isotropic g = 2.02 signal observed at
temperatures below 30°K. The oxidation state of the EPR active
nickel is proposed to be Ni(III). In the oxygenated form (unready)
the nickel center exhibits Ni signal A. In the deoxygenated form
(ready) it exhibits Ni signal B. The amount of ready state can be
increased drastically through anaerobic reoxidation [28]. EPR and
Mössbauer studies in the enzyme "as isolated" indicate that there
is no magnetic interaction between these four redox centers [28,48].

D. *baculatus* (DSM 1743) membrane-bound enzyme shows both Ni
signals A and B (Fig. 2) and a catalytic behavior similar to that
of the enzyme from D. *gigas*. D. *salexigens*, D. *baculatus* (DSM 1743)

(soluble forms), *D. desulfuricans* (Norway 4) (soluble), and *D. africanus* [57] hydrogenases are almost EPR silent as isolated and do not require a lag phase during the activation step, as maximal activity is observed from time zero. The soluble hydrogenase iso- lated from *D. desulfuricans* (Norway 4) requires an activation step only when its activity is measured at 0°C [58]. The activation of *D. desulfuricans* (ATCC 27774) hydrogenase is faster than that of the *D. gigas* enzyme, a phenomenon possibly associated with a reductive step [59].

The active state of the *D. gigas* periplasmic enzyme is EPR silent. It can be attained either from the inactive form through a complex and slow activation process (removal of oxygen followed by a reduction step) or it can be reached directly from the ready form (without a lag phase) (cf. Scheme). During this activation process, both the isotropic g = 2.02 and the nickel signals disappear. The loss of the g = 2.02 signal is attributed to the reduction of the [3Fe-xS] cluster. Furthermore, Mössbauer studies reveal that in the reduced state (under H_2 or poised at different redox potentials) the [3Fe-xS] center remains intact and does not convert into a [4Fe-4S] center [48]. However, the oxidation state of the nickel at this state is under discussion. We proposed previously [28] that one of the [4Fe-4S] clusters is reduced into a 1+ state (S = 1/2), spin- coupled with the Ni(III) center resulting in an EPR silent state. This proposal implies that the previously determined redox potential, E_0' = -220 mV, for the disappearance of Ni signal A is actually the midpoint redox potential for one of the [4Fe-4S] clusters. This assumption enables the proposal of a mechanism whereby the nickel center operates between the Ni(III) and Ni(II) states during activity, avoiding the alternative suggestion of a chemically less plausible mechanism that requires the nickel center to cycle between the Ni(III) and the Ni(0) states (see Refs. 15 and 28, and for an extended dis- cussion of the arguments involved see Ref. 3). However, the available data do not exclude other explanations. Preliminary Mössbauer studies show that approximately one [4Fe-4S] cluster is reduced at the EPR

silent state. Such a mechanism is also supported by optical studies
which indicate that the activation process involves the reduction of
iron-sulfur clusters [56]. Also, in the case of the hydrogenase from
Chromatium vinosum an EPR silent state (active form) was attributed
to a magnetic interaction between Ni(III) and a [4Fe-4S] cluster [60].

A correlation between the EPR spectral characteristics of the
hydrogenases in the native state and their need for an activation
step can thus be established: (1) enzymes showing EPR nickel(III)
signals that do not directly correlate with catalytic competent sites
require an activation step; and (2) enzymes that are EPR silent in
the native state are generally in a catalytic active state.

9.2. Catalytic Cycle

Since a hydride intermediate state was anticipated, a proton-binding
site is required in terms of the proposed mechanistic hypothesis.
Plausible candidates are either the [4Fe-4S] cluster or a ligand
(sulfur atom) at the nickel site [3]. Ni signal C is assumed to
represent the hydride-bound nickel center. Accumulated experimental
evidence supports this assignment for the D. gigas and Chromatium
vinosum hydrogenases. The development of the "g = 2.19" EPR signal
under hydrogen is concomitant with the activation of the enzyme [28].
This EPR signal is reversibly modified by illumination with visible
light in the frozen state originating a new rhombic EPR signal
(g values at 2.28, 2.12, and 2.03) [15,23]. The rate of conversion
was found to show a kinetic isotopic effect (slower in D_2O than in
H_2O) [15]. The light conversion of the EPR signal with $g_z \sim 2$ into
another signal with $g_z > 2$ led Albracht and coworkers [15] to propose
a Ni(I) transient state. Teixeira et al. [28] suggested that the
appearance of the Ni signal C indicates the breaking of the spin
coupling between the Ni(III) and the [4Fe-4S] cluster due to the
formation of the hydride. The lack of [57]Fe hyperfine broadening on
Ni signal C is in agreement with this hypothesis.

The nature of the complex "g = 2.21" EPR signal remains controversial. The midpoint redox potential for the development of

TABLE 4

Redox Processes Observed in *D. gigas* [NiFe] Hydrogenase

Redox center	E_0' (mV)[a]	pH dependence	Ref.
[3Fe-xS][b]	- 70	No	[18]
Ni-A[b]	-220	Yes	[18]
Ni-C[b]	-300 (appearance)	Yes	[28]
	-400 (disappearance)	Yes	[28]
Activation	-350	Yes	[52]
Deactivation	-220	Yes	[53]

[a]Around pH 8.0.
[b]Measured by EPR.

Ni signal C is consistent with a catalytic active species [28,52].
It is worth noticing that the midpoint redox potential associated
with the disappearance of the Ni signal A (-220 mV) and with the
appearance of Ni signal C are pH-dependent and that of the 3Fe
cluster (-70 mV) is pH-independent in agreement with the general
scheme proposed (see Table 4).

9.3. Role of Selenium on the Heterolytic Cleavage of the Hydrogen Molecule

Isotopic exchange between D_2 and H^+ and the ortho/para hydrogen
conversion are well suited for the study of the activation of the
hydrogen molecule at zero electronic balance. The data obtained
with both methods are consistent with the heterolytic cleavage of
the hydrogen molecule [22,61]. This last mechanism requires the
presence of a metal-hydride complex and of a proton acceptor site
as previously discussed (see Secs. 9.1 and 9.2). The stabilization
of the proton by a base (external or a metal ligand) is considered
to be a necessary requirement:

$$M + H_2 + B \longrightarrow M\text{-}H^- + B\text{-}H^+$$
$$M\text{-}X + H_2 \longrightarrow M\text{-}H^- \ XH^+$$

The first product of the D_2/H^+ exchange reaction is HD, as seen in the presence of whole cells, crude extracts, and purified enzymes. This result has been used in support of the heterolytic cleavage mechanism, assuming that one of the enzyme-bound H or D atoms exchanges more rapidly with the solvent than the other. Thus, HD is the initial product, but D_2 (or H_2) is nonetheless the final product of the total exchange process, since a secondary exchange step of the HD molecule occurs.

Assuming that the hydride and proton acceptor sites can exchange independently with the solvent, the amount of HD and D_2 (or H_2) produced depends on the relative exchange rates of both sites. According to this assumption, the ratio of products should be pH-dependent [61,62]. A change in the pK_a of the proton acceptor or active site can be viewed as responsible for attaining these isotope ratios. The comparison of the experimental data on the exchange reaction measured with different purified enzymes indicates that only the [NiFeSe] hydrogenases from D. baculatus (DSM 1743) and D. salexigens (British Guiana) have H_2/HD ratios greater than 1 and that the [NiFe] hydrogenases isolated from D. gigas, D. multispirans n.sp., and D. desulfuricans (ATCC 27774) show a ratio of H_2/HD smaller than 1 (0.22-0.40) at pH 7.6 [63] (Table 1). D. baculatus (DSM 1743) and D. gigas hydrogenases show pH-dependent H_2/HD ratios [22] indicating the protonation of the proton acceptor site.

The different exchange kinetics of the hydrogen-binding sites may reflect differences in the active centers. Selenium and nickel are present in equimolar amounts in the [NiFeSe] hydrogenases, suggesting that selenium is a ligand to the nickel site.

Substitution of one of the sulfur ligands at the nickel by the less electronegative selenium ligand may serve to destabilize the hydride form of this hydrogenase. Experiments using cells of ^{77}Se-enriched D. baculatus (DSM 1743) and selenium EXAFS experiments may help to determine the involvement of selenium in the nickel-binding site.

10. CONCLUSION

Among the nickel-containing hydrogenases isolated from the sulfate-reducing and methanogenic bacteria, multiple molecular forms of hydrogenase exist which exhibit different spectral properties in the "as-isolated" state; however, under reducing conditions several common spectral features emerge which are considered to reflect a common mechanism.

The [NiFeSe] hydrogenases clearly emerge as a distinct group of enzymes in terms of catalytic and active site composition, but the degree of structural homology between the [NiFe] and [NiFeSe] hydrogenases is yet to be determined. Selenium may play a role in modulating or fine-tuning the catalytic properties through an acid-base equilibrium at the proton acceptor site or at the hydride site.

11. ABSTRACT

Iron-sulfur centers have been implicated in the simplest known redox process, i.e., reduction of protons and oxidation of molecular hydrogen mediated by hydrogenases. Recently, nickel was also firmly established to be a required cofactor in a majority of cases. The metal in the native enzyme is described as a low-spin Ni(III) d^7 distorted octahedral complex, an unusual oxidation state in the light of nickel chemistry. Also, the reaction with the natural substrate, hydrogen, implies many relevant questions concerning the nature of the intermediate species and oxidation states utilized. The spectroscopic data available for the native and intermediate states of the enzyme are discussed in terms of the following topics: types of hydrogenases, nickel oxidation states and midpoint redox potentials involved, nickel ligation mode, definition of intermediate species, interaction between redox centers and catalytic/activation mechanism.

The hydrogenases isolated from sulfate-reducing bacteria of the genus *Desulfovibrio* are good representatives of the diversity and complexity observed within this group of nickel-containing enzymes, in terms of type of structure, metal content, stability, and reactivity. The actual data obtained for *Desulfovibrio gigas* hydrogenase are reviewed and complemented with experimental observations reported for hydrogenases isolated from other sulfate-reducers and related bacterial groups, in particular from methane-forming organisms.

ACKNOWLEDGMENTS

We would like to thank the following colleagues for valuable discussions and experimental contributions: Drs. B. H. Huynh, P. A. Lespinat, Y. M. Berlier, D. V. DerVartanian, and H. D. Peck, Jr. We are indebted to Ms. I. Ribeiro and Ms. M. Martinez for carefully typing this manuscript, and to I. Carvalho for skillful technical assistance. This research work was supported by grants from Instituto Nacional de Investigação, Junta Nacional de Investigação Cientifica e Tecnologica, AID 936-5542, G-SS-4003-00 (J.M.), National Science Foundation Grant DMB 8602789, Institute of Health 1-RO 1 GM 34903 (J. LeG.), and a grant from the NSF/CNRS fellowship program (G.D.F.).

NOTE ADDED IN PROOF

A nickel(III) iron containing hydrogenase has been isolated and partly characterized from the thermophilic strain of sulfate-reducing bacterium *D. thermophilus* DSM 1276 grown at 65°C (G. D. Fauque, M. H. Czechowski, J. J. G. Moura, and J. LeGall, 87th Annual Meeting of the American Society for Microbiology, Atlanta, 1-6 March, 1987, K84).

A [NiFe] hydrogenase containing FAD has also been purified from the thermophilic methanogenic bacterium *Methanosarcina* sp. (DSM 2905) (S. B. Woo and H. D. Peck, Jr., personal communication).

A [NiFeSe] hydrogenase containing FAD has been isolated and characterized from the methanogenic bacterium *Methanococcus voltae* (S. B. Woo and H. D. Peck, Jr., personal communication).

REFERENCES

1. R. K. Thauer, G. Diekert, and P. Schönheit, *Trends Biochem. Soc.*, *11*, 304 (1980).

2. A. J. Thomson, *Nature*, *298*, 602 (1982).

3. J. J. G. Moura, M. Teixeira, I. Moura, and J. LeGall, "Bioinorganic Chemistry of Nickel" (J. R. Lancaster, Jr., ed.), VCH, Deerfield Beach, Florida, 1987 (in press).

4. M. W. W. Adams, L. E. Mortenson, and J. S. Chen, *Biochim. Biophys. Acta*, *594*, 105 (1981).

5. H. G. Schlegel and K. Schneider, in "Hydrogenases: Their Catalytic Activity, Structure and Function" (H. G. Schlegel and K. Schneider, eds.), Enrich Goltze, KG, Göttingen, 1978, pp. 15-44.

6. J. LeGall, J. J. G. Moura, H. D. Peck, Jr., and A. V. Xavier, in "Iron-Sulfur Proteins" (T. Spiro, ed.), John Wiley, New York, 1982, pp. 177-246.

7. T. Lissolo, M. F. Cocquempot, D. Thomas, and J. LeGall, *Eur. J. Appl. Microb. Biotechnol.*, *171*, 158 (1983).

8. R. Hilhorst, C. Loane, and C. Veeger, *FEBS Lett.*, *159*, 225 (1983).

9. P. Schönheit, J. Moll, and R. K. Thauer, *Arch. Microbiol.*, *123*, 105 (1979).

10. J. M. Aggag and H. G. Schlegel, *Arch. Microbiol.*, *88*, 299 (1973).

11. B. Friedrich, E. Meine, A. Finck, and C. G. Friedrich, *J. Bacteriol.*, *145*, 1144 (1981).

12. J. R. Lancaster, *FEBS Lett.*, *536*, 165 (1980).

13. J. R. Lancaster, *Science*, *216*, 1324 (1982).

14. K. Schneider, H. G. Schlegel, and K. Joachim, *Eur. J. Biochem.*, *138*, 533 (1984).

15. J. W. Van der Zwaan, S. P. J. Albracht, R. D. Fontijn, and E. C. Slater, *FEBS Lett.*, *179*, 271 (1985).

16. R. Cammack, V. M. Fernandez, and K. Schneider, *Biochimie*, *68*, 85 (1986).

17. R. Cammack, D. Patil, R. Aguirre, and E. C. Hatchikian, *FEBS Lett.*, *142*, 289 (1982).

18. M. Teixeira, I. Moura, A. V. Xavier, D. V. DerVartanian, J. LeGall, H. D. Peck, Jr., B. H. Huynh, and J. J. G. Moura, *Eur. J. Biochem., 130,* 481 (1983).

19. F. Bossu, C. K. Murray, and D. K. Margerum, *Inorg. Chem., 17,* 1630 (1978).

20. E. Münck, in "Iron-Sulfur Proteins" (T. Spiro, ed.), John Wiley, New York, 1982, pp. 147-176.

21. S. P. J. Albracht, A. Kröger, J. W. Van der Zwaan, G. Unden R. Böcher, H. Mell, and R. D. Fontijn, *Biochim. Biophys. Acta, 874,* 116 (1986).

22. P. A. Lespinat, Y. Berlier, G. Fauque, M. H. Czechowski, B. Dimon, and J. LeGall, *Biochimie, 68,* 55 (1986).

23. M. K. Johnson, I. C. Zambrano, M. H. Czechowski, H. D. Peck, Jr., D. V. DerVartanian, and J. LeGall, in "Frontiers in Bioinorganic Chemistry" (A. V. Xavier, ed.), VCH, Deerfield Beach, Fl, 1986, pp. 36-44.

24. A. Scott, S. A. Wallin, M. H. Czechowski, D. V. DerVartanian, J. LeGall, H. D. Peck, Jr., and I. Moura, *J. Am. Chem. Soc., 106,* 6864 (1984).

25. B. H. Huynh, M. H. Czechowski, H. J. Krüger, D. V. DerVartanian, H. D. Peck, Jr., and J. LeGall, *Proc. Natl. Acad. Sci. USA, 81,* 3782 (1984).

26. H. M. Van der Westen, S. G. Mayhew, and C. Veeger, *FEBS Lett., 86,* 122 (1978).

27. J. LeGall, P. O. Ljungdahl, I. Moura, H. D. Peck, Jr., A. V. Xavier, J. J. G. Moura, M. Teixeira, B. H. Huynh, and D. V. DerVartanian, *Biochem. Biophys. Res. Commun., 106,* 610 (1982).

28. M. Teixeira, I. Moura, A. V. Xavier, B. H. Huynh, D. V. DerVartanian, H. D. Peck, Jr., J. LeGall, and J. J. G. Moura, *J. Biol. Chem., 260,* 8942 (1985).

29. J. J. G. Moura, M. Teixeira, I. Moura, A. V. Xavier, and J. LeGall, in "Frontiers in Bioinorganic Chemistry" (A. V. Xavier, ed.), VCH, Deerfield Beach, FL, 1986, pp. 3-11.

30. H. J. Krüger, B. H. Huynh, P. O. Ljungdahl, A. V. Xavier, D. V. DerVartanian, I. Moura, H. D. Peck, Jr., M. Teixeira, J. J. G. Moura, and J. LeGall, *J. Biol. Chem., 257,* 14620 (1982).

31. M. H. Czechowski, S. H. He, M. Nacro, D. V. DerVartanian, H. D. Peck, Jr., and J. LeGall, *Biochem. Biophys. Res. Commun., 108,* 1388 (1984).

32. R. Rieder, R. Cammack, and D. O. Hall, *Eur. J. Biochem., 145,* 637 (1984).

33. W. V. Lalla-Maharajh, D. O. Hall, R. Cammack, K. K. Rao, and J. LeGall, *Biochem. J., 209,* 445 (1983).

34. M. Teixeira, G. Fauque, I. Moura, P. A. Lespinat, Y. Berlier, B. Prickril, H. D. Peck, Jr., A. V. Xavier, J. LeGall, and J. J. G. Moura, *Eur. J. Biochem.*, *167*, 34 (1987).

35. M. Teixeira, I. Moura, G. Fauque, M. Czechowski, Y. Berlier, P. A. Lespinat, J. LeGall, A. V. Xavier, and J. J. G. Moura, *Biochimie*, *68*, 75 (1986).

36. S. P. J. Albracht, M. L. Kalkman, and E. C. Slater, *Biochim. Biophys. Acta*, *724*, 309 (1983).

37. G. Wang, M. J. Benecky, B. H. Huynh, J. F. Cline, M. W. W. Adams, L. E. Mortenson, B. H. Hoffman, and E. Münck, *J. Biol. Chem.*, *259*, 14328 (1984).

38. F. Rusnak, M. W. W. Adams, L. E. Mortenson, and E. Münck, *J. Biol. Chem.*, *262*, 000 (1987).

39. L. A. Gow, I. P. Pankhania, S. P. Ballantine, D. H. Boxer, and W. A. Hamilton, *Biochim. Biophys. Acta*, *851*, 57 (1986).

40. T. Yagi, K. Kimura, and H. Inokuchi, *J. Biochem.*, *97*, 181 (1985).

41. T. Lissolo, E. S. Choi, J. LeGall, and H. D. Peck, Jr., *Biochem. Biophys. Res. Commun.*, *139*, 701 (1986).

42. G. R. Bell, J. LeGall, and H. D. Peck, Jr., *J. Bacteriol.*, *120*, 994 (1974).

43. J. J. G. Moura, I. Moura, B. H. Huyhn, H. J. Krüger, M. Teixeira, R. C. DuVarney, D. V. DerVartanian, A. V. Xavier, H. D. Peck, Jr., and J. LeGall, *Biochem. Biophys. Res. Commun.*, *408*, 1388 (1982).

44. J. J. G. Moura, M. Teixeira, I. Moura, A. V. Xavier, and J. LeGall, *J. Mol. Cat.*, *23*, 303 (1984).

45. S. P. J. Albracht, E. G. Graf, and R. K. Thauer, *FEBS Lett.*, *140*, 311 (1982).

46. N. Kojima, J. Fox, R. P. Hausinger, L. Daniels, W. H. Orme-Johnson, and C. Walsh, *Proc. Nat. Acad. Sci. USA*, *80*, 378 (1983).

47. G. Fauque, M. Teixeira, I. Moura, P. A. Lespinat, A. V. Xavier, D. V. DerVartanian, H. D. Peck, Jr., J. LeGall, and J. J. G. Moura, *Eur. J. Biochem.*, *142*, 21 (1984).

48. B. H. Huynh, D. S. Patil, I. Moura, M. Teixeira, J. J. G. Moura, D. V. DerVartanian, M. H. Czechowski, B. C. Prickril, H. D. Peck, Jr., and J. LeGall, *J. Biol. Chem.*, *262*, 795 (1987).

49. S. H. Bell, D. P. E. Dickson, R. Rieder, R. Cammack, D. S. Patil, D. O. Hall, and K. K. Rao, *Eur. J. Biochem.*, *145*, 645 (1984).

50. M. Teixeira, I. Moura, A. V. Xavier, J. LeGall, and J. J. G. Moura, manuscript in preparation.

51. M. W. W. Adams, S. L. C. Jin, J. S. Chen, and L. E. Mortenson, *Biochim. Biophys. Acta*, *8*, 37 (1986).

52. T. Lissolo, S. Pulvin, and D. Thomas, *J. Biol. Chem.*, *259*, 11725 (1984).

53. R. M. Mege and C. Bourdillon, *J. Biol. Chem.*, *260*, 14701 (1985).

54. D. V. DerVartanian, H. J. Krüger, H. D. Peck, Jr., and J. LeGall, *Rev. Port. Quim.*, *27*, 70 (1985).

55. S. W. Ragsdale, H. G. Wood, and W. E. Antholine, *Proc. Natl. Acad. Sci. USA*, *82*, 6811 (1985).

56. Y. M. Berlier, G. Fauque, P. A. Lespinat, and J. LeGall, *FEBS Lett.*, *140*, 185 (1982).

57. V. Niviere, N. Forget, J. P. Gayda, and E. C. Hatchikian, *Biochem. Biophys. Res. Commun.*, *139*, 658 (1986).

58. V. M. Fernandez, K. K. Rao, M. A. Fernandez, and R. Cammack, *Biochimie*, *68*, 43 (1986).

59. H. J. Krüger, Masters thesis, Univ. of Georgia, Athens, 1983.

60. S. P. J. Albracht, J. W. Van der Zwaan, and R. D. Fontijn, *Biochim. Biophys. Acta*, *766*, 245 (1984).

61. H. F. Fischer, A. I. Krasna, and D. Rittenberg, *J. Biol. Chem.*, *20*, 569 (1954).

62. T. Yagi, M. Tsuda, and H. Inokuchi, *J. Biochem.*, *73*, 1069 (1973).

63. G. D. Fauque, Y. M. Berlier, M. H. Czechowski, B. Dimon, P. A. Lespinat, and J. LeGall, *J. Indust. Microbiol.*, *2*, 15 (1987). .

64. M. Teixeira, I. Moura, A. V. Xavier, J. J. G. Moura, G. Fauque, B. Prickril, and J. LeGall, *Rev. Port. Quim.* *27*, 194 (1985).

65. M. Czechowski, G. Fauque, Y. Berlier, P. A. Lespinat, and J. LeGall, *Rev. Port. Quim.*, *27*, 196 (1985).

66. M. Czechowski, G. Fauque, N. Galliano, B. Dimon, I. Moura, J. J. G. Moura, A. V. Xavier, B. A. S. Barata, A. R. Lino, and J. LeGall, *J. Indust. Microbiol.*, *1*, 139 (1986).

8

Nickel Ion Binding to Nucleosides and Nucleotides

R. Bruce Martin
Chemistry Department
McCormick Road
University of Virginia
Charlottesville, Virginia 22903

Volume 8 in this series, entitled *Nucleotides and Derivatives: Their Ligating Ambivalency*, contains chapters on both solid structure and solution properties with 34 entries for Ni^{2+} in the index [1]. That volume is highly recommended for general background coverage not focused on a single metal ion.

In all complexes discussed herein Ni^{2+} forms only hexacoordinate (octahedral) complexes. The transition to square-planar complexes described in Chap. 5 of this volume for tri- and higher peptides and for sulfhydryl-containing amino acids does not occur with nucleic bases or phosphates. Section 1 of Chap. 5 also compares properties of Ni^{2+} with those of neighboring metal ions of the first transition row [2].

1. BINDING TO BASES AND NUCLEOSIDES

Metal ions usually bind to the commonly occurring pyrimidine bases
at the same site, N3, as in the corresponding nucleosides. Without
a sugar moiety at N9, purine bases offer an effective metal ion
binding site at N9 in addition to the N1 and N7 sites available in
the purine nucleosides. Since availability of the N9 site greatly
increases the binding ambivalency, this section deals only with
9-substituted purines [3].

Weak binding makes difficult the determination of reliable
stability constants for first transition row metal ions to nucleic
base sites of low basicity. In the purine nucleosides binding may
occur at N1 or N7. The former nitrogen is protonated in neutral
solutions of inosine (pK_a = 8.7) and guanosine (pK_a = 9.2), so that
a metal ion may coordinate to the weakly basic N7 or compete with
the proton for the more basic N1. For weakly basic adenosine with
pK_a = 3.6 for N1, in neutral solutions both the N1 and N7 sites are
free to bind metal ions [3,4].

Table 1 lists stability constant logarithms (log K) for Ni^{2+}
binding to some nitrogen heterocycles. The first three ligands

TABLE 1

Ni^{2+} Binding to Bases, Stability Constant Logarithms

	pK_a	log K
Cytidine[a]	4.34	0.95
Pyridine[b]	5.47	1.88
Inosine (N1)[c]	8.7	2.8
Inosine (N7)[c]	1.4	1.1[d]
1-Methylinosine[c]	1.4	1.0
Guanosine[c]	2.33	1.4
1-Methylimidazole[a]	7.39	3.44

[a]Ref. 6.
[b]G. Faraglia, F. J. C. Rossotti, and H. S. Rossotti, *Inorg. Chim.
Acta, 4,* 488 (1970).
[c]H. Lonnberg and P. Vihanto, *Inorg. Chim. Acta, 56,* 157 (1981).
[d]Ref. c and Ref. 36.

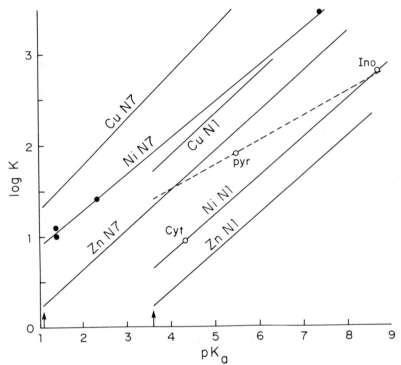

FIG. 1. Log stability constant vs. pK_a plots for nitrogen hetero-
cycle ligands with purine N7- and N1-type nitrogens and several
metal ions. Points for Ni^{2+} least squares lines are listed in
Table 1. For reference are drawn lines for Cu^{2+} and Zn^{2+}, all of
which have been established with more points than for Ni^{2+} [4,6].
The dashed line depicts the old line for Ni^{2+} binding to N1-type
nitrogens through eight substituted pyridines [4]. The new N1
line for Ni^{2+} excludes the point for pyridine in Table 1.

represent binding to a pyridine or N3 nitrogen in pyrimidine or
N1-type nitrogen in purine bases. The last four ligands represent
binding to an imidazole or N7-type nitrogen in purine bases. Plots
of log K vs. pK_a are shown in Fig. 1. Each type of nitrogen yields
its own straight least squares line, as already demonstrated for
(dien)Pd^{2+} [4,5], Cu^{2+}, and Zn^{2+} [4,6]. For comparison the least
square lines found for Cu^{2+} and Zn^{2+} binding also appear in Fig. 1.
Though for a given metal ion the slopes for the purine N1- and N7-
type nitrogens are similar, Figure 1 shows that the N7 line lies
0.9-1.3 log units higher on the log K axis.

TABLE 2

Slopes of Log Stability Constant vs. pK_a Plots

	Ni^{2+}	Cu^{2+}	Zn^{2+}	$(dien)Pd^{2+}$
Purine N1[a]	0.42[b]	0.45	0.42	0.66
Purine N7[a]	0.40	0.50	0.44	0.68
Amino acids[c]	0.35	0.55	0.47	
Iminodiacetates[d]	0.60	0.79	0.54	

[a]Ref. 6 for first three metal ions and Ref. 5 for $(dien)Pd^{2+}$.
[b]New in this chapter.
[c]Ref. 2.
[d]Ref. 7.

Table 2 tabulates the slopes corresponding to the least
squares lines in Fig. 1. For Ni^{2+}, Cu^{2+}, and Zn^{2+} the slopes for
both purine N1- and N7-type nitrogens span a narrow range from 0.40
to 0.50. Table 2 also presents slopes from least squares lines for
amino acid [2] and substituted iminodiacetate ligands [7]. The amino
acid slopes compare in value to those of the purine N1- and N7-type
nitrogens while slopes for the iminodiacetates are greater.

Stability constants reported recently for Ni^{2+} and other metal
ions with cytidine [8] and uridine [9] are often 100 times too strong,
and for Ni^{2+} the values for the two ligands differ by only 0.6 log
unit despite a 4.9 log unit difference in basicity. These titrimetric
results appear to be compromised by hydroxo complex formation and are
not considered further herein.

The straight lines of Fig. 1 permit estimates of log K_1 and
log K_7 for binding at purine N1- or N7-type nitrogens, provided a pK_a
is known. For adenosine the N1 site possesses $pK_1 = 3.6$ and for the
N7 site when the N1 site is not protonated, the intrinsic $pK_7 = 1.1$
[4,6]. These values are designated by arrows on the pK_a axis in
Fig. 1. All three N1 lines in Fig. 1 terminate at $pK_a = 3.6$, appropriate for pK_1 of adenosine. At these points the corresponding log
K_1 value for each of the three metal ions may be read off the log K
axis. Similarly, a log K_7 value may be estimated from the points

corresponding to an intersection of an N7 line with pK_a = 1.1, appropriate for proton binding to N7 when N1 is not protonated. This
value is the relevant reference for assessing metal ion division by
the N1 and N7 sites of adenosine in neutral solutions. For Ni^{2+} the
estimated stability constants are log K_7 = 0.9 and log K_1 = 0.6.
Thus this analysis indicates that in a neutral solution of adenosine
about two-thirds of the Ni^{2+} is at N7 and one-third at N1. For comparison the N1 percentage for Cu^{2+} is about 70% and for Zn^{2+} about
50% [4,6]. Therefore, all three transition metal ions bind competi-
tively to both the N1 and N7 sites of adenosine.

    The above estimates for Ni^{2+} and Cu^{2+} present problems where
experimentally determined stability constants for adenosine [10] are
less than the values estimated from reading the log K vs. pK_a plot of
Fig. 1. For Ni^{2+} experimentally determined stability constants (not
logs) for binding to adenosine range from 2.0 to 2.5 [10-12] with a
low value of 0.7 [13]. All but one [10] value were determined spec-
trophotometrically, and our attempts to determine the value by this
method resulted in an apparently greater stability constant at pH 1.5
where adenosine N1 is protonated than at pH 5 where it is not (S.-H.
Kim and R. B. Martin, unpublished experiments performed in 1982). We
conclude that the complexity of the system with two weak competitive
metal ion binding sites in experiments performed with up to a thousand
fold excess of metal ion lead to undiscovered complications in following small spectral perturbations on a strong absorption band. Possibly in neutral solutions the spectrophotometric method partly measures
binding of a second metal ion, which is present in great excess. It
is difficult to understand why Ni^{2+} binding to just the N1 site of
adenosine should be so much less than that to cytidine of comparable
basicity and no less steric hindrance. Similar discrepancies occur
with Cu^{2+} but nearly vanish with Zn^{2+} where observed [10] and graph-
ically estimated values for adenosine binding are in better accord
[4,6].

    Examination of titrimetrically determined stability constants
for Ni^{2+} and Cu^{2+} binding to a wide range of 9-methylpurines deepens
the confusion [14,15]. In making comparisons among ligands it is

essential to allow for basicity differences (Fig. 1) first, before invoking steric arguments. For both Ni^{2+} and Cu^{2+}, after allowing for basicity differences, substituents in position 6 that are $-CH_3$, $-NH_2$, or $-N(CH_3)_2$, (but not $-OCH_3$), lower the stability. As pointed out by the authors, this result provides another argument against metal ion chelation between N7 and the $-NH_2$ group at C6 in adenosine [14]. Steric hindrance by a C6 $-NH_2$ group would render suspect the above estimate of adenosine stability constants based on the Fig. 1 lines. No stability reduction appears with adenosine or AMP and $(dien)Pd^{2+}$ or even crowded $(pentamethyldien)Pd^{2+}$, where the stronger binding makes the determinations more reliable [5,16]. Cytidine, which contains two ortho substituents, falls on the N1 type binding straight lines for $(dien)Pd^{2+}$, $(pentamethyldien)Pd^{2+}$ [4,6], Cu^{2+}, and Zn^{2+} [4,6], but below the original Ni^{2+} line through eight substituted pyridines [4] shown as the dashed line in Fig. 1. Evidently the smaller and more rigorously octahedral Ni^{2+} has stricter steric requirements than the other metal ions [17]. The new line for Ni^{2+} in Fig. 1 is now drawn through the cytidine point rather than through points for eight substituted pyridines. The point for pyridine in Fig. 1 lies 0.45 log unit above the Ni^{2+} N1 line. The result of this switch is to convert Ni^{2+} to favoring the N7 from the N1-binding site in adenosine. In either case both sites must compete for Ni^{2+}.

    In conclusion we can state that adenosine binds Ni^{2+} weakly, so weakly that the stability constant is difficult to determine. Graphical stability correlation estimates suggest $\log K_1 = 0.6$ and $\log K_7 = 0.9$ while experiments yield $\log K = 0.3$ with the site undetermined. The range of these differences is a factor of 4 that for most purposes is inconsequential. For use in the next section we take an average $\log K_7 = 0.6$ or $K_7 = 4$ for the stability constant for Ni^{2+} binding at the N7 site of adenosine in neutral solutions. Selecting values at the ends of the range would not affect any conclusions.

2. BINDING TO NUCLEOTIDES

Despite the presence of phosphate groups on nucleotides some metal
ions such as Pd^{2+} and Pt^{2+} bind to the nucleic base nitrogens [16].
Like the other metal ions of the first transition row Ni^{2+} is affected
profoundly by the phosphates in nucleotides, a property accounting for
the sectional division in this chapter. All phosphates discussed
herein are 5'-phosphates. The phosphate pK_a remains nearly unaffected
by the nature of the base and from 0.1 to 0.2 ionic strength is 6.3 in
nucleoside monophosphates (NMP), 6.4 in nucleoside diphosphates (NDP),
and 6.5 in nucleoside triphosphates (NTP). There is little temperature dependence to phosphate deprotonations. The electronic spectral
properties of solid Ni^{2+} nucleotide complexes have been described [18].

Comparison of the Ni^{2+} stability constant of about 4 ($\log 4 =$
0.6) deduced at the end of the last section for Ni^{2+} binding to adenosine with the almost 2 log unit greater value for AMP [11,19], and
still greater differences with ADP and ATP stability constants in
Table 3, indicates that the presence of phosphate groups greatly
strengthens Ni^{2+} binding to adenosine nucleotides. A comparison of
Ni^{2+} stability constants with other phosphates [3,20] indicates that

TABLE 3

Ni^{2+} Binding to Nucleoside Phosphates,
Stability Constant Logarithms[a]

	NMP	NDP[b]	NTP[c]
Uridine			4.47
Cytidine	1.9[b]	3.5	4.52
Adenosine	2.5[d]	4.2	4.86
Inosine	3.0[e]		4.73

[a]At 25°C and 0.1 ionic strength unless otherwise stated.
[b]At 15°C from Ref. 20; corresponding AMP value is 2.6.
[c]Except for ITP from Ref. 21, ITP from Ref. 42.
[d]From Refs. 11 and 19.
[e]From Ref. 36.

the values for the uridine and cytidine entries in Table 3 represent binding only to the phosphates while the values for adenosine and inosine entries include additional contributions from base binding.

To what extent is a macrochelate formed in ATP by an interaction of the primarily triphosphate-bound metal ion with N7 of the adenine? There are two reactions: first, coordination of a metal ion at the triphosphate:

$$M^{2+} + ATP^{4-} \rightleftarrows (ATPM)^{2-}$$

with

$$K_p = \frac{[(ATPM)^{2-}]}{[M^{2+}][ATP^{4-}]}$$

potentially followed by isomerization to give a macrochelate:

$$(ATPM)^{2-} \rightleftarrows (A \cdot M \cdot TP)^{2-}$$

with $E = [(A \cdot M \cdot TP)^{2-}]/[(ATPM)^{2-}]$. The observed stability constant with ATP^{4-} is given by

$$K_s = \frac{[(ATPM)^{2-}] + [(A \cdot M \cdot TP)^{2-}]}{[M^{2+}][ATP^{4-}]} = K_p (1 + E)$$

The observed stability constant becomes enhanced beyond that for phosphate binding alone by the additional interaction of the metal ion at N7 of the nucleic base.

Metal ion binding to the phosphate only as represented by K_p corresponds to the equilibrium constant for binding to UTP, where no interaction with the base occurs [21]. Thus from Table 3 we have $4.86 - 4.47 = 0.39 = \log (K_s/K_p) = \log (1 + E)$ from which $E = 1.45$.

The fraction of closed or macrochelated complexes is given by

$$f = \frac{E}{1 + E} = 0.59$$

or by this calculation almost 60% of the ATP^{4-} complexes with Ni^{2+} bound primarily at the triphosphate contains a macrochelate with additional bonding of the same Ni^{2+} to N7 of the adenosine base. In about 40% of the complexes Ni^{2+} is bound only to the triphosphate group of ATP^{4-} in an open complex. (A similar treatment for ITP using the values in Table 3 indicates that almost half of the ITP·Ni^{2+} complexes are closed in a macrochelate.)

The 60% macrochelated or closed isomer for the Ni^{2+} and ATP^{4-} complex as determined by stability constant comparisons may in turn be compared with the percentage of the complexes that perturb the adenine ring as measured by ultraviolet absorption spectra. Two spectral perturbation studies yield 20% [13] and 29% [11] as the percentage of ATP^{4-} complexes in which Ni^{2+} perturbs the ultraviolet spectrum of the adenine ring. We continue with the 29% figure because the perturbant was not aqueous Ni^{2+} as in the former study but Ni^{2+}-tripolyphosphate (HOP$_3$O$_9$Ni^{2-}) analogous to the Ni^{2+} binding site in ATP^{4-}. This complex binds 11 times more strongly than aqueous Ni^{2+} to adenosine with an enhancement of binding at N7 over N1 [11]. It is the spectral perturbation due to binding at N7 that relates to the closed isomers of ATPNi^{2-} complexes. Significant spectral perturbations arise only when Ni^{2+} is bound to N7 in an inner sphere complex. Thus about 30% of the triphosphate bound Ni^{2+} in ATP^{4-} is also inner sphere-coordinated to N7 of the adenine base.

The difference between 60% closed isomers from the stability constant comparison and 30% closed isomers with inner sphere N7 coordination is assigned to outer sphere (via a water molecule) association of Ni^{2+} with the adenine base at N7 [11]. By difference the outer sphere associated closed isomers also amount to 30%. The 30% assigned to outer sphere coordination constitute isomers that must be closed but that do not perturb the adenine ring spectra.

Thus our final picture of Ni^{2+} complexing to ATP^{4-} becomes clear. Ni^{2+} is coordinated mainly at the triphosphate group. About 40% of the complexes are open with no additional involvement of the Ni^{2+}. Of the 60% of the complexes in a closed, macrochelate form, 30% contain Ni^{2+} inner sphere coordinated to N7 and 30% contain Ni^{2+} outer sphere coordinated to N7. Though Ni^{2+} represents the metal ion for which the above analysis is most sharply drawn, there is strong support for the concept from studies on other metal ions [21]. For example, for Zn^{2+} and ATP^{4-} two different methods—ultraviolet spectral perturbation and adenine base H8 chemical shift changes—

agree on 10-15% inner sphere coordination while the stability constant comparison designates 28% total closed isomers leading to
about 15% outer sphere isomers. For Mg^{2+} and ATP^{4-} ultraviolet
spectra [11] and NMR [22,23] results agree that there is no direct,
inner sphere Mg^{2+} to N7 interaction. Since ATP^{4-} enzymes use Mg^{2+},
they avoid having to cope with closed, macrochelate complexes that
occur in binary solutions with other metal ions such as Ni^{2+}.

Upon addition of imidazole to the $ATPNi^{2-}$ complex, macrochelate formation is markedly reduced or even eliminated [24,25].
This reduction further suggests that macrochelates are unlikely in
enzyme complexes.

Nucleoside triphosphate complexes with one proton also form
$NTPH \cdot Ni^{-}$. For the Ni^{2+} complexes of $UTPH^{3-}$, $CTPH^{3-}$, and $ATPH^{3-}$ the
pK_a values for proton loss are about 4.5, 4.7, and 4.5 [21]. In the
protonated complexes of UTP and ATP both the proton and the Ni^{2+}
occur on the triphosphate group with about half the complexes macrochelated through Ni^{2+} in the $ATPH^{3-}$ complex. For the Ni^{2+} complex
with $CTPH^{3-}$ there is no macrochelate, and about two-thirds of the
complexes are protonated on the triphosphate group and one-third on
the cytosine base [21].

From the differences between adenosine and cytosine values in
Table 3 for NMP^{2-} and NDP^{3-}, the above analysis leads to 80% total
closed isomers in both cases. The ultraviolet spectral perturbation
study indicates that all the closed isomers are inner sphere-coordinated at N7 in $AMP \cdot Ni$ and suggests that 66% of the complexes are
inner sphere-coordinated at N7 in $ADP \cdot Ni^{-}$ [11]. Thus for Ni^{2+} the
percentage of N7 inner sphere-coordinated isomers increases from ATP
(30%) through ADP (66%) to AMP (80%), while the N7 outer sphere-
coordinated percentages are 30%-14%-0%, and the open isomer percentages are 40%-20%-20% [11].

Crystal structures consistently show inner sphere binding of
hydrated metal ions to N7 of purine nucleoside monophosphates [26].
Examples for Ni^{2+} occur in AMP [27], GMP [28], deoxy-GMP [29], and
IMP [30,31]. Two phosphate oxygens are outer sphere-coordinated via
two Ni^{2+}-bound water molecules. In the 6-oxopurine derivatives GMP

and IMP there is no direct chelation between Ni^{2+} and O6, but a Ni^{2+}-coordinated water molecule hydrogen-bonds to O6. Chelation between a metal ion at N7 and O6 in 6-oxopurines is always indirect via a coordinated ligand on the N7-bound metal ion [3,26]. There are no crystal structures of nucleoside di- or triphosphates that show direct metal ion to N7 coordination.

The crystal structure determinations agree with the solution spectral studies that ascribe a high percentage of Ni^{2+} inner sphere-coordinated to N7 in AMP, but differ in that phosphate oxygens are only outer sphere-coordinated in the crystal while the calculation of percentage total closed forms uses inner sphere phosphate-based stability constants. To the extent that the N7 inner, phosphate outer sphere coordination in the crystal exists in solution, the percentage total closed forms calculation for AMP is invalid. The observed Ni^{2+} stability constants of 300 for AMP and 1,000 for IMP (Table 3) may be made consistent with the crystal structure determinations by multiplying the stability constants for Ni^{2+} binding to N7 of the nucleoside (4 for adenosine and 12 for inosine) by 12 for phosphate outer sphere complex formation [12], and by 6-7 for the chelate effect. For AMP we have 4 x 12 x 6.3 = 300 and for IMP 12 x 12 x 6.9 = 1,000.

Fast reaction methods have been used to investigate rate processes in Ni^{2+} complexation with CMP [32], CDP, CTP [33], adenosine [12], AMP [12,34], ADP, ATP [35], inosine, and IMP [36]. A general review of the kinetics of metal ion complexing to nucleic bases and nucleotides provides background reading on the method and data analysis [37]. No base interactions are found with any of the three cytidine nucleotides. With AMP, ADP, ATP, and IMP, there is an initial fast reaction with the phosphate group, followed by a slower reaction to form a macrochelate. In all the nucleotide cases the authors interpret the kinetics to suggest that there is inner sphere binding of both phosphate and base. This conclusion is inconsistent with the crystal structure determinations described above for Ni^{2+} and nucleoside monophosphates.

The summary scheme of binding to nucleotides by metal ions of the first transition row developed some years ago still seems capable of synthesizing most of the available information [3]. Individual metal ions may coordinate inner sphere to phosphates or to the nucleic base. In macrochelates, β and γ phosphates and N7 coordinate inner sphere to a single metal ion. Steric hindrance reduces the opportunity for both an α phosphate and N7 to be coordinated inner sphere in a macrochelate at the same time. One of the coordinate bonds is outer sphere: either N7 inner and α-phosphate outer sphere or N7 outer and α-phosphate inner sphere. These generalizations apply to NMP, NDP, and NTP.

It is easy to add a small amount of a paramagnetic metal ion to a solution containing a nucleotide and observe chemical shifts and line broadening in NMR spectroscopy. It is much more difficult to interpret correctly results of NMR experiments with paramagnetic metal ions and several pitfalls applicable to nucleotides have been described [3]. Compared to similar metal ions the electronic properties of Ni^{2+} are not advantageous for interpreting results, and of the many reported studies few use Ni^{2+}. Too often authors of NMR studies are unfamiliar with results obtained by other methods and reach untenable conclusions. For example, an often quoted multinuclear NMR investigation [38], performed at a single pH with a 10^4-fold excess of AMP over Mn^{2+}, proposes direct bonding of Mn^{2+} to the $C6-NH_2$ group, which is not a metal ion binding site [3]. In an elaborate multinuclear NMR study of Co^{2+} (similar to Ni^{2+}) binding to AMP it is suggested that in 80% of the complexes there is indirect phosphate and direct N7 coordination and in 20% of the complexes direct phosphate and indirect N7 coordination [39]. Except for the detail of the last 20% of indirect N7 coordination, these conclusions agree perfectly with that deduced from ultraviolet spectra and described above for the Ni^{2+} complex. Aside from the difference in the metal ion the NMR study was performed at -10°C at more than 150 times the AMP concentration of the ultraviolet spectral study. Under the conditions of high concentration, low temperature, and presence of a charge-neutralizing metal ion, almost all the AMP molecules will be stacked [22]. Thus, although the NMR spectra results are interpreted in terms of monomeric complexes, the observations were made on complexes with varying degrees of stacking. The observations probably refer more to intermolecular than to intramolecular interactions. It is well known that metal ions promote stack formation in nucleotides [22,40].

Sigel and collaborators provided extensive information on the ability of metal ions such as Ni^{2+} to promote the incipient stacking interaction between 2,2'-bipyridyl and the base moiety of nucleoside triphosphates [41-44]. Complexation of 2,2'-bipyridyl destroys macrochelate formation in favor of intramolecular stacking in the mixed complex. Ni^{2+} induced little stacking in a ternary complex of AMP and L-tryptophan [45].

High content of both ATP and catecholamines in chromaffin granules resulted in a popular hypothesis that a dipositive metal ion bridges the two molecules in a ternary complex [46]. This idea was never tenable because catecholamines such as dopamine and epinephrine do not form binary complexes with Mg^{2+} and Ca^{2+} in neutral solutions and complex only weakly with Ni^{2+} [2]. In the absence of metal ions, ATP and catecholamines form both 1:1 and 1:2 adducts stabilized mainly by stacking of aromatic rings. Addition of metal ions including Ni^{2+} destroys the 1:2 adduct and weakens association in the 1:1 adduct [47]. Rather than the metal ion serving as a bridge, it is the ATP phosphates that chelate the metal ion on one hand and the adenine ring that stacks with a catecholamine ring on the other.

Except in basic solutions Ni^{2+} is ineffective in promoting dephosphorylation of nucleoside triphosphates [48-50]. Compared to their high reactivity with some other substrates, metal ions are relatively unable to accelerate phosphate hydrolysis [51] and Ni^{2+} is one of the least effective ions. This ineffectiveness may be due to the slow rate of exchange of ligands out of the Ni^{2+} coordination sphere [3].

Base modified nucleotides often offer sites for metal ion binding. Because Ni^{2+} binds more strongly to the base N^6-ethenoadenosine than to adenosine, nucleoside phosphates of the former base exist in a higher percentage of closed forms than do the adenosine phosphates [52,53]. Formation of an N1-oxide in AMP [54], ADP, and IMP [55] provides an additional strong chelating site for Ni^{2+} that competes with phosphate binding [56].

Ni^{2+} binds weakly to adenylyl-3',5'-adenosine with a stability constant $K_S = 2.6$, similar to that with adenosine, but much more strongly to the polyelectrolyte polyadenylic acid with log $K_S = 3.9$ at 0.1 ionic strength [57]. Details of the binding to both ligands have been investigated by fast-reaction methods. Additional information on Ni^{2+} binding to polynucleotides appears in Chap. 9.

REFERENCES

1. H. Sigel (ed.), "Metal Ions in Biological Systems", Vol. 8, Marcel Dekker, New York, 1979.

2. R. B. Martin, *Met. Ions Biol. Syst.*, *23*, 123 (1988).

3. R. B. Martin and Y. H. Mariam, *Met. Ions Biol. Syst.*, *8*, 57 (1979).

4. R. B. Martin, *Accts. Chem. Res.*, *18*, 32 (1985).

5. S.-H. Kim and R. B. Martin, *Inorg. Chim. Acta*, *91*, 11 (1984).

6. S.-H. Kim and R. B. Martin, *Inorg. Chim. Acta*, *91*, 19 (1984).

7. R. B. Martin and H. Sigel, *Comments Inorg. Chem.*, in press.

8. P. R. Reddy and V. B. M. Rao, *Polyhedron*, *9*, 1603 (1985).

9. P. R. Reddy and V. B. M. Rao, *J. Chem. Soc. Dalton*, 2331 (1986).

10. H. Lonnberg and P. Vihanto, *Inorg. Chim. Acta*, *55*, 39 (1981).

11. Y. H. Mariam and R. B. Martin, *Inorg. Chim. Acta*, *35*, 23 (1979).

12. R. S. Taylor and H. Diebler, *Bioinorg. Chem.*, *6*, 247 (1976).

13. P. W. Schneider, H. Brintzinger, and H. Erlenmeyer, *Helv. Chim. Acta*, *47*, 992 (1964).

14. J. Arpalahti and E. Ottoila, *Inorg. Chim. Acta*, *107*, 105 (1985).

15. J. Arpalahti and H. Lonnberg, *Inorg. Chim. Acta*, *78*, 63 (1982).

16. K. H. Scheller, V. Scheller-Krattiger, and R. B. Martin, *J. Am. Chem. Soc.*, *103*, 6833 (1981).

17. L. G. Marzilli, *Adv. Inorg. Biochem.*, *3*, 47 (1981).

18. F. Walmsley and J. A. Walmsley, *J. Inorg. Nucl. Chem.*, *41*, 1711 (1979).

19. J. L. Banyasz and J. E. Stuehr, *J. Am. Chem. Soc.*, *96*, 6481 (1974).

20. C. M. Frey and J. E. Stuehr, *J. Am. Chem. Soc.*, *94*, 8898 (1972).

21. H. Sigel, R. Tribolet, R. Malini-Balakrishnan, and R. B. Martin, *Inorg. Chem.*, *26*, 2149 (1987).

22. K. H. Scheller, F. Hofstetter, P. R. Mitchell, B. Prijs, and H. Sigel, *J. Am. Chem. Soc.*, *103*, 247 (1981).

23. J. A. Happe and M. Morales, *J. Am. Chem. Soc.*, *88*, 1077 (1976).

24. R. Tribolet, R. B. Martin, and H. Sigel, *Inorg. Chem.*, *26*, 638 (1987).

25. N. Saha and H. Sigel, *J. Am. Chem. Soc.*, *104*, 4100 (1982).

26. R. W. Gellert and R. Bau, *Met. Ions Biol. Syst.*, *8*, 1 (1978).

27. A. D. Collins, P. DeMeester, D. M. L. Goodgame, and A. C. Skapski, *Biochim. Biophys. Acta*, *402*, 1 (1975).

28. P. DeMeester, D. M. L. Goodgame, A. C. Skapski, and B. T. Smith, *Biochim. Biophys. Acta*, *340*, 113 (1974).

29. R. W. Gellert, J. K. Shiba, and R. Bau, *Biochem. Biophys. Res. Commun.*, *88*, 1449 (1979).

30. K. Aoki, *Bull. Chem. Soc. Japan*, *48*, 1260 (1975).

31. G. R. Clark and J. D. Orbell, *J. Chem. Soc. Chem. Commun.*, 139 (1974).

32. J. C. Thomas, C. M. Frey, and J. E. Stuehr, *Inorg. Chem.*, *19*, 501 (1980).

33. C. M. Frey and J. E. Stuehr, *J. Am. Chem. Soc.*, *100*, 134 (1978).

34. J. C. Thomas, C. M. Frey, and J. E. Stuehr, *Inorg. Chem.*, *19*, 505 (1980).

35. C. M. Frey and J. E. Stuehr, *J. Am. Chem. Soc.*, *100*, 139 (1978).

36. A. Nagasawa and H. Diebler, *J. Phys. Chem.*, *85*, 3523 (1981).

37. C. M. Frey and J. Stuehr, *Met. Ions Biol. Syst.*, *1*, 51 (1974).

38. G. C. Levy and J. J. Dechter, *J. Am. Chem. Soc.*, *102*, 6191 (1980).

39. J. L. Leroy and M. Gueron, *J. Am. Chem. Soc.*, *108*, 5753 (1986).

40. K. H. Scheller and H. Sigel, *J. Am. Chem. Soc.*, *105*, 5891 (1983).

41. H. Sigel, *J. Inorg. Nucl. Chem.*, *39*, 1903 (1977).

42. P. Chaudhuri and H. Sigel, *J. Am. Chem. Soc.*, *99*, 3142 (1977).

43. Y. Fukuda, P. R. Mitchell, and H. Sigel, *Helv. Chim. Acta*, *61*, 638 (1978).

44. H. Sigel, B. E. Fischer, and B. Prijs, *J. Am. Chem. Soc.*, *99*, 4489 (1977).

45. J. B. Orenberg, B. E. Fischer, and H. Sigel, *J. Inorg. Nucl. Chem., 42*, 785 (1980).

46. K. S. Rajan, R. W. Colburn, and J. M. Davis, *Met. Ions Biol. Syst., 6*, 292 (1976).

47. J. Granot, *J. Am. Chem. Soc., 100*, 2886 (1978).

48. H. Sigel, F. Hofstetter, R. B. Martin, R. M. Milburn, V. Scheller-Krattiger, and K. H. Scheller, *J. Am. Chem. Soc., 106*, 7935 (1984).

49. H. Sigel and P. E. Amsler, *J. Am. Chem. Soc., 98*, 7390 (1976).

50. P. E. Amsler and H. Sigel, *Eur. J. Biochem., 63*, 569 (1976).

51. B. S. Cooperman, *Met. Ions Biol. Syst., 5*, 79 (1976).

52. H. Sigel and K. H. Scheller, *Eur. J. Biochem., 138*, 291 (1984).

53. H. Sigel, K. H. Scheller, V. Scheller-Krattiger, and B. Prijs, *J. Am. Chem. Soc., 108*, 4171 (1986).

54. H. Sigel and H. Brintzinger, *Helv. Chim. Acta, 47*, 1701 (1964).

55. H. Sigel, *Helv. Chim. Acta, 48*, 1519 (1965).

56. H. Sigel, *Met. Ions Biol. Syst., 8*, 125 (1979).

57. M. J. Hynes and H. Diebler, *Biophys. Chem., 16*, 79 (1982).

9

Interactions between Nickel and DNA: Considerations about the Role of Nickel in Carcinogenesis

E. L. Andronikashvili, V. G. Bregadze,
and J. R. Monaselidze
Institute of Physics
Academy of Sciences of the Georgian SSR
Tbilisi 77, USSR

1. INTRODUCTION

It was shown in 1970 [1] that tumors contain significant amounts of
endogenously bound ions of the first transition series. Some of
these ions are redistributed rather efficiently among subcellular
structures in the processes of proliferation and malignant growth.

The investigation of the thermodynamic properties of dilute
solutions of DNA isolated from such tumors (mice and rat sarcoma,
and hepatoma of various origins) has shown that the double helix of
tumor DNA is damaged with respect to the norm. These results were
repeatedly confirmed using different methods [2,3], and the type and
location of the damages was also established [4]. The significance
of such investigations is beyond doubt, since in recent years special
attention is paid to defects in DNA structure in studies devoted to
elucidate malignant growth mechanisms and biological cell aging.

It has been shown that the DNA melting temperature represents
a rather sensitive parameter for detecting slight changes as observed
at the malignant transformation. Based on these and related investigations [5], it was concluded that the amount of defects in the DNA
double helix may serve as a test to distinguish the genetic material
of normal cells from that of tumorous ones [6].

The presence of defects in DNA, i.e., locally unwound sites,
modifications, wrong base pairings, does not contradict the two
existing principal concepts of carcinogenesis. According to these,
tumorous transformation of cells is due to (1) the change of genetic
material as a result of mutations [7] and (2) redundant expression

of a potential oncogene of viral or cellular origin as a result of stable and nonspecific genome rearrangements due to physical, chemical, or biological factors.

In this chapter we present results supporting the "mutational" concept and make an attempt to establish a relationship between these two concepts.

1.1. DNA as a Trap for Metal Ions

Table 1 shows the metal content of calf thymus DNA and mammalian serum, as well as the accumulation indices, as we call them, representing the ratios of the amounts of atoms endogenously bound with DNA to their concentration in serum, and carcinogenicity indices taken from the work of Flessel et al. [8].

It is seen that a real DNA molecule contains no less than 10^4 metal atoms per 10^6 base pairs, i.e., 1 atom per 100 base pairs.

TABLE 1

DNA as a Trap for Metal Ions: Metal Content
of Calf Thymus DNA and Mammalian Serum

Metal	Number of metal atoms per 10^6 DNA base pairs	Metal concentration in serum (μM)	Accumulation index	Carcinogenicity index [8]
Mg	3500	1000^a	3.5	—
Ca	5000	2500^a	2	—
Cr	320	0.09^b	3560	21
Fe	710	20.3^b	30	11
Ni	230	0.044^b	5180	21
Cu	230	19.3^b	14	3
Zn	210	18.5^b	11	7

[a]Yu. I. Moskalev, "Mineral Exchange", Moscow, 'Meditsina', 1985.
[b]E. J. Underwood, "Trace Elements in Human and Animal Nutrition", Academic, 1977.

Table 1 clearly shows a greater accumulation of Cr (3,560) and especially of Ni (5,180), as compared to that of Fe, Cu, and Zn, whose
accumulation indices are between 10 and 30.

The comparative investigation of such ions as Fe(II), Ni(II),
Cu(II), and Zn(II) leads nearly always to the Irving-Williams series.
At least, Cu(II) is always more active than Ni(II). The indicated
elements reveal their activity, i.e., causing errors in the poly-
nucleotide chain synthesis [9]. It is well known that living systems
can control the concentrations of necessary substances, including
trace elements, at the required level. This is shown, for instance,
by the mechanism of copper regulation: gastrointestinal tract/liver/
ceruloplasmin/blood serum. As far as we know, there is no analog of
this mechanism in the case of Cr and Ni. Hence, the investigation
of the interaction of chromium and nickel ions with DNA is of special
interest; especially that of nickel, which is, according to some data
[10], essential as well (in fact, there is a specific protein called
nickelplasmin in blood serum).

1.2. Peculiarities of the Interaction of Divalent Metal Ions with DNA

It is well known that the principal binding sites for the metal ions
of the first transition series are negatively charged phosphate
groups, N-7 of adenine, N-7 and O-6 of guanine, O-2 of cytosine, and
O-4 of thymine. In addition, in some special cases the deprotonated
NH_2 groups of adenine and cytosine are also discussed in the literature as potential binding sites.

Divalent metal ions, M(II), may coordinate directly to the
above mentioned electron donor groups of DNA; they may be bound via
a water molecule, and they are also affected by the electrostatic
potential of the mentioned binding sites. Indeed, all these complexes will coexist to some extent in equilibrium, the equilibrium
constants being dependent on the DNA nucleotide composition, the kind
of M(II), and the ionic strength (i.e., the background electrolyte,
e.g., NaCl).

Experimental data show that hydrated M(II) ions may interact with nucleotides [11] and DNA [12] in solution by means of hydrogen bonding. This is also directly observable in nucleotide complexes in the crystalline state [13].

2. NICKEL(II) INTERACTION WITH DNA

Ni(II) and other metal ions of the first transition series when interacting with DNA in solution change the electronic properties of DNA, which leads with some probability to tautomeric base changes. This in turn causes changes in the double helix stability and leads to various defects of its structure, in particular depurination. Finally, M(II) interactions with DNA will manifest themselves by various physical and chemical methods. In particular, these interactions change the absorption and circular dichroism spectra as well as the thermostability and viscosity of DNA. Besides, as it will be shown in Sec. 5.1, tautomeric base changes in vivo due to a direct interaction of the M(II) ion with the DNA guanine base may lead to a point mutation of the transition type. The interactions of square-planar complexes capable of intercalation may be the cause of other point mutations.

2.1. Structure of the Nickel(II)-DNA Complex: Influence of Ionic Strength and Nucleotide Composition

We have investigated the change in the electronic properties occurring upon the Ni(II)-DNA interaction by observing the absorption spectra in the visible and ultraviolet regions.

Figure 1 shows that the octahedral surrounding of Ni(II) is preserved in the formation of the $Ni(H_2O)_6^{2+}$ complex with DNA. We have interpreted the slight increase in the absorption band intensity as an additional distortion of the Ni(II)-hexaqua ion by the formation of the complex with DNA. Thus $Ni(H_2O)_6^{2+}$ seems to interact

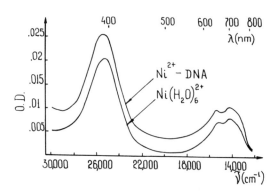

FIG. 1.   Absorption spectra of $Ni(H_2O)_6^{2+}$ and its complex with DNA from calf thymus (Serva) in the visible spectral region.   Concentrations: Ni(II)—0.8 mM; DNA—2.4 mM; NaCl—20 mM.   The spectra were recorded with the double-beam spectrophotometer "Specord M40" (Carl Zeiss, Jena) in 5-cm cells.

with DNA as a hydrated ion, i.e., it forms an outer sphere complex with DNA.

It is noteworthy that according to the degree of their influence on the DNA absorption spectrum, M(II) ions may be arranged in the so-called Irving-Williams series [14]:

Cu(II) > Ni(II) > Co(II) > Zn(II) > Mn(II)

In addition, we have also investigated UV difference absorption spectra (UDS) of DNA complexes with various Ni(II) compounds.   Figure 2 presents three of the most characteristic UV difference spectra of DNA complexes with $NiCl_2$, $Ni[(en)(H_2O)_4]^{2+}$, and $cis\text{-}Ni[(en)_2(NO_2)(NCS)]^0$, where en is ethylenediamine.

Comparison of the UDS of DNA with $NiCl_2$ and $Ni[(en)(H_2O)_4]^{2+}$ (Fig. 2) shows that substitution of ethylenediamine for two water molecules in the cis position within the xy plane decreases the difference spectrum twice.   Substitution of the remaining water molecules in the Ni(II)-hexaqua ion leads to a sharp UDS decrease.   This indicates that DNA interacts with the Ni(II)-hexaqua ion by forming a chelate in the xy plane.   The z axis of the octahedral aqua complex is parallel to the axis of the DNA double helix.

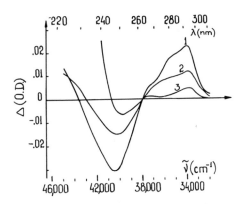

FIG. 2. UV difference spectra of the complexes of thymus DNA with
various nickel(II) compounds: $NiCl_2$ (1), $Ni[(en)(H_2O)_4]^{2+}$ (2), and
cis-$Ni[(en)_2(NO_2)(NCS)]^0$ (3). The Ni(II) concentration is in all
cases 0.25 per mol DNA. The DNA concentration is 0.2 mM in 10 mM
NaCl solution. In all cases the DNA-Ni(II) complex is compared with
DNA in the absence of Ni(II).

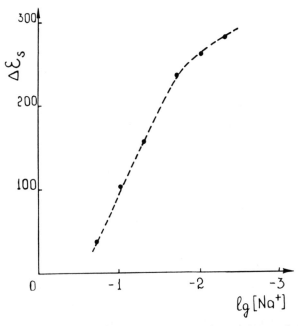

FIG. 3. UDS investigation of the DNA-Ni(II) complex. Influence of
the ionic strength. $\Delta\varepsilon_s = |\Delta\varepsilon\tilde{\nu}=\tilde{\nu}_{min}| + |\Delta\varepsilon\tilde{\nu}=\tilde{\nu}_{max}|$ in dependence on
Na^+ concentration. The Ni(II) concentration is 0.25 per mol DNA.
The DNA concentration per phosphorus is 0.2 mM.

TABLE 2

Magnitudes of the Sums $\Delta\varepsilon_s$ of Absolute Values of Molar
Absorption Coefficient Differences at $\tilde{v} = \tilde{v}_{min}$ and
$\tilde{v} = \tilde{v}_{max}$ for Ni(II) Complexes of DNA with Various
Nucleotide Compositions (% GC-pair)[a]

E. coli (51%)	Salmon sperm (44%)	Calf thymus (40%)	Clostridium perfringens (27%)
360	260	230	100

[a]$\Delta\varepsilon_s = |\Delta\varepsilon_{\tilde{v}=\tilde{v}_{min}}| + |\Delta\varepsilon_{\tilde{v}=\tilde{v}_{max}}|$. The Ni(II) concentration per DNA phosphorus equals 0.25; the NaCl concentration was 10 mM.

The strong influence of the NaCl concentration (ionic strength) on the Ni(II) interaction with DNA points to the presence of chelate complexes of the nitrogen base-phosphate group type. Figure 3 shows the dependence on NaCl concentration of the spectral changes in the DNA absorption bands due to the Ni(II) action on DNA.

From investigations of UDS of Ni(II) and DNA with various contents of GC pairs as carried out by Bregadze and coworkers follows that the amount of Ni(II) bound to DNA increases with an increasing ratio of GC pairs to AT pairs. Thus, Ni(II) preferentially interacts with guanine. The results of these measurements are given in Table 2 for DNA from *Clostridium perfringens*, calf thymus, salmon sperm, and *E. coli*.

2.2. Stability Constants for the Interaction of Nickel(II) with DNA

An excellent method for the investigation of M(II) interactions with DNA is the equilibrium dialysis through cellophane membranes coupled with pulse atomic emission spectroscopy using UHF inductively coupled plasma of reduced pressure as the light source in the optical spectral analysis of metals [15]. It should be emphasized that 3-5 µl of a solution with a metal ion concentration of about 10^{-8} M is quite

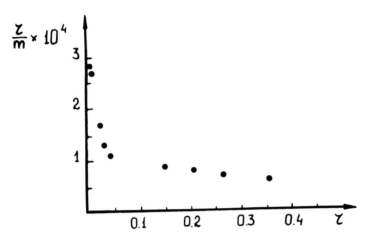

FIG. 4. Isotherm of Ni(II) adsorption on thymus DNA in the coordinates of a Scatchard plot as obtained by the joint use of equilibrium
dialysis and plasma emission analysis of nickel. Concentrations:
DNA—0.1 mM; NaCl—10 mM; the Ni(II) concentration varied from 0.001
to 0.1 mM. The Ni concentration was measured with a pulse multi-
channel spectrometer using VHF inductively coupled plasma at reduced
pressure in helium as the light source in the optical atomic emission
spectral analysis [15]. The spectrometer built at the Institute of
Physics, Academy of Sciences of the Georgian SSR, was tuned for the
Ni line λ = 341.48 nm.

sufficient for the analysis. The isotherm of the Ni(II) adsorption
on DNA in the coordinates of a Scatchard plot is shown in Fig. 4.
The isotherms of Co(II), Cu(II), and Zn(II) adsorption on DNA are of
a similar shape.

Since we have no possibility to consider such an interesting
problem as the nonlinear character of Scatchard isotherms for the
metal interaction of DNA, we shall use below the magnitude K; i.e.,
the equilibrium (stability) constant obtained by means of r/m extrapolation to r $\to$ 0, where m is the Ni(II) concentration in the free state
and r is the number of bound ions per mol DNA. Bregadze and Gelagutashvili [16] established the "thermodynamic" equilibrium constants K^0 for
DNA complexes with Co(II), Ni(II), Cu(II), and Zn(II) by measuring the
stability constants at three different values of ionic strength with
their subsequent extrapolation to an ionic strength equal to 1 Na^+ per

DNA phosphorus. pK^0 for the equilibrium of the Ni(II)-DNA (calf thymus) complex proved to be equal to 5.56 at 20°C.

2.3. Thermodynamic Properties of the Nickel(II)-DNA Complex

As early as 1973, Andronikashvili, Monaselidze, and coworkers [2] investigated the process of intramolecular melting of DNA isolated from normal tissue (healthy animal liver) and from tumors of certain transplanted sarcomas and hepatomas. In all cases they have recorded the decrease of the temperature corresponding to the melting curve maximum by 0.7°K and the broadening of the melting interval (ΔT_{melt}) by 0.5°K at the tumor growth with respect to the norm. A similar pattern is observed in the process of intramolecular melting of DNA isolated from the ascitic cells of mice (C3HA line). However, $NiCl_2$

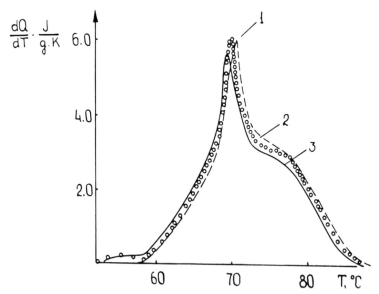

FIG. 5. Heat absorption curves of DNA solutions recalculated per gram of dry matter in 0.1 SSC; pH 7.2; C_{DNA} = 0.12%; the volume of the measuring cell was 0.2 ml. (1) DNA from mice liver (C3HA line); (2) DNA from ascitic cells; (3) DNA from ascitic cells of mice (C3HA line) injected with $NiCl_2$ (see text).

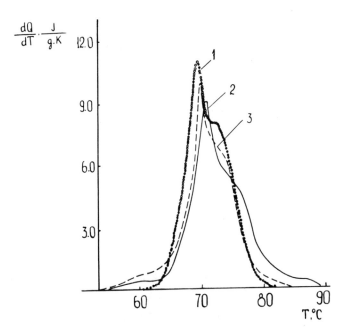

FIG. 6. Heat absorption curves of the solutions of BALB/c mice spleen DNA in 0.1 SSC; pH 7.2; C_{DNA} = 0.12%; cell volume 0.2 ml. (1) Spleen DNA; (2) DNA from leukemic mice spleen; (3) DNA from the spleen of leukemic mice injected with $NiCl_2$ (see text).

injection to the same mice leads to a decrease of T_{melt} of the ascitic DNA by 1.0°K as compared to the same DNA of mice without the injection of Ni(II) (Fig. 5). In case of the melting process of DNA from the spleen of BALB/c mice with Rauscher leukemia, the melting curve shifts toward higher temperatures as compared to the norm, which is in contrast to the transplanted tumors (Fig. 6). However, the opposite process takes place when $NiCl_2$ is injected to leukemic mice: the melting curve of spleen DNA shifts toward lower temperatures by 0.5°K as compared to sarcoma DNA.

The above results testify that Ni(II) incorporation into the DNA molecule distorts its structure, i.e., Ni(II) ions are mutagenous. This conclusion is also confirmed by the results presented in Fig. 7. The figure shows that T_{melt} of the alternating polymer poly d(A-T)poly d(A-T) does not change within the limits of τ_t = 0.005-0.01, where τ_t

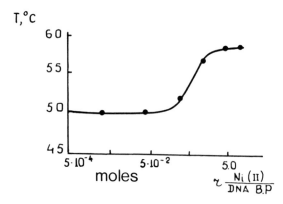

FIG. 7. The dependence of T_{melt} of poly d(A-T)poly d(A-T) on the molar ratio of Ni(II)/base.

is the molar ratio of Ni(II) ions per base. T_{melt} increases with increasing Ni(II) concentration (τ_t = 0.01-5) reaching its maximum value at 58.3°C. A further increase of Ni(II) ion concentration does not change T_{melt} of the polymer. These data show that Ni(II) interacting with DNA in dilute solutions causes over a wide range of $NiCl_2$ concentration only a stabilization of the DNA structure (see also the data of Eichhorn [17]).

The determination of the Ni(II) concentration in DNA by pulse plasma emission analysis gave values of 16.9 µg/g of DNA for ascites DNA, 17.2 µg/g of DNA for DNA isolated from the ascitic cells of C3HA mice injected with Ni(II), 12.7 µg/g of DNA for DNA from BALB/c mice spleen, and 18.0 µg/g of DNA for DNA from the spleen of leukemic BALB/c mice injected with Ni(II). These amounts of Ni(II) are approximately 3 orders of magnitude lower than the amount of Ni(II) ions capable to affect the thermostability of poly d(A-T)poly d(A-T) in solution (the polymer concentration in the calorimetric experiments was equal to 0.04%) (Fig. 6). Therefore, to explain the observed decrease of DNA thermostability by the introduction of Ni(II) ions in vivo, we have used the concept of Andronikashvili and Esipova [18] according to which divalent ions of the transition metals, including Ni(II), bind with N-7 of guanine and make its electronic structure

similar to that of adenine: $G \cdot M(II) \rightarrow A$ [19]. Hence, the amount
of AT pairs increases and consequently the intramolecular melting
temperature decreases. According to this pattern, the distortion of
the DNA structure under the influence of divalent metal ions mani-
fests itself by the appearance of noncomplementary sites, as in the
experiments of Sirover and Loeb [9], according to which the most probable disturbance of the DNA fidelity synthesis is the formation of
noncomplementary GT pairs and, to a lesser degree, of AC pairs. A
similar effect is observed, for instance, at aging [20], when the
error rates in copying the matrix increases two to four times. These
differences are even more marked in tumor DNA; the rate of noncomplementary base introduction increases about 10 times [20]. Such noncomplementary pairings may be detected using highly sensitive scanning microcalorimetry.

Figure 8 presents microcalorimetric curves of dilute solutions
of poly d(G-T) in 1.2 M NaCl and poly d(G-C) in 1.7 M $NaClO_4$. The
thermodynamic characteristics of the melting process of these polymers
are T_{melt} = 34.5°C, ΔT_{melt} = 2.9°C, ΔH_{melt} = (2.4 ± 0.4) kcal/mol pair,
and T_{melt} = 108.2°C, ΔT_{melt} = 2.0°C, ΔH_{melt} = (8.4 ± 0.4) kcal/mol
pair, respectively.

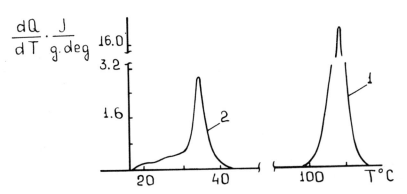

FIG. 8. Microcalorimetric melting curves of solutions of poly d(G-T)
(pH 7.0, C = 0.22%) at 1.2 M NaCl and poly d(G-C) (pH 7.0, C = 0.05%)
at 1.7 M $NaClO_4$.

Thus, T_{melt} decreases by 73.5°C and ΔH_{melt} decreases 3.5 times as compared to T_{melt} and ΔH_{melt} of poly d(G-C), respectively. This allows us to conclude that even one substitution of T for C in a DNA chain consisting of 1,000 bases during the mutation should lead to changes in thermostability which may be recorded by means of scanning microcalorimetry.

2.4. Defects in DNA Caused by Nickel(II): Depurination

In this section we investigate the defects arising at the interaction of M(II), in particular, nickel(II), with DNA in vitro in the system DNA-water-metal ion.

The increase in the tautomerization constant of nitrogen DNA bases due to the interaction with M(II) ions explains not only the change in DNA thermostability, but also their involvement in the adenine and guanine depurination of DNA. The stronger the action is upon N-7, the higher should be the depurination percentage; this has been confirmed by the experiments of Loeb and Tkeshelashvili [21], who have kindly given us their results, some of which are presented in Table 3. It is evident that the depurination of adenine proceeds

TABLE 3

DNA Depurination Caused by M(II) Ions [21]

Metal	[8-^{3}H]adenine DNA % depuration	[8-^{3}H]guanine DNA % depuration
Without metal	<0.01	<0.02
$CuCl_2$	7.4	0.10
$NiCl_2$	0.42	0.05
$CaCl_2$	<0.01	0.10
$MnCl_2$	0.14	<0.02
$MgCl_2$	<0.01	<0.02
$ZnCl_2$	<0.20	<0.20

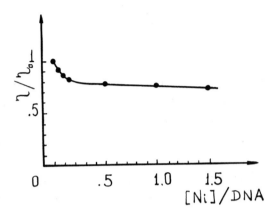

FIG. 9. Dependence of the viscosity of calf thymus DNA solution on $NiCl_2$ concentration. The measurements were carried out using a micro-viscosimeter with a magnetic suspension of the rotor as built at the Institute of Physics, Academy of Sciences of the Georgian SSR [22]. The solution contained 0.01 M NaCl; the P-DNA concentration was $2 \cdot 10^{-4}$ M. The experiments were carried out at 30°C. $[NiCl_2]$ varied from 10^{-5} M to $8 \cdot 10^{-3}$ M. η_0 = specific viscosity of the DNA solution at $[Ni^{2+}] = 0$; η = the specific viscosity of the DNA solution at various $[Ni^{2+}]/[phosphate]$ ratios. The increase of nickel concentration up to $[Ni^{2+}]/[P] = 40$ does not cause any significant viscosity changes.

much easier than that of guanine, and that the adenine depurination degree under the action of nickel is at least 17 times lower than in the case of copper. According to the degree of their action upon DNA regarding adenine depurination, the M(II) ions of Table 3 may be arranged in the following order:

Cu(II) > Ni(II) > Zn(II) > Mn(II) > Mg(II)

Viscometric investigations carried out by Kiziria and Gamtsem-lidze on nickel(II)-DNA solutions (Fig. 9) indicate the ability of Ni(II) to induce defects in the DNA structure. It is evident that the role of nickel(II) cannot be reduced to a simple increase in the ionic strength of the solution; it is clear that a stronger inter-action of nickel(II) with DNA takes place, leading to an increase of its flexibility. These changes are so pronounced that they may be explained by the appearance of significant local disturbances of the DNA double helix, such as depurination or one-strain breakages.

3.   NICKEL AND NUCLEOPROTEINS

3.1. Model Investigations

Nickel(II) significantly affects the binding character and the structure of the DNA complexes formed with the antibiotics distamycin and
netropsin, which represent good models of protein molecules due to
the presence of peptide groups in their composition and which are
widely used in investigations of DNA-protein interactions.

Kharatishvili and Esipova have shown by circular dichroism,
that nickel(II) ions, as well as cobalt(II) and zinc(II) ions, change
the structure of DNA complexed with antibiotics much more than the

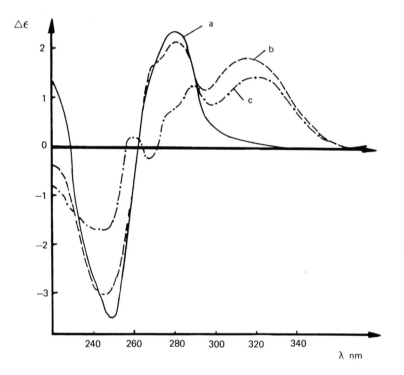

FIG. 10.   CD spectra of calf thymus DNA (a); DNA complexes with DM-2
before (b) and after (c) the addition of Ni(II).   Concentrations:
[DNA] = $5 \cdot 10^{-6}$ M P/2, bound [DM-2] = $3.1 \cdot 10^{-7}$ M, [Ni^{2+}] = $2.4 \cdot 10^{-5}$ M;
pH 5.9; μ = 10^{-3} M.

structure of free DNA. The effect of M(II) ions and an antibiotic (represented in this case by an analog of distamycin containing two pyrrolcarboxamide cycles and three peptide groups) on the structure of DNA is not simply the sum of their separate actions; here synergistic effects are occurring; this is illustrated in Fig. 10.

In this figure the CD spectra of DNA (a), the DNA-distamycin-2 complex (b), and the DNA-distamycin-2-nickel(II) complex (c) are presented. The authors interpret spectrum (c) as a result of superhelix nuclei formation. A large increase of this effect under the action of nickel(II) is (probably) connected with the defects in DNA caused by the metal ion [the nickel(II) concentration per P/2 DNA is ~5].

3.2. Nickel(II) Interaction with DNA and Mononucleosomes of Normal and Tumor Cells. Stability Constants

We have investigated in collaboration with Gelagutashvili and Sapozhnikova the stability constants of nickel(II) complexes with DNA and mononucleosomes isolated from hepatoma 22a cell nuclei and from C3HA line mice liver.

These stability constants of Ni(II) complexes with ascitic and normal DNA and mononucleosomes have been determined by the joint use of equilibrium dialysis and plasma emission analysis as described in Sec. 2.2. The stability constants for the mentioned complexes are given in Table 4. A difference between the stability constants of

TABLE 4

Stability Constants of Ni(II) with DNA and Mononucleosomes of Mice Liver (C3HA line) and of the Ascitic Hepatoma 22a Cells (expressed in mol^{-1})

DNA		Mononucleosome	
Liver	Hepatoma 22a	Liver	Hepatoma 22a
$4.40 \cdot 10^4$	$4.75 \cdot 10^4$	$2.13 \cdot 10^4$	$2.15 \cdot 10^4$

TABLE 5

Metal Content[a] in Soluble Chromatin of Mice Liver
(C3HA line) and in Ascitic Hepatoma 22a Cells,
Expressed as Number of Metal Atoms per
Million of DNA Base Pairs

Element:	Mg	Ca	Cr	Fe	Ni	Cu	Zn	Cd	Pb
Liver	2060	3020	420	580	400	150	290	18	66
Hepatoma 22a	6820	13400	1820	2280	560	480	950	38	260

[a]The precision of the Cd and Pb determinations is estimated as 20%,
that of the other elements as 10%.

ascitic and normal DNA, determined in a simultaneous experiment, is
indicated. The stability constants of the nickel(II) complexes with
DNA and mononucleosome are significantly different.

3.3. Some Metal Ions in Normal and Cancerous Chromatin

We have established in collaboration with Belokobyl'ski and Gelagutashvili the contents of Mg, Ca, Cr, Fe, Ni, Cu, Zn, Cd, and Pb in
the so-called soluble chromatin (chromatin treated with micrococcal
nuclease and a subsequent clarification by centrifugation) isolated
from hepatoma 22a cell nuclei and mice liver (C3HA line). The results
are presented in Table 5. It is evident that the nickel concentration
in chromatin (as well as that of other metals) increases with malignant growth, though not quite as significantly as in case of other
elements.

4. IN VIVO INFLUENCE OF NICKEL(II) ON THE THERMAL PROPERTIES OF CHROMATIN INSIDE THE NUCLEI, AND ON CELLS AND TISSUES OF NORMAL AND TUMOROUS ORIGIN

Recently, the leading role of the 3d transition metal ions on the
conformational rearrangements of the genetic material at the sub-

cellular level, which plays the principal role in the regulation of
genome activity, has become increasingly evident. Such changes in
the genetic material may include rearrangements at the highest levels
of DNA organization in the eukaryotic chromatin:

1. Solenoidal (condensation-decondensation of fibers dependent
    on Mg(II) concentration [23])
2. Domain (Zn(II), Mn(II), Cu(II) chelation from supertwisted
    loops linked with the nuclear matrix [24]).

In this connection, the investigation of the conformational
rearrangements of the genetic material in chromatin using the in vivo
introduction of carcinogenous metal ions seems rather important and
promising. The interest in these investigations is growing, since
the highly sensitive method of differential scanning microcalorimetry
devised at the Institute of Physics, Academy of Sciences of the
Georgian SSR, makes it possible to observe the conformational changes
in chromatin, proteins, and the DNP complex in the composition of
cells, tissues, and organs.

Figure 11 shows microcalorimetric heat absorption curves of
hepatoma 22a ascitic cells. Curve 1 corresponds to the denaturation
of cells isolated from mice (C3HA line) on the sixth day, while curve
2 shows the denaturation of cells which were isolated on the sixth day
as well, but in this case the mice were 24 hr before intraperitoneally

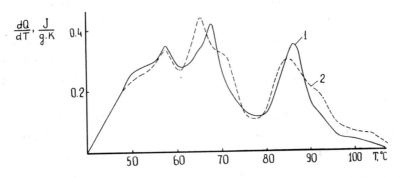

FIG. 11. Heat absorption curves recalculated per gram of dry matter:
(1) ascitic cell suspension; (2) suspension of the ascitic cells of
mice (C3HA line) injected with $NiCl_2$ (see text).

injected with 0.3 ml of a phosphate-buffered solution (pH 7.0) containing 0.65 mg of $NiCl_2$.

Comparison of the profiles of the heat absorption curve shows
that considerable differences are observed in the temperature intervals 60-70°C and 75-110°C. However, we shall consider here only the
higher temperature interval reflecting the chromatin denaturation.

Evidently, the peak of the heat absorption corresponding to
the chromatin denaturation in the temperature interval 75-110°C
(curve 2 in Fig. 11) is shifted toward lower temperatures by 1.5°C,
and the half-width of its temperature interval (ΔT_d) increases by
4.0°C. Evidently, such changes in the chromatin denaturation parameters indicate directly that Ni(II) ions damage the DNA in the
chromatin composition.

The differences in the profiles of the heat absorption curves
are also observed when the denaturation processes of spleen tissues
of healthy BALB/c mice and of those with Rauscher leukemia are compared (see Fig. 12). However, significant differences are only
observed between the sixth and tenth day after the intraperitoneal
injection of 0.4 ml of blood plasma taken from BALB/c mice with
Rauscher leukemia on the 24th day. We shall not discuss the problem
of heat redistribution between the peaks with maxima at 73 and 57°C
(Fig. 12; curves 2, 3, 4); we shall discuss here only the changes
concerning the chromatin complex (temperature interval 75-110°C).
The presented data show that the thermodynamic parameters (T_d, ΔT_d,
Q_d) and peak profiles corresponding to the normal and leukemic
chromatin denaturation are identical from the 10th to the 40th day.
The differences in the denaturation processes are observed only in
case of a comparison of the heat absorption curves of spleen tissues
of BALB/c mice injected with Ni(II) ions and without (Fig. 12; curves
5, 6). One can see that the difference in chromatin T_d is 0.8-1.0°C.

It was also of interest to study the structural changes of
chromatin after the in vivo administration of Ni_3S_2, which is one of
the well-known mutagens. The investigation of the thermal properties
of ascitic cells isolated from C3HA line mice has not revealed any
substantial changes in the structural transitions of proteins and

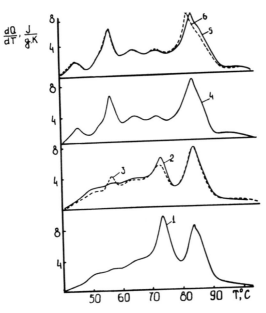

FIG. 12. Heat absorption curves of spleen tissue isolated from BALB/c mice on various days after the animal was injected with Rauscher virus: (1) healthy mouse spleen tissue; (2) on the 4th day after the injection; (3) on the 6th day; (4) on the 18th day; (5) on the 27th day; (6) on the 27th day, 24 hr before the spleen isolation the mice were injected with $NiCl_2$ (see text).

nucleoprotein complexes. This is not surprising, since the life duration of the ascitic mice did not exceed 8-10 days after Ni_3S_2 injection. Consequently, Ni_3S_2 does not dissolve in such quantities [25] as to influence the thermal characteristics of cells. However, if Ni_3S_2 is injected intraperitoneally to healthy rats and the thermal properties of the spleen are investigated on the 20th to 25th day, drastic changes are observed. In particular, if the process of the intact chromatin denaturation is characterized by an intensive heat absorption peak at 81.2°C and a shoulder at about 86°C, then two pronounced peaks with maxima at about 79 and 85°C are observed in the chromatin of rats injected with Ni_3S_2. Thus, we observe substantial changes in the chromatin structure at Ni_3S_2 in vivo.

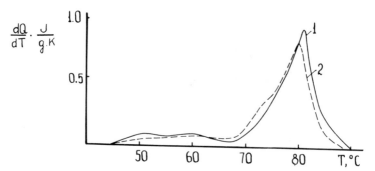

FIG. 13. Heat absorption curves of nuclei of BALB/c mice spleen cells recalculated per gram of dry matter (pH 7.8). Buffering solution: saccharose 250 mM, tris-HCl 10 mM NaCl, $MgCl_2$ 1 mM. (1) healthy spleen nuclei; (2) leukemic mice spleen nuclei.

The influence of Ni(II) ions on the chromatin structure is revealed at the nuclei level as well (Fig. 13). The shift of the heat absorption curve of the suspension of nuclei of mice with Rauscher leukemia toward lower temperatures is quite marked and equals $0.8°C$.

5. NICKEL(II) INCORPORATION IN DNA AND THE ROLE OF NICKEL IN CARCINOGENESIS

5.1. M(II) Binding by DNA-Guanine and G·M(II) → A Transition

It is well known that the necessary condition for the cell transformation is the mutation of bases in DNA, e.g., point mutation, as in case of the ras oncogene. However, it is also necessary for the fixation of a mutation that the repair system be incapable of restoring the primary DNA structure. On the other hand, metal ions in vivo and in particular Ni(II) can cause a disturbance of the electronic structures of the nitrogen bases only if the metal ion interaction with DNA leads to the formation of strong complexes. Under the in vivo conditions of DNA existence and functioning (0.15 M of monovalent cation), a strong complex with DNA may be formed by the interaction of a soft metal ion, i.e., a metal ion preferring covalent binding to DNA-guanine bases. This is well illustrated by the example of Cu(II) → Cu(I) reduction in copper(II) complexes of DNA.

As we know, Cu(II) interacts with DNA preferentially at N-7 of guanine and the phosphate group of the same nucleotide using at least one H_2O molecule [12]. At Cu(II) → Cu(I) reduction by, e.g., ascorbic acid [26], UV [14] or γ irradiation [26], instead of the mentioned complex we obtain the chelate N-7/O-6* (soft interaction). This then leads to the fact that, as Ivanov and Minchenkova showed [19], the guanine electronic structure (copper being preferentially bound to guanine) becomes similar to that of adenine, which should finally lead to a point mutation, i.e., substitution of G·M(II) as shown by Andronikashvili and Esipova [18].

5.2. Ni_3S_2 Solubilization in DNA by the Presence of Microsomal Proteins

In coordination chemistry the phenomenon of symbiosis is well known; it consists of the fact that an electron donor group with high polarizability coordinating to a M(II) ion of the first transition series makes this ion softer. For instance, Ni in Ni_3S_2 may be solubilized by soft imidazole groups of histidine residues in a protein [27], forming then a strong complex with DNA [25]. Clearly, protein molecules may further also play the role of "catalysts" in Ni(II) binding to DNA, especially those proteins which are characterized by a high affinity for DNA.

The matters mentioned above are confirmed by Lee et al. [25], who have observed a high Ni(II) concentration (1-10 mM) in 5 mM tris-HCl solution at pH 7.4 containing the water-insoluble subsulfide Ni_3S_2, DNA, and microsomal proteins, i.e., these authors have managed to solubilize Ni(II) using soft bases. The latter is confirmed by the absorption spectra over the interval from 350 to 700 nm (d-d

---

Editors' note: Regarding this type of chelate formation see also the comments in Chap. 8.

transitions) as recorded by the authors of [25]. (They show a short wavelength shift and a considerable intensity change of the absorption spectrum compared to the $Ni(H_2O)_6^{2+}$ spectrum).

Costa and Mollenhauer showed in 1980 [28] that the subsulfide Ni_3S_2 in the crystalline, and not in the amorphous form (suspension of 1- to 2-μm particles) is actively phagocytized by hamster culture cells (CHO line), increasing their morphologic transformation rate. In another study of Sen and Costa [29], the transformations of the cells of the same line were induced not only by Ni_3S_2, but also by the monosulfide NiS, as well as by $NiCl_2$. In all these cases the authors [29] conclude that Ni(II) ions are responsible for the cell transformation inducing, in their opinion, the breakages of DNA chains. Although the latter are rapidly repaired, the protein extraction ability of DNA, for nonhistonic, chromosomal, and DNA-binding proteins, decreases. This indicates that Ni(II) is a DNA-protein crosslink. This is in good agreement with the above mentioned role of protein molecules as catalysts in M(II) binding with DNA.

6. CONCLUSION

According to the results presented in this chapter, the key mechanism of the transformation of tumorous cells is the DNA damage as a result of a mutation. The mutation arises due to the fact that M(II) binding, for instance, with one of the guanines in a DNA chain changes its electronic structure so that it becomes similar to the electronic structure of adenine. As a result, the altered guanine is recognized in the process of replication as adenine, which gives rise to a site of noncomplementary guanine pairing with thymine leading to the formation of an AT pair instead of a GC pair in the DNA double helix in the subsequent generations already in the first replication cycle.

In the proposed scheme, the fixation of the mutation in the genome is of primary importance since it determines the degree of malignancy. In fact, if a single mutation is not fixed, then the malignancy will be defined by the population of the daughter cell

containing an AT pair instead of a GC pair. However, if the mutation
is fixed, i.e., M(II) remains bound with guanine, then both daughter
cells will contain DNA with altered sequences and the malignancy will
manifest itself to a higher degree.

The fixation of the mutation may occur as a result of the formation of a strong protein-M(II)-DNA complex. For instance, in case of
Ni_3S_2 Ni(II) is solubilized by the microsomal proteins [25], and it
is the Ni(II)-protein which forms a strong complex with DNA.

Taking into account the possible transitions, e.g., G·M(II) → A,
the malignant growth caused by chemical substances (including metals)
may be approximated to the malignant growth caused by oncogenous
viruses. Let us give two examples:

1. Suppose, a divalent ion of the first transition series has
been bound to one of the guanines in the regulatory gene. Then, due
to the arisen transition, an amino acid substitution arises in the
repressor protein, and this protein will not be recognized by the
operator, and the respective structural gene will become uncontrol-
lable, which leads to its higher expression.

2. On the other hand, it is well known that in the twelfth
triplet of the cellular gene c-ras, a point mutation GGC → GTC may
arise which represents the sign of the oncogene of human bladder
carcinoma of chemical origin. However, the same triplet GGC of c-ras
gene may mutate in another way: GGC → AGC, and this mutation is the
sign of the oncogene of the Kirsten sarcoma of mice (v-ki-ras) of
viral origin. It is sufficient for the point mutation GGC → AGC that
a divalent ion of the first transition series would bind with the
first guanine of this triplet making the electronic structure of
guanine similar to that of adenine.

In our opinion, these examples are quite sufficient to confirm
the absence of a sharp boundary between the malignant growth caused
by a chemical substance on the one hand and by a virus on the other.

ACKNOWLEDGMENT

The authors would like to express deep gratitude to Dr. L. Tkeshelashvili, Dr. A. Belokobyl'ski, Dr. E. Kiziria, Dr. G. Majagaladze, Prof. A. Shvelashvili, Miss Z. Chanchalashvili, Miss E. Gelagutashvili, Miss N. Sapozhnikova, Miss E. Lomidze, Mr. M. Kharatishvili, Miss G. Chitadze, and Miss I. Khutsishvili for their assistance in this work.

REFERENCES

1. E. L. Andronikashvili, L. M. Mosulishvili, V. P. Manjgaladze, A. I. Belokobyl'ski, N. E. Kharabadze, and E. Yu. Efremova, *Dokl. Akad. Nauk SSSR, 195,* 979 (1970).

2. E. L. Andronikashvili, Z. I. Chanchalashvili, G. V. Majagaladze, G. M. Mrevlishvili, and J. R. Monaselidze, in "Conformational Changes of Biopolymers in Solutions", Proc. 2nd All-Union Conf. in Tbilisi in 1973 (E. L. Andronikashvili, ed.), Metsniereba, Tbilisi, 1975, p. 161 ff.

3. E. L. Andronikashvili, J. R. Monaselidze, Z. I. Chanchalashvili, E. M. Lomidze, and E. M. Lukanidin, *Biofizika, 31,* 256 (1986).

4. T. P. Zhizhina, S. I. Bovochich, A. V. Alesenko, O. E. Petrov, and K. E. Kruglyakova, *Dokl. Akad. Nauk SSSR, 222,* 973 (1975).

5. A. V. Shugalii, *Experimental'naya Onkologiya, 5,* 31 (1983).

6. E. L. Andronikashvili, J. R. Monaselidze, Z. I. Chanchalashvili, and G. V. Majagaladze, *Biofizika, 28,* 528 (1983).

7. T. P. Zhizhina, *Dokl. Akad. Nauk SSSR, 265,* 1268 (1982).

8. C. P. Flessel, A. Furst, and S. B. Radding, in "Metal Ions in Biological Systems", Vol. 10 (H. Sigel, ed.), Marcel Dekker, New York, 1980, p. 23 ff.

9. M. A. Sirover and L. A. Loeb, *Science, 194,* 1434 (1976).

10. F. H. Nielsen, in "Biochemistry of the Essential Ultratrace Elements", Plenum, New York, 1984, p. 293 ff.

11. G. P. P. Kuntz, T. A. Glassman, C. Cooper, and T. J. Swift, *Biochemistry, 11,* 538 (1972).

12. (a) V. G. Bregadze, *Biofizika, 19,* 179 (1974). (b) V. G. Bregadze and E. Yu. Efremova, in "Conformational Changes of Biopolymers in Solutions", Metsniereba, Tbilisi, 1975, p. 96 ff.

13. R. Gellert and R. Bau, in "Metal Ions in Biological Systems", Vol. 8 (H. Sigel, ed.), (a) Marcel Dekker, New York, 1979, p. 1 ff.; (b) "Mir", Moscow, 1982, p. 9 ff.

14. V. G. Bregadze, *J. Quantum Chem.*, *17*, 1213 (1980).

15. V. G. Bregadze, in "New Physical Methods in Biological Investigations", Nauka, Moscow, 1987, p. 33 ff.

16. E. S. Gelagutashvili and V. G. Bregadze, *Bull. Acad. Sci. Georgian SSSR*, 1987, in press.

17. G. L. Eichhorn and Y. A. Shin, *J. Am. Chem. Soc.*, *90*, 7323 (1968).

18. E. L. Andronikashvili and N. G. Esipova, *J. Mol. Catalysis*, *23*, 195 (1984).

19. V. I. Ivanov and L. E. Minchenkova, *Biopolymers*, *5*, 615 (1967).

20. V. Murray and R. Holliday, *J. Mol. Biol.*, *146*, 55 (1981).

21. L. K. Tkeshelashvili and L. A. Loeb, "Gordon Conference on Mutagenesis", Plymouth, New Hampshire, 1986.

22. E. L. Kiziria, in "New Physical Methods in Biological Investigations", Nauka, Moscow, 1987, p. 204 ff.

23. J. T. Finch and A. Klug, *Proc. Natl. Acad. Sci. USA*, *73*, 1897 (1976).

24. J. S. Lebkowski and U. K. Laemmli, *J. Mol. Biol.*, *156*, 309 (1982).

25. J. E. Lee, R. B. Ciccarelli, and K. W. Jennette, *Biochemistry*, *21*, 771 (1982).

26. (a) V. G. Bregadze and M. G. Kharatishvili, *Biofizika*, *25*, 615 (1980). (b) V. G. Bregadze, E. S. Gelagutashvili, and M. G. Kharatishvili, *Studia Biophysica* (Berlin), *67*, 25 and Microfiche 2/17-26 (1978).

27. F. W. Sunderman, Jr., *Prev. Med.*, *5*, 279 (1976).

28. M. Costa and H. H. Mollenhauer, *Cancer Res.*, *40*, 2688 (1980).

29. P. Sen and M. Costa, *Cancer Res.*, *45*, 2320 (1985).

10

Toxicology of Nickel Compounds

Evert Nieboer, Franco E. Rossetto,
and C. Rajeshwari Menon
Department of Biochemistry and Occupational Health Program
Health Sciences Centre, McMaster University
1200 Main Street West
Hamilton, Ontario L8N 3Z5 Canada

1. INTRODUCTION

Nickel toxicology dates back to the acute animal toxicity studies conducted by Stuart in 1880-1883 at the Pharmacology Institute in Strassbourg, Germany (now in France) [1-3]. The systematic study of the toxicity of nickel compounds began to gain momentum in the 1960s and early 1970s, and since then has blossomed into a vibrant research area. This may be judged from the large number of related research papers, reviews, and monographs that have been published during the past decade [3-12], as well as from the frequency of international conferences highlighting and promoting nickel toxicology [2]. The catalysts for this development were the discovery in 1939 of excess lung and nasal cancers among workers of the Mond Nickel Company, Clydach, Wales, U.K. (first reported in published form in 1958) [13, 14], and the frequent reports of human poisoning and death after inhalation of nickel tetracarbonyl [3].

The objective of this review is to present a concise overview of the known toxic effects of nickel compounds in animals and humans.

As in Chap. 4, human data are highlighted. Animal and cell culture
studies will serve a supporting role, especially in discussions
of toxicologic mechanisms. At the outset, the types of exposure
encountered in the human experience are examined. Nickel carbonyl
poisoning and its treatment by chelation therapy is then reviewed,
followed by a consideration of the clinical, cellular, and molecular
aspects of the occurrence and pathogenesis of nickel hypersensitivity
expressed as nickel dermatitis and occupational asthma. A summary of
the epidemiologic evidence for nickel-related nasal and lung cancers
is subsequently presented, and the genotoxicity of nickel compounds
is discussed in the context of mechanisms of nickel carcinogenesis.
To complete the review, the evidence for a number of less prominent
toxic effects are assessed, including: renal toxicity, reproductive
and developmental deficiencies, immunotoxicity, cardiotoxicity, and
nickel-related iatrogenic hazards associated with certain medical
treatments.

2. HUMAN EXPOSURES

2.1. Nonoccupational Exposures

Even though ambient nickel concentrations are low (e.g., 0.008 $\mu g/m^3$
was the 1982 US average [11]), they appear to account for the observed
lung tissue nickel contents at necropsy (see Chap. 4). Based on an
inhalation rate of 20 m^3/day [11], 50% deposition and a lung tissue
wet/dry weight ratio of 10 [15], 30 mg/kg dry weight of nickel is
predicted for a 70-year-old compared to an average value of 200 $\mu g/kg$
observed [15]. Perhaps this comparison can be interpreted to indicate
that about 1% of inhaled nickel is not cleared nor absorbed and is
retained on a long-term basis. Speciation assignments of nickel in
ambient air support this perspective. The major contribution to
ambient nickel originates from the combustion of fossil fuels and the
predominate forms in air are nickel sulfate (water-soluble), nickel
oxide, and complex metal oxides containing nickel [11]. As summarized

in Chap. 4 (Sec. 2.1), animal studies have shown that green nickel(II) oxide is relatively inert in the lung in contrast to water-soluble nickel salts which are rapidly cleared by absorption.

Nickel concentrations in drinking water are low (estimated US municipal average was estimated as 5 µg/liter in 1964) [3,11]. Since this assessment predated the concern about rigorous control of nickel contamination during sample collection, storage, and handling (see Chap. 4, Sec. 3; [16]), it likely constitutes an overestimate. Utilizing appropriate contamination control, McNeeley et al. in 1972 reported a mean value of 1.1 ± 0.3 µg/liter (range 0.8-1.5 µg/liter) for tap water in Hartford, Connecticut [17]. Thus, under normal circumstances, drinking water should make only a minor contribution to the daily dietary intake of 160 µg Ni/day. As already indicated (Chap. 4), certain foods are relatively rich in nickel, such as oatmeal, dried legumes, and chocolate products.

Nickel leaches from some nickel alloy and stainless steel objects and thus inadvertent exposure sources include jewelry, coins, cooking utensils, cutlery, and surgical implants. Contact through other household items such as detergents and cosmetics is also known (see Sec. 4).

2.2. Occupational Exposures

Occupational exposure to nickel compounds is strongly dependent on the industrial processes considered [18-20]. In electroplating shops and electrorefining plants, workers are exposed to aerosols of dissolved nickel salts. Ambient nickel concentrations of up to 0.2 mg/m^3 have been reported [19-22]. Exposures associated with pyrometallurgical refining processes are to nickel oxides and sulfides, as well as to nickel powder as the final product of secondary refining. Ambient concentrations of nickel are known to exceed 1 mg/m^3 in some of these working areas, although an average concentration of 0.2 mg/m^3 is more typical [12,19-22]. Recent toxicologic work with nickel oxides suggests that the chemical and biological reactivities depend strongly on the thermal pretreatment [23]. In nickel user industries, the main exposure is to the metal itself or its alloys [19,20]. Miscellaneous occupational exposures are numerous and include exposure to nickel hydroxides (e.g., in nickel/cadmium battery production [24]); water-soluble and insoluble nickel in stainless steel welding fumes [25,26]; nickel tetracarbonyl, $Ni(CO)_4$, in the production of highly purified nickel [18,20]; and water-soluble nickel salts in hydrometallurgical refining and nickel salts production [19,20].

3. NICKEL CARBONYL POISONING

3.1. Carbonyl Refining of Nickel

$Ni(CO)_4$ is formed at temperatures between 40 and 100°C from carbon monoxide and activated nickel (obtained from reduction of nickel oxide). Typically, the "carbonylation reaction extracts 97.5% of the nickel and 30% of the iron leaving copper, cobalt, precious metals and other impurities behind as a porous granular residue" [27]. Separation of $Ni(CO)_4$ (b.p. 43°C) from $Fe(CO)_5$ (b.p. 103°C) is by fractionation. The purified $Ni(CO)_4$ is decomposed at temperatures between 200 and 250°C to yield pure nickel powder [18,27]. The carbonyl process has been employed in the commercial refining of nickel since 1902 [20]; plants are currently operated in Canada [27], the U.K. [18], and China [28]. Uses of $Ni(CO)_4$ in non-nickel-producing industries is difficult to assess. Interestingly, some of the early gassing accidents occurred in plants employing nickel catalysts.

3.2. Clinical Manifestations of Exposure and Treatment

The number of accidental gassings have been reduced significantly since the early 1970s and no fatalities have been reported in recent years [21,28]. By contrast, of the 245 cases reviewed in 1975 by the Panel on Nickel, U.S. National Academy of Sciences [3], 19 resulted

in death. In one early report in 1954 in which 100 workers suffered
accidental exposure to $Ni(CO)_4$, 31 required hospitalization with two
fatalities [29].

Both clinical experience and experimental work with animals
have attested to the highly toxic nature of inhaled nickel carbonyl.
The clinical signs and symptoms can be divided into two stages:
initial and delayed. The initial symptoms are usually mild and transitory and may include dyspnea, fatigue, nausea, vertigo, headache,
vomiting, insomnia, irritability, and the characteristic sooty odor
of $Ni(CO)_4$ in exhaled breath [3,29-31]. The delayed effects develop
gradually during the 1- to 5-day period following exposure and may
consist of dyspnea with tachycardia, coughing, muscular weakness and
pain, excessive sweating, substernal pain, visual disturbances, and
diarrhea. The physical signs are compatible with pneumonitis or
bronchopneumonia and resemble those of a viral influenzal pneumonia.
In the most severe cases, death results from respiratory failure
(diffuse interstitial pneumonitis), although cerebral edema and
hemorrhage may contribute in some patients [3]. Convalescence in
patients recovering from acute nickel carbonyl poisoning is usually
very protracted and is characterized particularly by fatigue on
slight exertion.

Nickel uresis occurs after exposure to nickel carbonyl and the
urinary nickel concentration correlates with the clinical severity.
Sunderman and coworkers concluded [3,32,33] that if the concentration
of nickel in urine collected during an 8-hr interval immediately after
exposure is less than 100 µg/liter (normal background value is 2 µg/
liter), the exposure may be classified as mild. Delayed symptoms, if
they appear at all, will be relatively mild. Urinary nickel concentrations between 100 and 500 µg/liter of urine have been observed to
result from moderately severe exposure with delayed symptoms. It was
recommended further that if the nickel in urine exceeds 500 µg/liter,
the delayed symptoms are likely to be severe and warrant hospitalization and immediate chelation therapy (see below). Nieboer et al.
[21] reviewed all of the available data on the relationship between
clinical symptoms and urinary nickel levels. For reasons explained

in Sec. 3.3, they question the wisdom of the delay inherent in the
8-hr urine sample and conclude that the exact "action levels" suggested by Sunderman and his colleagues may not be appropriate in all
instances. Furthermore, they express the opinion that urinary nickel
levels in spot samples of more than 100 μg/liter on the day of exposure suggest that careful monitoring of the patient would be judi-
cious even in the absence of initial symptoms. Clinical judgment
must of course prevail.

In addition to the use of corticosteroids and prophylactic
treatment of infection [28,30], chelation therapy with diethyldithiocarbamate (DDC) or tetraethylthiuram disulfide (TETD; known as Antabuse) has been routinely employed in the treatment of nickel carbonyl
poisoning since it was recommended by the Sundermans in 1958 [3,32].
The prevailing opinion is that chelating agents, and especially DDC,
are of unquestionable value in improving prognosis by reducing the
body burden of nickel [31-34]. After a critical review of the published case reports, Nieboer et al. [21] expressed a number of reservations concerning this practice. They concluded that it is not known
with certainty if the rate of nickel ion excretion after exposure to
$Ni(CO)_4$ is significantly enhanced by administering chelating agents
or if the rate of such assumed drug-induced nickel uresis correlates
with an accelerated destruction or elimination of the pathogenic
agent. In addition, they noted that a proper clinical evaluation
of the effectiveness of DDC or TETD has not been performed and experimental evidence and considerations of plausible toxicologic mechanisms
suggest that the timing of administration might be crucial (as discussed in Sec. 3.3).

3.3. Pathogenesis and Toxicologic Mechanisms

"The most significant pathologic changes in acute poisoning from the
inhalation of nickel carbonyl are found in the lungs. The lungs show
a severe degree of congestion, pulmonary edema, hemorrhage, and interstitial pneumonia with hyaline membrane formation. Concomitant

hepatic and adrenal cortical degeneration and brain and renal congestion are also usually observed" [35]. The pathogenesis in humans and
animals is almost indistinguishable. Deposition of nickel occurs in
the lungs, brain, kidney, bladder, heart muscle, diaphragm, and ovary
[36-38]. Because of its lipid solubility, $Ni(CO)_4$ can cross the
alveolar membrane in either direction without undergoing metabolic
alterations, suggesting it does not decompose immediately [36,39].

A number of reactions can be postulated for the decomposition
of $Ni(CO)_4$ in tissues [21,39].

$$Ni(CO)_4 \rightarrow Ni^\circ + 4CO\uparrow \tag{1}$$
$$Ni^\circ + Ox \rightarrow Ni^{2+} + Ox^{2-} \tag{2}$$
$$3Ni(CO)_4 + 2L \rightarrow NiL_2^{2+} + 6CO\uparrow + Ni_2(CO)_6^{2-} \tag{3}$$
$$\text{unstable}$$

In these equations, Ox denotes an oxidizing agent and L an endogenous
ligand. Studies in rats have shown that in blood, hemoglobin appears
to facilitate this dissociation as it serves as a receptor for the
released CO. Presumably myoglobin acts similarly in tissues. It is
further postulated that the metallic Ni° is subsequently oxidized to
Ni^{2+} by a suitable oxidizing agent [Eq. (2)], perhaps aided by enzyme
catalysis. Ligand-promoted disproportionation, such as in Eq. (3),
also seems plausible. It is not clear whether the direct interaction
of $Ni(CO)_4$ with tissues [e.g., Eq. (3)] or the resulting Ni° or Ni^{2+}
is responsible for the biological lesions.

The extent to which each of the above processes actually occurs
has implications for chelation therapy. Animal studies have indicated that DDC is not a very efficient antidote to $NiCl_2$ poisoning
[40]. In fact, it enhanced the deposition of nickel in such organs
as lung and brain [41]. This may be attributed to the lipid solubility of the $Ni(DDC)_2$ complex. It is also of interest that DDC was
most effective in protecting rats when administered prior to (-10 min)
or just after (+10 min) exposure to $Ni(CO)_4$ [34,42]. This is especially striking since DDC administered 30 min after exposure reduced
significantly the lung nickel level at 1 hr [38]. These observations
suggest that the action of DDC may be directly on $Ni(CO)_4$ [cf., Eq.
(3)] or that its role is not primarily to lower tissue Ni(II) levels.

Recent reports have shown that DDC is an effective in vitro and in vivo radical scavenger, and is also radioprotective [43-45]. Since the reactions depicted in Eqs. (2) and (3) involve oxidation-reduction processes, it is conceivable that the tissue damage is mediated by radicals. In addition, it was recently concluded that lipid peroxidation constitutes a molecular mechanism of acute nickel toxicity in $NiCl_2$-treated rats [46]. This concurs with the suggestion that the Ni(III)/Ni(II) redox couple is biologically active and potentiates dioxygen species (see Chap. 4, Sec. 5.2). It is also relevant to point out that the delayed pulmonary cellular damage and edema induced by $Ni(CO)_4$ are shared by other agents suspected of acting by radical pathways such as ozone, nitrogen dioxide, phosgene, and the pesticide paraquat [47]. These new insights and developments make imperative a careful clinical evaluation of the use of DDC and TETD in the treatment of nickel carbonyl poisoning and of the suggestion that the time of administration may be crucial to their efficacies.

4. NICKEL HYPERSENSITIVITY

4.1. Occurrence and Clinical Aspects of Nickel Contact Dermatitis

Nickel is one of the most common skin allergens in the general population. Nickel contact dermatitis was estimated in 1979 to occur in approximately 5% of the general population with a 10-fold higher incidence in women than men [e.g., 48-50; 21]. Among contact dermatitis patients (all causes), the reported incidence of nickel-related responses is dependent on the proportion of men and women and is in the range of 5-20% [e.g., 51-54]. The greater incidence in women most likely reflects exposure to domestic nickel-containing objects and substances. Numerous nonoccupational sources of nickel such as jewelry, coins, clothing fasteners, tools, cutlery, cosmetics, detergents, surgical implants, and medical instruments are known to be responsible for the licitation by nickel of contact dermatitis [3,54-58]. Occupational outbreaks have been documented [e.g., 59], although nickel dermatitis is currently not regarded as an occupa-

tional problem in the nickel-producing industry [21]. Current expe-
riences appear to agree with the findings at the Finsen Institute in
Denmark during the period 1936-1955; only 4% of the cases could be
assigned to nickel-plating operations, 9.5% to other employment, and
86.5% to contact with domestic articles [60]. In a more recent study
of occupational dermatitis, one-fifth of the women who developed
nickel allergy did so in work described as "cleaning" [61]. It was
suspected that work involving contact with water facilitates the
occurrence of eczema.

Clinically, nickel dermatitis tends to be widespread, chronic,
and persistent and many different types of lesions occur [54,62].
Hand eczema is a common finding in patients with contact allergy to
nickel (e.g., in 16-60% of cases). According to Calnan [63], a characteristic feature of nickel dermatitis is the spreading from primary
areas in direct contact with nickel (primary site) to secondary sites.
Primary sites are very numerous, while secondary eruptions often are
localized to the elbow flexures, the eyelids, and sides of the neck
[54,63]. Prognosis is generally poor. Oral challenges with nickel
salts exacerbate nickel dermatitis and this observation serves as
evidence for some contribution of internal nickel to the development
of the dermatitis [64,65]. Similarly, the consumption of tap water
with elevated nickel levels (>50 μg/liter) has been associated with
the internal provocation of nickel hand eczema [67]. A low-nickel
diet has also been correlated with improvement in the condition of
patients [66]. Treatment of nickel dermatitis with Antabuse (TETD)
in an effort to reduce the body nickel burden has only met with partial success [68,69]. It is conceivable that TETD may enhance gastrointestinal absorption in addition to mobilizing and redistributing
tissue nickel [21].

Nickel contact dermatitis can be diagnosed by three conventional
methods employed for other contact allergens: (1) the patch test;
(2) the leukocyte migration test (LMT); and (3) the lymphocyte transformation test (LTT). The patch test is the most convenient and most
commonly used although patients may also become sensitized as a result
of the secondary challenge with the antigen [70]. LMT is a somewhat

specialized and erratic in vitro test requiring about 72 hr. It is
based on the release of a soluble mediator (macrophage-inhibiting
factor) from sensitized lymphocytes on contact with the specific
antigen such as a nickel(II)-protein complex [71,72]. The in vitro
LTT is effective in detecting nickel sensitivity by culturing peripheral lymphocytes with the appropriate nickel compound which stimulates
transformation into lymphoblasts [73,74]. As illustrated in Fig. 1,
the extent of this transformation is measured by the incorporation
of radiolabeled thymidine.

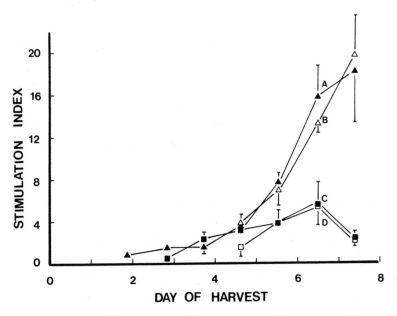

FIG. 1. Dependence on incubation time and antigen concentration of
the nickel(II)-induced antigenic response in the lymphocyte transformation test. Curves A (25 μg NiSO$_4$·6H$_2$O/ml; 9.5 x 10^{-5} M) and
B (12.5 μg NiSO$_4$·6H$_2$O/ml; 4.75 x 10^{-5} M) are for a nickel-sensitive
donor, while curves C (9.5 x 10^{-5} M NiSO$_4$) and D (4.75 x 10^{-5} M NiSO$_4$)
are for a nonsensitized individual. The stimulation index, defined
as the ratio of cell-associated ^{3}H counts due to the uptake of [^{3}H-
methyl]thymidine in the presence of NiSO$_4$ to that in its absence,
expresses the extent of cell transformation. (Reproduced with permission from [75].)

4.2. Immunological Mechanism of Nickel
        Contact Dermatitis

Nickel contact dermatitis is a delayed reaction (type IV cell-
mediated immune response) and flares up approximately 24-28 hr
following exposure to a secondary challenge. Four different developmental phases have been recognized: refractory, induction, elicitation, and persistence [76,77]. In the refractory state (duration
of days to lifetime), an individual is exposed to the allergen but
remains unaffected. An inductive phase of 4 days to several weeks
following exposure is required for full sensitization. To commence
the activation process, the hapten [e.g., Ni(II)] must penetrate the
skin [78], conjugate with epidermal protein to form the antigen, and
then activate antigen-inexperienced effector T-lymphocytes. The
activated cells migrate to draining lymph nodes where proliferation
ensues. The resultant immunoblasts differentiate into effector and
memory T-lymphocytes for general circulation. During elicitation or
secondary response, the process of penetration and recognition by
effector cells (previously activated) is repeated. Upon contact, the
effector cells release chemotactic substances causing cellular inva-
sion and skin inflammation. As illustrated in Fig. 2, T-cell activation is a cooperative effect requiring the presence of macrophage-
like cells (Langerhans cells) and keratinocytes (epidermal cells
synthesizing keratin). Langerhans cells present the antigen to a
specific helper T-cell ("signal 1") and secrete a peptide cytokine
(interleukin-1) which serves to activate the effector cell fully
("signal 2") [79]. Alternatively, the second signal may be provided
in the form of an epidermal cell-derived thymocyte-activating factor
(ETAF) which is, or resembles, interleukin-1. After activation, one
subset of helper T-cells produces a second soluble helper factor
(interleukin-2) which promotes T-cell proliferation [77,80]. Recently,
Ni(II)-reactive T-cell populations have been isolated from nickel dermatitis patients and characterized [81,82]. The continued presence
of effector cells capable of recognizing specific haptens like Ni(II)
and producing inflammation is responsible for the persistence phase
which may last from months to a lifetime.

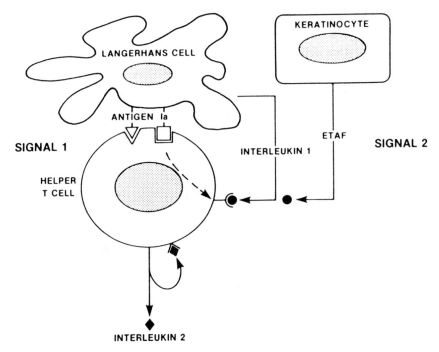

FIG. 2. T-cell activation by epidermal cells (see text). (Reproduced with permission from [77].)

Tolerance and desensitization may explain the somewhat surprising observation that nickel dermatitis is not a severe occupational problem. Loss of the immune response to an antigen or hapten to which an individual has already been sensitized is termed "desensitization" [72]. The evidence for its existence in man is weak. By contrast, "tolerance" to the development of contact sensitivity occurs in nonsensitized (naive) subjects on exposure to the hapten. Tolerance appears to depend on the presence and action of suppressor T-lymphocytes which modulate the induction or elicitation phase [77]. Recently, it has been reported that tolerance to nickel is characterized by defective interleukin-2 production and receptor induction during nickel stimulation in vitro [83]. Interestingly, the removal of histamine receptor-bearing suppressor lymphocytes from human peripheral mononuclear leukocytes caused a 63% enhancement of nickel-induced lymphoblast transformation [84].

4.3. Nickel-Induced Occupational Asthma

Occupational asthma from nickel sensitivity is uncommon, although
reports of isolated and even case series are becoming more frequent
[85-91]. As for other respiratory allergens [92], three types of
responses are known: (1) immediate with the development of bronchoconstriction in minutes, which is reversible with bronchodilators
[86,87]; (2) late with a response time of 2-12 hr [89], which may
not reverse completely with bronchodilators; and (3) dual (immediate
and late) [88].

The mechanisms of the immediate and late asthmatic responses
are quite different. The late asthmatic response involves the
release of a neutrophil chemotactic factor by epithelium (surface)
cells of the airway following stimulation by the antigen [presumably
Ni(II) bound to protein] [93,94]. Induced airway hypersensitivity
is also characteristic of delayed asthma. The release of a biochemical mediator has been implicated in this enhancement of bronchial
sensitivity by way of lowering the levels of histamine or methacholine
required for the induction of bronchoconstriction [93,95]. Both the
inflammatory process and the hyperresponsiveness appear to require
the activation of the arachidonic acid pathway.

It is generally accepted that the immediate asthmatic response
is antibody-mediated. Nickel-reactive antibodies have been identified
in sera of immediate asthmatics [85,86,88,91], while they are absent
in late asthmatics [89,91]. In one male worker, the nature of the
antigenic determinant of serum antibodies has been characterized
[96,97]. The antibody recognized Ni(II) bound at the natural Cu(II)/
Ni(II) transport site of human serum albumin (HSA), but not Cu(II)
nor Co(II). This interpretation was deduced from metal ion blocking
experiments and from the good agreement obtained between the pH dependence of the formation of the Ni(II)-HSA complex and the antigen-
antibody complex. The blocking experiments showed that only Ni(II),
Cu(II), and Co(II) inhibited the formation of the ternary ^{63}Ni(II)-
HSA-antibody complex, while metal ions known not to bind to the primary HSA Cu(II)-Ni(II) site did not (i.e., Zn(II), Mn(II), and

Cr(III); see Chap. 4, Sec. 5.1). Nonradiolabeled Ni(II) was most
effective in these competition experiments, suggesting that the
antibody interaction depended on a special structural feature of
the interaction of Ni(II) with HSA. Perhaps the ability to form an
octahedral complex affords one explanation (see Chap. 4, Sec. 5.1).

5. NICKEL CARCINOGENESIS

5.1. Epidemiologic and Animal Studies

Nickel-related human cancer constitutes an industrial disease. It
is virtually restricted to the nickel-producing industry and is
largely historical [13]. The epidemiologic and animal data have
recently been expertly reviewed in considerable detail by the U.S.
Environmental Protection Agency (EPA) [11]. Less elaborate summaries
or progress reports are also available [9,12,98,99]. Of the toxicologic effects of nickel, carcinogenesis has been the most intensively
studied. The coverage in this subsection is limited to a review of
those nickel-containing substances for which either human or animal
evidence is deemed sufficient for designation as human or suspected
human carcinogens. Determinants and a model of nickel carcinogenesis
are formulated in Sec. 5.2 on the basis of the animal and human expe-
riences and the known genotoxicity of nickel compounds is reviewed
in Sec. 6.

Evidence of a cancer hazard is strongest for nickel refinery
dust associated with pyrometallurgical processes. Excess risk of
nasal and lung cancer has been reported for nickel refinery workers
in Canada [100-102], United Kingdom [103,104], and Norway [105,106].
Agreement in findings by different researchers on the epidemiologic
results obtained for similar categories of workers in different coun-
tries is an important epidemiologic criterion. The nickel data fulfil this condition. A typical set of results is summarized in Tables
1 and 2. It is evident from these data that excess risk of lung
cancer is dependent on the duration of exposure, which is a surrogate
measure of total exposure. (The weakness of most historical prospec-

TABLE 1

Lung Cancer Deaths by Exposure Duration
(more than 15 years since first exposure)[a]

Study subgroup[b] (refining process)	<5-yr exposure[c]			>5-yr exposure[c]		
	O	E	O/E	O	E	O/E
Sudbury (S)	18	6.5	2.8*	24	2.6	9.4*
Copper Cliff (S)	17	5.8	2.9*	20	1.9	10.7*
Coniston (S)	1	0.7	1.4	4	0.7	5.8*
Port Colborne (L, C, & S)	17	9.2	1.8*	32	7.2	4.5*

[a]Based on the results of an historical prospective study of 55,000 nickel refinery workers, followed up between 1949 and 1976 and of which 5,280 had died [100,101].
[b]Abbreviations: S, sintering; L, leaching; C, calcining. Sintering and calcining are high-temperature oxidation processes to remove sulfur. At Port Colborne, L, C, and S were all carried out in separate buildings, although materials were cycled between them.
[c]Abbreviations: O, observed number of deaths; E, expected deaths based on Ontario mortality rates; O/E, standardized mortality ratio or relative risk; *$p < 0.001$. Risks for nonsinter operations in the plants were not elevated.

TABLE 2

Nasal Cancer Deaths by Exposure Duration
(more than 15 years since first exposure)[a,b]

Study subgroup (refining process)	<5-yr exposure			>5-yr exposure		
	O	E	O/E	O	E	O/E
Sudbury (S)	1	0.07	14	1	0.02	45
Port Colborne (L, C, & S)	3	0.09	33*	13	0.08	170*

[a]Based on the results of an historical prospective study of 55,000 nickel refinery workers [100,101].
[b]See notes to Table 1 for abbreviations and explanation.

tive studies is the absence of reliable, quantitative exposure
data.) Since nasal cancer is very rare, the number of expected
cases is small and the increased risk extremely high (Table 2). The
"15 years since first exposure" restriction applied in the study in
question allowed appropriate time for the cancer to develop (the
latency period). While no risk of lung and nasal cancer was associated with most nonsintering/calcining processes at the Canadian and
U.K. plants, the Norway studies do associate cancer with electrorefining of nickel. There is good evidence that in the Norway plant
there was mixed exposure; in part because of the closeness of the
smelter and tank house (electroplating) facilities. This complicates
the search for causation.

The majority of published animal nickel cancer studies involved
injection as the route of administration [11,12]. The relationship
of injection-site-only tumors to human carcinogenic hazards by inhalation is uncertain and thus must be qualified as providing only limited
evidence. However, as illustrated in Sec. 5.2, such studies are useful for mechanistic considerations.

Only three nickel compounds or mixtures of nickel compounds
have been classified by the U.S. EPA as either known human carcinogens
(group A) or probable human carcinogens (group B) [11]. The group A
compounds include nickel refinery dust from pyrometallurgical nickel
sulfide refineries and nickel subsulfide; nickel carbonyl is assigned
to group B. The refinery dust classification is based largely on epidemiologic data. The dust is a complex mixture of nickel oxides and
sulfides, but contains nickel subsulfide (Ni_3S_2) as a major component
at certain stages of the refining process. Since some animal inhalation data are available for Ni_3S_2 [107] and because of its presence
in the refinery air, the combined evidence is considered positive
proof for the human carcinogen classification. By contrast, there
is inadequate evidence for nickel carbonyl acting as a human carcinogen [13,104] and the animal inhalation data are limited [11,108].
The assessment of the carcinogenic potential of other nickel compounds
is an area of intense debate and research. Biochemical and in vitro
cell culture work suggest that the Ni(II) ion may be the ultimate

carcinogen—the "nickel ion hypothesis". In Sec. 5.2 it is concluded that, if verified, this hypothesis does not predict that all nickel compounds are carcinogenic.

Based on epidemiologic data, the unit risk of excess respiratory cancer resulting from a lifetime exposure to 1 μg Ni/m^3 as nickel refinery dust is estimated as 2 x 10^{-4} (μg/m^3)$^{-1}$ [11], compared to 6 x 10^{-2} (μg/m^3)$^{-1}$ for chromium(VI) compounds [109]. This places refinery dust in the third quartile of 55 substances evaluated as suspect carcinogens (i.e., <50% are less potent); Cr(VI) is in the first quartile and thus is in the top 25% [11].

5.2. Determinants and Model of Nickel Carcinogenesis

"Nickel compounds that have been shown to be carcinogenic for rodents generally share the following properties: (1) they are crystalline substances that may dissolve slowly in serum or cytosol, but are essentially insoluble in water; (2) when present as small particles, they undergo phagocytosis in vitro by tissue culture and peritoneal macrophages; (3) they cause morphological transformation of cultured Syrian hamster embryo (SHE) and baby hamster kidney cells; and (4) they induce erythrocytosis in rats following intrarenal (i.r.) injection, owing to increased renal production of erythropoietin" [23]. In fact, "the cell transformation and erythropoiesis stimulation assays have been proposed as screening tests for carcinogenic nickel compounds" [23]. These quotes from a recent paper by Sunderman and his colleagues concisely summarize the current state of affairs in experimental nickel carcinogenesis. The reader is referred to this article [23] for original references.

Chromosomal abnormalities and mutational changes in DNA are believed to be responsible for the transformation of SHE cells [110, 111] (subsequent transplantation into mutant nude mice results in malignant tumors [110,112]). This interpretation is consistent with the general view that many forms of neoplastic transformation are a consequence of DNA damage which permanently alter gene expression by, for example, mutation [113]. From biological and epidemiologic perspectives, chemical carcinogenesis is a multistage process [114-116]. Animal studies have identified the following stages [116-118]: (1) initiation (a single or brief exposure to a carcinogen induces a permanent change in tissues which is recognized after promotion as local proliferations, such as papillomas, nodules, or polyps); (2) promotion (the process involving cell proliferation whereby tumor formation is accelerated or encouraged in a tissue that has been initiated); and (3) progression (focal proliferative lesions resulting from a promoting environment become precancerous lesions before malignant behavior is expressed). Stages 2 and 3 are often reversible.

The strong association observed between erythrocytosis and carcinogenesis in rats [119,120] does not imply that these phenomena are linked. Sunderman et al. [119] provide the following two explanations: (1) similar dependence on intracellular bioavailability of Ni(II), which is likely influenced by a complex interplay of chemical reactivities, surface properties, and extracellular or intracellular dissolution kinetics of the nickel compounds; and (2) altered expression of the erythropoietin gene during renal carcinogenesis. In light of the potentiation by Ni(II) complexes of dioxygen species at physiologic pH values (Chap. 4, Sec. 5.2), molecular oxygen depletion in tissues may provide an alternative explanation. Tissue hypoxia is the main stimulus for erythropoietin synthesis and this hormone controls erythrocyte production [121,122].

The collective evidence derived from in vitro genotoxicity, animal carcinogenicity, and erythropoiesis studies can be interpreted to indicate that the Ni(II) ion is the putative agent—the ultimate carcinogen. It is of interest therefore to review this hypothesis and its ramifications and limitations. The summary and personal perspective [123] provided below affords a convenient avenue for bringing together a number of features of the available animal and genotoxicity (Sec. 6) studies, as well as the nickel biochemistry discussed in Chap. 4.

Intramuscular and intrarenal injection experiments with rats clearly indicate that solid nickel compounds differ in carcinogenic

potency [120]. Such properties of the solid nickel compounds as
particle size, crystallinity, surface charge, and solubilization
kinetics are believed to determine respiratory deposition and clearance, as well as cellular uptake; they are said to determine the bioavailability of Ni(II) [123]. In humans and animals, intracellular
compartmentalization of Ni(II) must necessarily be balanced by extracellular transport and excretion. Endogenous ligands at physiologic
concentrations bind Ni(II) tightly and usually maintain low cellular
levels. They facilitate its mobilization, transport and subsequent
renal excretion. Presumably, extracellular and intracellular (uptake
is by phagocytosis) pools of particulate nickel compounds in the
lung and nasal passages overwhelm this removal mechanism and can
provide a continuous inward cellular flux of Ni(II). A half-life of
3-4 years has been assigned to such pools in retired workers (Chap. 4,
Sec. 2). By contrast, exposure to water-soluble nickel salts (such
as aerosols of dissolved $NiCl_2$ and $NiSO_4$) results in more rapid turn-
over (half-lives of of 1-2 days in humans). Consequently, the probability of intracellular accumulation of Ni(II) in individuals exposed
to water-soluble salts is lowered, as is presumably the opportunity
to express fully its carcinogenic potential. Animal experiments concur with this, as the results with the water-soluble salts $NiCl_2$ and
$NiSO_4$ are negative [11]. Potency differences are also implied by the
epidemiologic data. As already indicated, it assigns the highest risk
to exposure to refinery dust containing solid nickel oxides and sulfides of relatively low solubility. The absence of renal cancers in
nickel workers further attests to this since the kidney is a major
route of excretion of nickel as Ni(II) (likely in the form of a
complex).

From this brief overview it may be concluded that the "nickel
ion hypothesis", if proven correct, predicts a range of carcinogenic
potencies. The observed effect will depend on a number of physicochemical properties of nickel compounds, their biological residence
time, the magnitude of pools built up in exposed organs, and the
effective intracellular load of Ni(II) in affected tissue. It is
obvious that the "nickel ion hypothesis" as stated leaves a number of

unresolved issues in nickel carcinogenesis which severely limit our understanding of this process. (1) There is a lack of understanding concerning the dynamics of nickel compounds in the body (e.g., uptake by and solubilization in tissue). (2) The precise events at the molecular level involving Ni(II) that induce mutations and/or affect gene expression are unknown. And (3), the exact meaning of the strong correlation between erythrocytosis and carcinogenesis induced by nickel compounds is perplexing and needs to be further explored. Obviously, additional research is needed.

6. GENETIC TOXICOLOGY

6.1. Objectives

The main objective of this section on the genotoxicity of nickel compounds is to summarize the known effects at the cellular, chromosomal, and DNA levels. The coverage is not comprehensive and only the major developments are considered. Additional details may be sought in recent reviews [11,99,124-127] (see also Chapter 9).

6.2. Mutagenicity in Prokaryotic and Eukaryotic Cell Systems

Positive mutagenic responses with $NiSO_4$ or $NiCl_2$ have been observed in a limited number of mammalian cells, e.g., for the yeast *Saccharomyces cerevisiae* D7 [128,129], mouse lymphoma L5178Y [130], and Chinese hamster V79 cells [131]. Results for prokaryotes with water-soluble Ni(II) salts have been mostly negative (e.g., for the bacteria *Salmonella typhimurium* [132,133], *Bacillus subtilis* [134], and *Escherichia coli* [135]). A dose-dependent response has been reported for *Corynebacterium* [136], as well as the potentiation of the mutagenicity of known alkylating agents by $NiCl_2$ in bacteria [137]. All but two of the studies [130,137] have deficiencies such as weak responses, insufficient experimental data, or are preliminary reports. Consequently, it is not possible to conclude with certainty that Ni(II) is mutagenic in prokaryotic and eukaryotic in vitro cell systems [11]. It has been suggested that components of cell culture medium circumvent, by complex formation, the cellular uptake of Ni(II) [125,133], although internalization of nickel by *Pseudomonas tabaci* has been confirmed [146].

6.3. Transformation of Cultured Mammalian Cells

The data summarized in Table 3 clearly illustrate the transforming power of nickel compounds, especially when in crystalline form. Good correlations have been observed between morphologic cell transformation and phagocytotic activity [23,147] or erythropoiesis [23], suggesting that cellular uptake is the common feature. As indicated earlier (Sec. 5.2), transformation in vitro is a characteristic of nickel compounds that are carcinogenic in rodents. In fact, clones of Ni_3S_2-transformed SHE cells when transplanted subcutaneously into nude mice developed injection site undifferentiated sarcomas [112]. Di Paolo emphasized the difficulty of transforming human cells in comparison to animal cells. He suggests that there are three reasons for this: "our inexperience in culturing human material, a central mechanism that is responsible for the relatively stable human phenotype, and the heterogeneity of humans" [110]. It is noteworthy, therefore, that $NiSO_4$ was not able to transform fully human bronchial epithelial cells as indicated in Table 3. Some support for the "nickel ion hypothesis" can be drawn from the work by Hansen and Stern [144]. As noted in Table 3, at the LD_{50} (measured by survival rates) all of the nickel compounds tested demonstrated equal transformation ability. They reasoned that equal toxicity implied equal intracellular concentrations of Ni(II). This study has a number of limitations and this hypothesis needs to be tested further, including an assessment of the actual intracellular nickel levels and its subcellular distribution.

TABLE 3

Transformation of Cultured Mammalian Cells[a]

Cells	Compound tested	Comments	Ref.
SHE	Cryst. Ni_3S_2, amorph. NiS	Ni_3S_2, +; dose response	112,138
SHE	Cryst. Ni_3S_2, $NiSO_4$	Both +; Ni_3S_2 most effective	139
SHE	Cryst. Ni_3S_2, cryst. NiS, cryst. Ni_3Se_2, amorph. NiS, Ni powder	Cryst. compounds, +; dose response	140
SHE	$NiSO_4$	Synergism with organic carcinogens	141
SHE	$NiSO_4$	Enhanced transf. by SA7	142
$C3H/10T\frac{1}{2}$	Cryst. Ni_3S_2	+; dose response	143
BHK-21	Ni, NiO, Ni_3S_2, Ni_2O_3, $Ni(C_2H_3O_2)_2$	All +; equal transf. at LD_{50}	144
Human bronchial epithelial	$NiSO_4$	Partial transf.	145

[a]Abbreviations: SHE, Syrian hamster embryo; BHK, baby hamster kidney; $C3H/10T\frac{1}{2}$, mouse embryo fibroblast; SA7, simian adenovirus; cryst., crystalline; amorph., amorphous; transf., transformation; +, positive response.

6.4. Chromosomal and DNA Damage and Related Effects

There is little doubt that in mammalian cells internalized Ni(II) can
interact with the cellular genetic material [99,126]. To facilitate
the discussion of the known genotoxic effects of nickel compounds,
the uptake model presented in Fig. 3 is considered first.

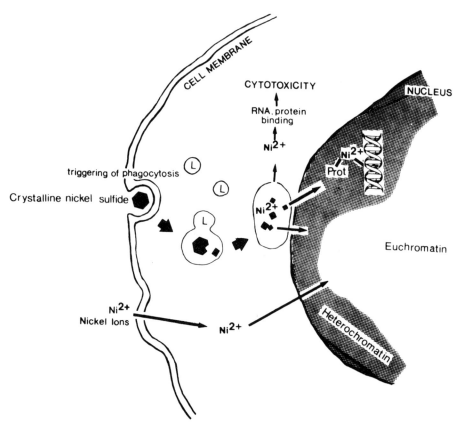

FIG. 3. Model of cellular uptake of Ni(II) by diffusion and crystalline NiS by phagocytosis. Dissolution of the particulate NiS inside
the phagocytic vesicles is believed to be aided by the attachment of
lysosomes (L), which are digestive vesicles. Phagocytic vesicles
with internalized NiS have been observed to aggregate around the
nuclear membrane and thus may act as vehicles of nuclear delivery
of Ni(II). The toxicologic consequences of the release of Ni(II)
to the cytosol and nucleus are discussed in the text. (Reproduced
with permission from [148].)

As indicated in Fig. 3, uptake of Ni(II) can occur by diffusion whereas that of particulate crystalline NiS is by phagocytosis. It is also conceivable (not shown) that the Ni(II) ion could be taken up by pinocytosis, which is an endocytotic process similar to phagocytosis; in this instance the internalized vesicle contains only extracellular fluid. Lysosome interaction with cytoplasmic particles has been confirmed [149], and the nuclear membrane appears to be impermeable to particulates of nickel compounds [140,149]. How the encapsulated nickel particles dissolve is unknown, but the resultant accumulation of Ni(II) in the nucleus has been assessed [140]. Nuclear Ni(II) appears to be associated strongly with protein [140], as depicted in Fig. 3. As is evident from the data in Table 3, water-soluble nickel salts are less effective than crystalline particulates in the transformation assay. The same can be concluded when X-chromosome fragmentation in Chinese hamster ovary cells (CHO) is considered [148,150]; $NiCl_2$ was ineffective while crystalline NiS exhibited a dose response. By contrast, the delivery and uptake by phagocytosis of $NiCl_2$ artificially encapsulated in liposomes also induced fragmentation. Therefore, the pathway of delivery seems to be an important determinant in the genotoxicity of nickel compounds in vitro. These observations are consistent with the "nickel ion hypothesis" outlined in Sec. 5.2.

Studies with cultured cells and biochemical studies have demonstrated that Ni(II) compounds detrimentally affect chromosomes, DNA, DNA synthesis, and transcription. Some of the effects observed are induction in CHO cells of DNA strand breaks by $NiCl_2$ and crystalline NiS [151] and of DNA-protein crosslinks by $NiCl_2$ [152]; sister chromatid exchanges (SCEs) in CHO cells by $NiCl_2$ and crystalline NiS [153] and in cultured human lymphocytes by $NiCl_2$, $NiSO_4$, and Ni_3S_2 [111,143,154]; chromosomal aberrations in FM3A murine carcinoma cells by dissolved NiS [155] and X-chromosome fragmentation in CHO cells as already mentioned [148,150]; blocking of cell growth in S phase in CHO cells with $NiCl_2$, crystalline Ni_3S_2, NiS, Ni_3Se_2, and NiO [e.g., 156]; induction of DNA repair synthesis in CHO cells by $NiCl_2$ and crystalline NiS [157]; reduction by Ni(II) in the fidelity of DNA synthesis [158,159]; weak inhibition by Ni(II) and slight loss

of fidelity in DNA transcription [160,161]; apparent chromosome
length shortening by $NiSO_4$ in cultured human peripheral lymphocytes
[162]; and conversion by Ni(II) of oligonucleotides from right-
handed conformations (B form) to left-handed conformers (Z form)
[123,163,164].

A number of comments about, or relevant to, the compiled genotoxic effects are warranted. (1) $NiCl_2$ is surprisingly genotoxic
which suggests that some of the outcomes assessed have low intracellular Ni(II) concentration requirements (cf. earlier discussion
of Fig. 3). (2) Both histone (required for DNA condensation) and
nonhistone proteins were involved in the binding of Ni(II) to chromatin [165,166]. (3) As indicated in Fig. 3, Ni(II) prefers to
associate with heterochromatin rather than the genetically active
euchromatin. Heterochromatic DNA has few actively transcribed genes.
(4) Single-strand breaks, DNA-protein crosslinks, and DNA interstrand
crosslinks have also been induced in rat tissues following intraperitoneal injection of nickel(II) carbonate [167]. (5) Interestingly, an increased level of chromosomal gaps and breaks, but not
SCEs, have been observed for workers employed or retired from a
nickel refinery [168,169]. (6) The induction of reversible conformational changes in DNA is not an exclusive property of Ni(II) and
other ions share this ability [123,163]. It has been suggested that
since nucleotide sequences with Z-forming potential are widely found
in evolutionarily diverse eukaryotic genomes [170], and left-handed
oligonucleotides can serve as templates for RNA synthesis but with
reduced activity compared to the B form [163,171,172], they may serve
a role in the control of gene expression. Although speculative,
"they could be hot spots for gene recombination or rearrangement, or
they could be especially reactive with chemical reagents such as
mutagens and carcinogens" [170] as they are rich in guanine.

6.5. Implications

The genotoxic ability of nickel compounds is clearly established,
although they are only weakly mutagenic when compared to chromium(VI)

compounds [173]. Ultimately, the aim is to understand the relevance
of these genotoxic outcomes to cytotoxicity, carcinogenesis, and
reproductive and developmental deficits. Nickel compounds are not
recognized as reproductive hazards in humans (Sec. 7). Their specific role in the development of cancer remains unresolved because of
our limited understanding of chemical carcinogenesis. In addition,
other uncertainties exist. What is the relevance of phagocytosis to
human epithelial cells of the respiratory tract? Or is this capability limited to mobile phagocytes? Since renal cancers are not
associated with exposure to nickel compounds, what is the significance of the strong relationship between erythropoietic ability and
carcinogenic potency in rodents? There are no simple answers.

The somatic cell mutation model of cancer postulates that
carcinogens initiate neoplastic transformation by interacting with
the DNA of cells. If DNA repair is absent, or misrepair occurs,
mutations may arise in the descendents of the affected cells. Such
mutations may affect a variety of genes controlling normal cellular
growth and differentiation [113,116,174]. A number of viral genes
(oncogenes) have been shown to induce cancers in a variety of animal
species [175]. It has been suggested that genes similar in nucleotide sequence to these viral oncogenes (protooncogenes) exist in all
cells [176]. Furthermore, a role for certain of these genes in the
regulation of the mammalian cell cycle has been proposed [177]. It
is hypothesized that protooncogenes are "activated" by mutations
induced by chemical mutagens and/or ionizing radiation [176,178].
Alternatively, alterations at the chromosomal level (e.g., abnormal
condensation, deletions, and translocations) may result in altered
gene expression (of a protooncogene?) crucial for neoplastic transformation [179,180]. A loss of the normal expression of tumor suppressor genes or antioncogenes may arise in a similar fashion [181].
Another interesting development concerns chemical promoters of cancer
which facilitate the promotion stage of cancer (see Sec. 5.2). Such
chemicals appear to have the ability to produce radicals derived from
molecular oxygen (O_2), namely, the superoxide anion (O_2^-), the hydroxyl
radical $(\cdot OH)$, and the peroxy radical $R\dot{O}_2$ [116]. In vitro, dioxygen

radicals can damage DNA. It is known that tumor promoters induce the formation of peroxide hormones derived from arachidonic acid. These hormones are intimately involved in cell division, differentiation, and apparently also in tumor growth [182].

Within the context of the above theory of chemical carcinogenesis [116], the proven genotoxicity and ability to potentiate dioxygen species (see Chap. 4, Sec. 5.2) of nickel compounds provide plausible mechanisms for the development of cellular growth independence via the uncoupling of cell multiplication and differentiation (i.e., the neoplastic state). Epidemiologic analysis has suggested that nickel refinery dust may act as both an initiator and promoter of cancer [115]. Furthermore, orally administered $NiCl_2$ has been shown to have an apparent promoting effect on chemically induced renal tumorigenesis in rats [183].

7. MISCELLANEOUS HEALTH EFFECTS

7.1. Renal Toxicity

Only one published study reports on the renal toxicity among electroplaters and electrolytic refinery workers exposed to aerosols of dissolved nickel salts [184]. Although there was some correlation between proteinuria (measured by β_2-microglobulin) and urinary nickel concentrations, the elevation of this low molecular mass protein was mild in the 48 workers (10-450 µg/liter compared to 5-229 µg/liter in controls). Cadmium workers with nephrotoxicity have urinary β_2-microglobulin levels >>1,000 µg/liter [185,186]. A recent survey in the authors' laboratory of 26 electrolytic nickel refinery workers, which included those with long-term employment, indicated very few excursions of urinary β_2-microglobulin outside the normal range [187]. Aminoaciduria and proteinuria can be induced in rats with a single dose of $NiCl_2$ (4-6 mg Ni/kg intraperitoneally) or $Ni(CO)_4$ (acute toxic dose by inhalation) [31,188,189]. In the former case, amino acid and protein excretions consistently returned to normal by day 5 after exposure.

7.2. Reproductive and Developmental Effects

No detrimental reproductive and developmental effects have been documented in humans [190]. A range of effects can be induced in rodents and this is common for compounds of most metals [191]. Nickel is known to cross the placenta in rodents [192]. A number of extensive reviews are available which summarize the animal reproductive studies [11,12,193,194]. When applying a number of criteria to a critical appraisal of the animal experiments with nickel compounds, no firm conclusion can be derived concerning the likely risk to humans. In evaluation of teratogenic and developmental hazards, it is common practice to consider the route of exposure, male and maternal toxicity, the nature and susceptibility of the test animal to specific effects, and some evidence for mechanism. The predictive value of animal experiments for effects in human pregnancy is generally poor and complex; fertility effects in rodents are generally better indicators for effects in humans [195,196]. With this perspective in mind, a small number of nickel-related studies are examined.

Sunderman et al. [197] observed excess embryonic mortality (up to 24%) when administering $NiCl_2$ intramuscularly to rats on day 6-10 of gestation at total doses near the LD_{50} (15-20 mg Ni/kg). Mas et al. [198] found excess fetal abnormalities in rats under conditions toxic to the dam such as hemorrhage and hydrocephalus (administration of 2-4 mg Ni/kg was intraperitoneally). Injection of mice intraperitoneally with a similar dose (4 mg Ni/kg) did not affect implantation but produced a small increase in the total number of nonspecific abnormalities and stillborn offspring (3% compared to 0.3% in controls) [199]. A more specific occular effect (anophthalmia and microphthalmia) in the absence of other abnormalities has been induced in rats on administration of $Ni(CO)_4$ by inhalation at doses one-half of the apparent LD_{50} (93% dams survived) on day 7 or 8 of gestation [200]. Injection studies have reproduced this result [194]. By contrast, in Syrian golden hamsters, anophthalmia was not observed (only in one fetus of 17) when $Ni(CO)_4$ was administered by inhalation near the LD_{50} on day 4 or 5 (the most sensitive days for other nonspecific malformations) [201].

As an aside to our discussion of chelation therapy in $Ni(CO)_4$ poisoning (Secs. 3.2 and 3.3), it is of interest to mention that in pregnant mice treated orally with $^{63}NiCl_2$, coadministration of the chelating drug DDC (or thiuram sulfides) enhanced the deposition of the radiolabel in the brains and other tissues of both adults and fetuses [202].

7.3. Immunotoxicity

Inflammatory alterations characterized by a loss of cilia and gradual modification of epithelial cell shape and arrangement have been observed in the nasal mucosa of active and retired nickel workers. The changes ranged from normal to squamous, dysplastic, and ultimately carcinomatous in a number of individuals. Epithelial keratinization was also in evidence [203-205]. Not surprisingly, a number of effects on the respiratory tract of animals have produced changes that may be termed immunosuppressive. The observed immunotoxic responses are numerous [11,12] and include: reduced ciliary activity by inhalation of $NiCl_2$ (Syrian golden hamsters) and depression of phagocytic and bacteriocidal function of alveolar macrophages in mice [206,207]; distinct changes in alveolar macrophage size and function and the cellular association of viscous substances in inhalation studies with $NiCl_2$, Ni metal, or NiO in a number of animals [208-210]; reduced specific antibody-producing spleen cells in mice and rats following inhalation of $NiCl_2$ and NiO, respectively [211,212]; a suppression of T-lymphocyte-mediated reactions and a reduced natural killer cell activity in mice after injection of a high dose of $NiCl_2$ (18 mg Ni/kg) intramuscularly [213,214]. The implication of these immunosuppressive effects in animals for humans is not known, especially since ambient exposure levels were generally low (e.g., 100-250 μg Ni/m^3) in the experiments summarized.

7.4. Cardiotoxicity

As discussed in relation to the essentiality of nickel, hypernickelemia occurs in a number of diseased states (Chap. 4, Sec. 7.4). Of
major concern is its frequent occurrence in patients with acute
myocardial infarction or unstable angina pectoris [215]. Because
coronary vasoconstriction in dog heart can be induced in vivo and
in perfused isolated hearts by injection of Ni(II) [216], hypernickelemia is surmised to be related to the pathogenesis of ischemia.
It is obvious that this raises a genuine concern, especially since
nickel workers are known to have elevated body fluid nickel levels
and iatrogenic exposures to nickel have been identified in isolated
instances. For example, elevated nickel levels occur in association
with hemodialysis treatments [22], prostheses and implants [22,217],
and with medications such as DDC and Antabuse (see Secs. 3.2 and 3.3)
[22,218]. The knowledge that Ni(II) can assume an agonistic or
antagonistic relationship with respect to Ca(II) functions in excitable tissues (Chap. 4, Sec. 5.1) adds to the concern. The publication of research papers which report extremely high body fluid nickel
levels in healthy and unhealthy individuals unnecessarily complicates
the problem. For example, a recent study [219] links posttraumatic
myocardial damage to endogenous nickel release. The analytical data
on which this is based assigns normal serum nickel levels at between
40 and 80 µg/liter, corresponding to 100-300 times the accepted level.
Values up to 2105 µg/liter are reported in patients. This study must
be discounted simply on analytical grounds; contamination of the
specimens is the obvious deficiency.

In a 1983 editorial, F. W. Sunderman, Jr. recommends that the
maximum permissible level of nickel in diluted solutions used in
hospitals be set at 5 µg/liter for common intravenous solutions,
and 10 µg/liter for solutions that contain albumin or amino acids
such as histidine which have a high affinity for Ni(II) [220].
Severe contamination of such solutions with Ni(II) have indeed been
documented [221].

To gain some perspective, reference needs to be made to the
epidemiologic data concerning the potential cardiotoxic effects in
the manner suggested by Doll [13]. Although the epidemiologic data
relate to broad categories of cardiovascular disease, they are
encouraging since the fractional standardized mortality ratio (e.g.,
the relative risk) is near unity. Of course a small hazard of specific diseases could exist unnoticed within the broad categories
examined. Additional correlations between laboratory research and
clinical chemistry measurements are to be encouraged to better document the fluctuations in body fluid nickel levels with heart disease.
It is crucial that laboratory personnel adopt sound protocols for
avoiding contamination of specimens during collection, storage,
handling, and the analytical procedures [22,222]. Of course, a
formal analysis of existing epidemiologic data and the initiation
of new epidemiologic studies are necessary in clarifying the potential seriousness of the human cardiotoxicity of nickel compounds.

8. CONCLUDING REMARKS AND SUMMARY

It is evident from the material reviewed in this chapter and its
companion on nickel metabolism (Chap. 4) that organisms require
trace amounts of nickel for optimum function; by contrast, large
doses or inappropriate forms are toxic. Nonoccupational environmental exposures are minimal. The major source of nickel is dietary,
with drinking water providing a minor component. Nickel leachates
from certain nickel alloys and stainless steels constitute a major
inadvertant source in dermal exposure. From the toxicologic perspective, the cancer story is the one most extensively documented.
Nickel-related human cancer constitutes an industrial disease. It
is virtually restricted to the nickel-producing industry and is
largely historical. The epidemiologic evidence is considered unequiv-
ocal, while the elucidation of the biological mechanisms involved have
proven more difficult and controversial. Collectively, the evidence
(epidemiologic, animal studies, and short-term in vitro assays) imply

a range of potencies for different compounds. The current hypothesis
is that the nickel ion is the ultimate carcinogen and the bioavailability, compartmentalization, and perhaps the mode of intracellular
delivery are determinants.

The volatility of nickel carbonyl is the basis for its use in
the purification of nickel. It is lipophilic and this, in part,
explains its severe toxicity. The lung is the major target organ
although neurological and gastrointestinal disturbances contribute
to the clinical manifestations (resemble viral influenzal pneumonia).
It is noted that the induced pulmonary damage and edema are characteristic of agents known to act by a radical pathway. Diethyldithiocarbamate (DDC) or Antabuse serve as therapeutic chelating agents and
a number of reservations concerning this practice are itemized.

Nickel hypersensitivity is expressed as contact dermatitis or
asthma. The dermatological response is a delayed, cell-mediated
occurrence. More women are affected than men and this is rationalized in terms of higher domestic exposures. Nickel-induced occupational asthma involves either an immediate or delayed response to
the nickel-containing allergen with distinct underlying immunological
mechanisms. The asthmatic response to nickel compounds is not widespread.

No detrimental developmental and reproductive effects or
serious renal toxicity have been documented in humans. In animals,
nickel compounds act as immunosuppressants, especially when the lung
is the route of uptake. For example, exposed animals are more suscep-
tible to infection. Histopathological changes are known to occur in
nasal mucosa of active and retired nickel workers.

Nickel(II) salts are cardiotoxic in dogs and hypernickelemia
is known to occur in patients with acute myocardial infarction or
unstable angina pectoris. The relationship of elevated plasma nickel
levels to the pathogenesis of ischemia is unknown. An analysis of
the available epidemiologic data on nickel refinery workers suggest
that cardiovascular disease is not a major risk associated with
occupational exposure to nickel compounds.

ABBREVIATIONS

CHO   — Chinese hamster ovary

DDC   — diethyldithiocarbamate

ETAF — epidermal cell-derived thymocyte-activating factor

HSA   — human serum albumin

i.r. — intrarenal

LMT   — leukocyte migration test

LTT   — lymphocyte transformation test

SCE   — sister chromatid exchange

SHE   — Syrian hamster embryo

TETD — tetraethylthiuram disulfide (known as Antabuse)

ACKNOWLEDGMENTS

Financial support from the Natural Sciences and Engineering Research Council of Canada and from the Occupational Health and Safety Division of the Ontario Ministry of Labour is gratefully acknowledged.

REFERENCES

1. T. P. A. Stuart, *Arch. Exp. Pathol. Pharmakol.*, *18*, 151 (1884).

2. F. W. Sunderman, Jr., and S. S. Brown, in "Progress in Nickel Toxicology" (S. S. Brown and F. W. Sunderman, Jr., eds.), Blackwell Scientific, Oxford, UK, 1985, pp. 1-6.

3. F. W. Sunderman, Jr. (ed.-in-chief), "Nickel", National Academy of Sciences, Washington, D.C., 1975.

4. DHEW (NIOSH), "Criteria for a Recommended Standard, Occupational Exposure to Inorganic Nickel", U.S. Department of Health, Education and Welfare, Washington, D.C., 1977.

5. S. S. Brown and F. W. Sunderman, Jr., "Nickel Toxicology", Academic, London, 1980.

6. NRCC, "Effects of Nickel in the Canadian Environment", NRCC Document 18568, National Research Council of Canada (Environmental Secretariat), Ottawa, Canada, 1981.

7. J. O. Nriagu, "Nickel in the Environment", John Wiley, New York, 1980.

8. J.-P. Rigaut, "Rapport Préparatoire sur les Critères de Santé pour le Nickel", Commission of the European Community Directorate General, Luxembourg, 1983.

9. F. W. Sunderman, Jr. (ed.-in-chief), "Nickel in the Human Environment", IARC Sci. Publ. 53, Oxford Univ. Press, Oxford, UK, 1984.

10. S. S. Brown and F. W. Sunderman, Jr. (eds.), "Progress in Nickel Toxicology", Blackwell Scientific, Oxford, UK, 1985.

11. EPA, "Health Assessment Document for Nickel and Nickel Compounds", EPA/600/8-83/012FF, U.S. Environmental Protection Agency, Research Triangle Park, NC, 1986.

12. P. Grandjean, "Health Effects Document on Nickel", Ontario Ministry of Labour Document, Toronto, Canada, 1986.

13. R. Doll, in "Nickel in the Human Environment", IARC Sci. Publ. 53 (F. W. Sunderman, Jr., ed.-in-chief), Oxford Univ. Press, Oxford, UK, 1984, pp. 3-21.

14. J. G. Morgan, *Br. J. Indust. Med.*, *15*, 224 (1958).

15. W. N. Rezuke, J. A. Knight, and F. W. Sunderman, Jr., *Am. J. Indust. Med.*, *11*, 419 (1987).

16. S. S. Brown, S. Nomoto, M. Stoeppler, F. W. Sunderman, Jr., *Pure Appl. Chem.*, *53*, 773 (1981).

17. M. D. McNeely, M. W. Nechay, and F. W. Sunderman, Jr., *Clin. Chem.*, *18*, 992 (1972).

18. L. G. Morgan, *J. Soc. Occup. Med.*, *29*, 33 (1979).

19. J. S. Warner, in "Nickel in the Human Environment", IARC Sci. Publ. 53 (F. W. Sunderman, Jr., ed.-in-chief), Oxford Univ. Press, Oxford, UK, 1984, pp. 419-437.

20. E. Mastromatteo, *Am. Indust. Hyg. Assoc. J.*, *47*, 589 (1986).

21. E. Nieboer, A. Yassi, A. A. Jusys, and D. C. F. Muir, "The Technical Feasibility and Usefulness of Biological Monitoring in the Nickel Producing Industry", special document, McMaster University, Hamilton, Canada, 1984. (Available from the Nickel Producers Environmental Research Association, Tribune Tower, 435 North Michigan Avenue, Chicago, IL 60611.)

22. F. W. Sunderman, Jr., A. Aitio, L. G. Morgan, and T. Norseth, *Toxicol. Indust. Health*, *2*, 17 (1986).

23. F. W. Sunderman, Jr., S. M. Hopfer, J. A. Knight, K. S. McCully, A. G. Cecutti, P. G. Thornhill, K. Conway, C. Miller, S. R. Patierno, and M. Costa, *Carcinogenesis*, *8*, 305 (1987).

24. E. Hassler, B. Lind, B. Nilsson, and M. Piscator, *Ann. Clin. Lab. Sci.*, *13*, 217 (1983).

25. P. Grandjean, I. J. Selikoff, S. K. Shen, and F. W. Sunderman, Jr., *Am. J. Indust. Med.*, *1*, 181 (1980).

26. R. M. Stern, in "Biological and Environmental Aspects of Chromium", Topics in Environmental Health, Vol. 5 (S. Langård, ed.), Elsevier, Amsterdam, 1982, pp. 5-47.

27. R. A. Bale and E. T. Chapman, paper presented at the 25th Annual Conference of Metallurgists, Canadian Institute of Mining and Metallurgy, Toronto, Canada, Aug. 17-20, 1986.

28. S. Zhicheng, Br. J. Indust. Med., 43, 422 (1986).

29. F. W. Sunderman, Sr., and J. F. Kincaid, J. Am. Med. Assoc., 155, 889 (1954).

30. U. Vuopala, E. Huhti, J. Takkunen, and M. Huikko, Ann. Clin. Res., 2, 214 (1970).

31. F. W. Sunderman, Jr., Ann. Clin. Lab. Sci., 7, 377 (1977).

32. F. W. Sunderman, Sr., and F. W. Sunderman, Jr., Am. J. Med. Sci., 236, 26 (1958).

33. F. W. Sunderman, Sr., Ann. Clin. Res., 3, 182 (1971).

34. F. W. Sunderman, Sr., Ann. Clin. Lab. Sci., 9, 1 (1979).

35. F. W. Sunderman, Sr., Ann. Clin. Lab. Sci., 11, 1 (1981).

36. F. W. Sunderman, Jr., and C. E. Selin, Toxicol. Appl. Pharmacol., 12, 207 (1968).

37. A. Oskarsson and H. Tjälve, Br. J. Indust. Med., 36, 326 (1979).

38. H. Tjälve, S. Jasim, and A. Oskarsson, in "Nickel in the Human Environment", IARC Sci. Publ. 53 (F. W. Sunderman, Jr., ed.-in-chief), Oxford Univ. Press, Oxford, UK, 1984, pp. 311-320.

39. K. S. Kasprzak and F. W. Sunderman, Jr., Toxicol. Appl. Pharmacol., 15, 295 (1969).

40. E. Horak, F. W. Sunderman, Jr., and B. Sarkar, Res. Commun. Chem. Pathol. Pharmacol., 14, 153 (1976).

41. A. Oskarsson and H. Tjälve, Arch. Toxicol., 45, 45 (1980).

42. R. C. Baselt, F. W. Sunderman, Jr., J. Mitchell, and E. Horak, Res. Commun. Chem. Pathol. Pharmacol., 18, 677 (1977).

43. R. G. Evans, C. R. Engel, C. L. Wheatley, J. R. Nielson, and L. J. Ciborowski, Int. J. Radiat. Oncol. Biol. Phys., 9, 1635 (1983).

44. R. G. Evans, Int. J. Radiat. Oncol. Biol. Phys., 11, 1163 (1985).

45. A. W. Maners, E. M. Walker, Jr., M. Baker, and A. Pappas, Ann. Clin. Lab. Sci., 15, 343 (1985).

46. F. W. Sunderman, Jr., Toxicol. Environ. Chem., 15, 59 (1987).

47. C. D. Klaassen, M. O. Amdur, and J. Doull (eds.), "Casarett and Doull's Toxicology", 3rd ed., Macmillan, New York, 1986, pp. 351, 556.

48. S. D. Prystowsky, A. M. Allen, R. W. Smith, J. H. Nonomura, R. B. Odom, and W. A. Akers, *Arch. Dermatol.*, *115*, 959 (1979).

49. L. Peltonen, *Contact Dermatitis, 5*, 27 (1979).

50. M. Kieffer, *Contact Dermatitis, 5*, 398 (1979).

51. J. Oleffe, M. J. Nopp-Oger, and G. Achten, *Berufs-Dermatosen, 20*, 209 (1972).

52. North American Contact Dermatitis Group, *Arch. Dermatol.*, *108*, 537 (1973).

53. C. C. Sun, *J. Formosan Med. Assoc.*, *79*, 220 (1980).

54. E. Cronin, "Contact Dermatitis", Churchill Livingstone, Edinburgh, UK, 1980, pp. 338-366.

55. J. M. Lachapelle and D. Tennstedt, *Contact Dermatitis, 8*, 193 (1982).

56. T. Menné and N. Hjorth, *Sem. Dermatol.*, *1*, 15 (1982).

57. C. Cavelier, J. Foussereau, and M. Massin, *Contact Dermatitis, 12*, 65 (1985).

58. T. Fischer, S. Fregert, B. Gruvberger, and I. Rystedt, *Contact Dermatitis, 10*, 23 (1984).

59. L. M. Wall and C. D. Calnan, *Contact Dermatitis, 6*, 414 (1980).

60. P. V. Marcussen, *Br. J. Indust. Med.*, *17*, 65 (1960).

61. S. Fregert, *Contact Dermatitis, 1*, 96 (1975).

62. O. B. Christensen and H. Möller, *Contact Dermatitis, 1*, 129 (1975).

63. C. D. Calnan, *Br. J. Dermatol.*, *68*, 229 (1956).

64. O. B. Christensen and H. Möller, *Contact Dermatitis, 1*, 136 (1975).

65. E. Cronin, A. D. DiMichiel, and S. S. Brown, in "Nickel Toxicology" (S. S. Brown and F. W. Sunderman, Jr., eds.), Academic, London, 1980, pp. 149-152.

66. K. Kaaber, N. K. Veien, and J. C. Tjell, *Br. J. Dermatol.*, *98*, 197 (1978).

67. K. E. Andersen, G. D. Nielsen, M.-A. Flyvholm, S. Fregert, and B. Gruvberger, *Contact Dermatitis, 9*, 140 (1983).

68. T. Menné, K. Kaaber, and J. C. Tjell, *Ann. Clin. Lab. Sci.*, *10*, 160 (1980).

69. K. Kaaber, T. Menné, N. Veien, and P. Hougaard, *Contact Dermatitis, 9*, 297 (1983).

70. D. Spruit, P. J. M. Bongaarts, and K. E. Malten, in "Nickel in the Environment" (J. O. Nriagu, ed.), John Wiley, New York, 1980, pp. 601-609.

71. H. Thulin, *Acta Derm. Venereol. (Stockholm)*, *56*, 377 (1976).

72. L. Polak, in "Chromium Metabolism and Toxicity" (D. Burrows, ed.), CRC, Boca Raton, FL, 1983, pp. 51-136.

73. N. G. Al-Tawil, J. A. Marcusson, and E. Möller, *Acta Derm. Venereol. (Stockholm)*, *61*, 511 (1981).

74. N. G. Al-Tawil, G. Berggren, L. Emtestam, J. Fransson, R. Jernselius, and J. A. Marcusson, *Acta Derm. Venereol. (Stockholm)*, *65*, 385 (1985).

75. C. R. Menon and E. Nieboer, *J. Inorg. Biochem.*, *28*, 217 (1986).

76. R. M. Adams, "Occupational Skin Disease", Grune and Stratton, New York, 1983, pp. 1-26.

77. D. N. Sauder, in "Pathogenesis of Skin Disease" (B. H. Thiers and R. L. Dobsen, eds.), Churchill Livingstone, Edinburgh, UK, 1986, pp. 3-12.

78. A. Fullerton, J. R. Andersen, A. Hoelgaard, and T. Menné, *Contact Dermatitis*, *15*, 173 (1986).

79. S. Sjöborg, A. Andersson, and O. B. Christensen, *Acta Derm. Venereol. (Stockholm)*, Suppl. 111, 1 (1984).

80. R. L. Edelson and J. M. Fink, *Sci. Am.*, *252(6)*, 46 (1985).

81. F. Sinigaglia, D. Scheidegger, G. Garotta, R. Scheper, M. Pletscher, and A. Lanzavecchia, *J. Immunol.*, *135*, 3929 (1985).

82. S. Silvennoinen-Kassinen, H. Jakkula, and J. Karvonen, *J. Invest. Dermatol.*, *86*, 18 (1986).

83. S. Silvennoinen-Kassinen, R. Karttunen, and J. Karvonen, *Scand. J. Immunol.*, *22*, 452 (1985).

84. S. Silvennoinen-Kassinen, *Acta Dermat. Venereol. (Stockholm)*, *62*, 251 (1982).

85. L. H. McConnell, J. N. Fink, D. P. Schleuter, and M. G. Schmidt, *Ann. Intern. Med.*, *78*, 888 (1973).

86. J.-L. Malo, A. Cartier, N. Doepner, E. Nieboer, S. Evans, and J. Dolovich, *J. Allergy Clin. Immunol.*, *69*, 55 (1982).

87. G. T. Block and M. Yeung, *J. Am. Med. Assoc.*, *247*, 1600 (1982).

88. H. S. Novey, M. Habib, and I. D. Wells, *J. Allergy Clin. Immunol.*, *72*, 407 (1983).

89. J.-L. Malo, A. Cartier, G. Gagnon, S. Evans, and J. Dolovich, *Clin. Allergy*, *15*, 95 (1985).

90. J. E. Davies, *J. Soc. Occup. Med.*, *36*, 29 (1986).

91. A. M. Cirla, F. Bernabeo, F. Ottoboni, and R. Ratti, in "Progress in Nickel Toxicology" (S. S. Brown and F. W. Sunderman, Jr., eds.), Blackwell Scientific, Oxford, UK, 1985, pp. 165-168.

92. J. Pepys, *J. Allergy Clin. Immunol., 66,* 179 (1980).

93. P. M. O'Byrne, E. H. Walters, H. Aizawa, L. M. Fabbri, M. J. Holtzman, and J. A. Nadel, *Am. Rev. Respir. Dis., 130,* 220 (1984).

94. J. G. R. De Monchy, H. F. Kauffman, P. Venge, G. H. Koeter, H. M. Jansen, H. J. Sluiter, and K. de Vries, *Am. Rev. Respir. Dis., 131,* 373 (1985).

95. F. E. Hargreave, E. H. Ramsdale, and S. O. Pugsley, *Am. Rev. Respir. Dis., 130,* 513 (1984).

96. J. Dolovich, S. L. Evans, and E. Nieboer, *Br. J. Indust. Med., 41,* 51 (1984).

97. E. Nieboer, S. L. Evans, and J. Dolovich, *Br. J. Indust. Med., 41,* 56 (1984).

98. F. W. Sunderman, Jr., *Environ. Health Persp., 40,* 131 (1981).

99. F. W. Sunderman, Jr., *Ann. Clin. Lab. Sci., 14,* 93 (1984).

100. R. S. Roberts, J. A. Julian, and D. C. F. Muir, "The JOHC-INCO Mortality Study: An Analysis of Mortality from Cancer", McMaster Univ., Hamilton, Canada, 1982.

101. R. S. Roberts, J. A. Julian, D. C. F. Muir, and H. S. Shannon, in "Nickel in the Human Environment", IARC Sci. Publ. 53 (F. W. Sunderman, Jr., ed.-in-chief), Oxford Univ. Press, Oxford, UK, 1984, pp. 23-55.

102. H. S. Shannon, J. A. Julian, and R. S. Roberts, *J. Natl. Cancer Inst., 73,* 1251 (1984).

103. R. Doll, L. G. Morgan, and F. E. Speizer, *Br. J. Cancer, 24,* 623 (1970).

104. R. Doll, J. D. Mathews, and L. G. Morgan, *Br. J. Indust. Med., 34,* 102 (1977).

105. E. Pedersen, A. C. Høgetveit, and A. Andersen, *Int. J. Cancer, 12,* 32 (1973).

106. K. Magnus, A. Andersen, and A. C. Høgetveit, *Int. J. Cancer, 30,* 681 (1982).

107. A. D. Ottolenghi, J. K. Haseman, W. W. Payne, H. L. Falk, and H. N. MacFarland, *J. Natl. Cancer Inst., 54,* 1165 (1974).

108. F. W. Sunderman, Sr., and A. J. Donnelly, *Am. J. Pathol., 46,* 1027 (1965).

109. A. Yassi and E. Nieboer, in "Chromium in the Natural and Human Environments", Vol. 19, Advances in Environmental Science and Technology (J. O. Nriagu and E. Nieboer, eds.), John Wiley, New York, 1988, in press.

110. J. A. DiPaolo, *J. Natl. Cancer Inst., 70,* 3 (1983).

111. M. L. Larramendy, N. C. Popescu, and J. A. DiPaolo, *Environ. Mutagen., 3,* 597 (1981).

112. M. Costa, J. S. Nye, F. W. Sunderman, Jr., P. R. Allpass, and B. Gondos, *Cancer Res., 39,* 3591 (1979).

113. E. C. Friedberg, "DNA Repair", W. H. Freeman, New York, 1985, pp. 505-574.

114. N. E. Day and C. C. Brown, *J. Natl. Cancer Inst., 64,* 977 (1980).

115. J. Kaldor, J. Peto, D. Easton, R. Doll, C. Hermon, and L. Morgan, *J. Natl. Cancer Inst., 77,* 841 (1986).

116. U.S. Interagency Staff Group on Carcinogens, *Environ. Health Persp., 67,* 201 (1986).

117. E. Farber, *N. Engl. J. Med., 305,* 1379 (1981).

118. E. Boyland, *Br. J. Indust. Med., 42,* 716 (1985).

119. F. W. Sunderman, Jr., K. S. McCully, and S. M. Hopfer, *Carcinogenesis, 5,* 1511 (1984).

120. F. W. Sunderman, Jr., in "Nickel in the Human Environment", IARC Sci. Publ. 53 (F. W. Sunderman, Jr., ed.-in-chief), Oxford Univ. Press, Oxford, UK, 1984, pp. 127-142.

121. W. Jelkmann, *Rev. Physiol. Biochem. Pharmacol., 104,* 139 (1986).

122. A. C. Guyton, "Textbook of Medical Physiology", W. B. Saunders, Philadelphia, 1986, pp. 42-50.

123. E. Nieboer, R. I. Maxwell, F. E. Rossetto, A. R. Stafford, and P. I. Stetsko, in "Frontiers in Bioinorganic Chemistry" (A. V. Xavier, ed.), VCH Verlagsgesellschaft, Weinheim, FRG, 1986, pp. 142-151.

124. J. D. Heck and M. Costa, *Biol. Trace Element Res., 4,* 71 (1982).

125. J. D. Heck and M. Costa, *Biol. Trace Element Res., 4,* 319 (1982).

126. N. T. Christie and M. Costa, *Biol. Trace Element Res., 5,* 55 (1983).

127. N. T. Christie and M. Costa, *Biol. Trace Element Res., 6,* 139 (1984).

128. I. Singh, *Mut. Res., 117,* 149 (1983).

129. I. Singh, *Mut. Res., 137,* 47 (1984).

130. D. E. Amacher and S. C. Paillet, *Mut. Res., 78,* 279 (1980).

131. M. Miyaki, N. Akamatsu, T. Ono, and H. Koyama, *Mut. Res., 68,* 259 (1979).

132. N. W. Biggart and M. Costa, *Mut. Res., 175,* 209 (1986).

133. A. Arlauskas, R. S. U. Baker, A. M. Bonin, R. K. Tandon, P. T. Crisp, and J. Ellis, *Environ. Res., 36,* 379 (1985).

134. N. Kanematsu, M. Hara, and T. Kada, *Mut. Res., 77,* 109 (1980).

135. M. H. L. Green, W. J. Muriel, and B. A. Bridges, *Mut. Res., 38,* 33 (1976).

136. P. Pikálek and J. Nečásek, *Folia Microbiol.*, *28*, 17 (1983).

137. J. S. Dubins and J. M. LaVelle, *Mut. Res.*, *162*, 187 (1986).

138. M. Costa and H. H. Mollenhauer, *Cancer Res.*, *40*, 2688 (1980).

139. J. A. DiPaolo and B. C. Casto, *Cancer Res.*, *39*, 1008 (1979).

140. M. Costa, J. Simmons-Hansen, C. W. M. Bedrossian, J. Bonura, and R. M. Caprioli, *Cancer Res.*, *41*, 2868 (1981).

141. E. Rivedal and T. Sanner, *Cancer Res.*, *41*, 2950 (1981).

142. B. C. Casto, J. Meyers, and J. A. DiPaola, *Cancer Res.*, *39*, 193 (1979).

143. H. J. K. Saxholm, A. Reith, and A. Brogger, *Cancer Res.*, *41*, 4136 (1981).

144. K. Hansen and R. M. Stern, *Environ. Health Persp.*, *51*, 223 (1983).

145. J. F. Lechner, T. Tokiwa, I. A. McClendon, and A. Haugen, *Carcinogenesis*, *5*, 1697 (1984).

146. R. H. Al-Rabaee and D. C. Sigee, *J. Cell Sci.*, *69*, 87 (1984).

147. M. Costa and J. D. Heck, *Adv. Inorg. Biochem.*, *6*, 285 (1984).

148. P. Sen and M. Costa, *Toxicol. Appl. Pharmacol.*, *84*, 278 (1986).

149. M. Costa, J. D. Heck, and S. H. Robison, *Cancer Res.*, *42*, 2757 (1982).

150. P. Sen and M. Costa, *Cancer Res.*, *45*, 2320 (1985).

151. S. H. Robison and M. Costa, *Cancer Lett.*, *15*, 35 (1982).

152. S. R. Patierno, M. Sugiyama, J. P. Basilion, and M. Costa, *Cancer Res.*, *45*, 5787 (1985).

153. P. Sen and M. Costa, *Carcinogenesis*, *7*, 1527 (1986).

154. S. M. Newman, R. L. Summitt, and L. J. Nunez, *Mut. Res.*, *101*, 67 (1982).

155. M. Umeda and M. Nishimura, *Mut. Res.*, *67*, 221 (1979).

156. M. Costa, O. Cantoni, M. de Mars, and D. E. Swartzendruber, *Res. Commun. Chem. Pathol. Pharmacol.*, *38*, 405 (1982).

157. S. H. Robison, O. Cantoni, J. D. Heck, and M. Costa, *Cancer Lett.*, *17*, 273 (1983).

158. M. A. Sirover and L. A. Loeb, *Science*, *194*, 1434 (1976).

159. L. A. Loeb and A. S. Mildvan, in "Metal Ions in Genetic Information Transfer" (G. L. Eichhorn and L. G. Marzilli, eds.), Elsevier, New York, 1981, pp. 125-142.

160. S. K. Niyogi, R. P. Feldman, and D. J. Hoffman, *Toxicology*, *22*, 9 (1981).

161. S. K. Niyogi and R. P. Feldman, *Nucl. Acid. Res.*, *9*, 2615 (1981).

162. O. Andersen, *Res. Commun. Chem. Pathol. Pharmacol.*, *50*, 379 (1985).

163. J. H. van de Sande and T. M. Jovin, *EMBO J.*, *1*, 115 (1982).

164. J. Liquier, P. Bourtayre, L. Pizzorni, F. Sournies, J.-F. Labarre, and E. Taillandier, *Anticancer Res.*, *4*, 41 (1984).

165. R. B. Ciccarelli and K. E. Wetterhahn, *Chem.-Biol. Interact.*, *52*, 347 (1985).

166. R. B. Ciccarelli and K. E. Wetterhahn, *Cancer Res.*, *44*, 3892 (1984).

167. R. B. Ciccarelli and K. E. Wetterhahn, *Cancer Res.*, *42*, 3544 (1982).

168. H. Waksvik and M. Boysen, *Mut. Res.*, *103*, 185 (1982).

169. H. Waksvik, M. Boysen, and A. C. Høgetveit, *Carcinogenesis, 5,* 1525 (1984).

170. H. Hamada, M. G. Petrino, and T. Kakunaga, *Proc. Natl. Acad. Sci. USA, 79,* 6465 (1982).

171. R. Durand, C. Job, D. A. Zarling, M. Teissère, T. M. Jovin, and D. Job, *EMBO J.*, *2*, 1707 (1983).

172. J. J. Butzow, Y. A. Shin, and G. L. Eichhorn, *Biochemistry, 23,* 4837 (1984).

173. E. Nieboer and S. L. Shaw, in "Chromium in the Natural and Human Environments", Vol. 19, Advances in Environmental Science and Technology (J. O. Nriagu and E. Nieboer, eds.), John Wiley, New York, 1988, in press.

174. L. Sachs, *Sci. Am.*, *254(1)*, 40 (1986).

175. J. M. Bishop, *Ann. Rev. Biochem.*, *52*, 301 (1983).

176. G. M. Brodeur, *Prog. Hematol.*, *14*, 229 (1986).

177. D. T. Denhardt, D. R. Edwards, and C. L. J. Parfett, *Biochim. Biophys. Acta, 865,* 83 (1986).

178. D. G. Scarpelli, *Ann. Clin. Lab. Sci.*, *13*, 249 (1983).

179. J. C. Eissenberg, I. L. Cartwright, G. H. Thomas, and S. C. R. Elgin, *Ann. Rev. Genet.*, *19*, 485 (1985).

180. J. J. Yunis, *Science, 221,* 227 (1983).

181. R. Sager, *Cancer Res.*, *46*, 1573 (1986).

182. B. N. Ames, *Science, 221,* 1256 (1983).

183. Y. Kurokawa, M. Matsushima, T. Imazawa, N. Takamura, M. Takahashi, and Y. Hayashi, *J. Am. Coll. Toxicol.*, *4*, 321 (1985).

184. F. W. Sunderman, Jr., and E. Horak, in "Chemical Indices and Mechanisms of Organ-Directed Toxicity" (S. S. Brown and D. S. Davies, eds.), Pergamon, Oxford, UK, 1981, pp. 52-64.

185. R. R. Lauwerys, A. Bernard, H. A. Roels, J.- P. Buchet, and C. Viau, *Environ. Health Persp.*, *54*, 147 (1984).

186. C. G. Elinder, C. Edling, E. Lindberg, B. Kågedal, and A. Vesterberg, *Br. J. Indust. Med.*, *42*, 754 (1985).

187. W. E. Sanford, "Renal Clearance and Nephrotoxicity of Nickel", Ph.D. thesis, Univ. of Surrey, UK, December 1987.

188. P. H. Gitlitz, F. W. Sunderman, Jr., and P. J. Goldblatt, *Toxicol. Appl. Pharmacol.*, *34*, 430 (1975).

189. E. Horak and F. W. Sunderman, Jr., *Ann. Clin. Lab. Sci.*, *10*, 425 (1980).

190. T. W. Clarkson, G. F. Nordberg, and P. R. Sager, *Scand. J. Work Environ. Health*, *11*, 145 (1985).

191. J. G. Wilson, in "Handbook of Teratology, Vol. 1, General Principles and Etiology" (J. G. Wilson and F. C. Fraser, eds.), Plenum, New York, 1977, pp. 357-370.

192. C.-C. Lu, N. Matsumoto, and S. Iijima, *Toxicol. Appl. Pharmacol.*, *59*, 409 (1981).

193. A. Leonard, G. B. Gerber, and P. Jacquet, *Mut. Res.*, *87*, 1 (1981).

194. F. W. Sunderman, Jr., M. C. Reid, S. K. Shen, and C. B. Kevorkian, in "Reproductive and Developmental Toxicity of Metals" (T. W. Clarkson, G. F. Nordberg, and P. R. Sagar, eds.), Plenum, 1983, pp. 399-416.

195. S. M. Barlow and F. M. Sullivan, "Reproductive Hazards of Industrial Chemicals", Academic, London, 1981.

196. J. G. Wilson, in "Handbook of Teratology, Vol. 1, General Principles and Etiology" (J. G. Wilson and F. C. Fraser, eds.), Plenum, New York, 1977, pp. 47-74.

197. F. W. Sunderman, Jr., S. K. Shen, J. M. Mitchell, P. R. Allpass, and I. Damjanov, *Toxicol. Appl. Pharmacol.*, *43*, 381 (1978).

198. A. Mas, D. Holt, and M. Webb, *Toxicology*, *35*, 47 (1985).

199. R. Storeng and J. Jonsen, *Toxicology*, *20*, 45 (1981).

200. F. W. Sunderman, Jr., P. R. Allpass, J. M. Mitchell, R. C. Baselt, and D. M. Albert, *Science*, *203*, 550 (1979).

201. F. W. Sunderman, Jr., S. K. Shen, M. C. Reid, and P. R. Allpass, *Teratogenesis Carcinog. Mutagen.*, *1*, 223 (1980).

202. S. Jasim and H. Tjälve, *Toxicology*, *32*, 297 (1984).

203. W. Torjussen, L. A. Solberg, and A. C. Høgetveit, *Br. J. Cancer*, *40*, 568 (1979).

204. M. Boysen, R. Puntervold, B. Schüler, and A. Reith, in "Nickel Toxicology" (S. S. Brown and F. W. Sunderman, Jr., eds.), Academic, London, 1980, pp. 39-42.

205. M. Boysen, L. A. Solberg, I. Andersen, A. C. Høgetveit, and W. Torjussen, *Scand. J. Work Environ. Health, 8,* 283 (1982).

206. D. Adalis, D. E. Gardner, and F. J. Miller, *Am. Rev. Res. Dis., 118,* 347 (1978).

207. D. E. Gardner, in "Nickel Toxicology" (S. S. Brown and F. W. Sunderman, Jr., eds.), Academic, London, 1980, pp. 121-124.

208. R. C. Murthy and W. J. Niklowitz, *J. Submicrosc. Cytol., 15,* 655 (1983).

209. P. Camner, A. Johansson, and M. Lundborg, *Environ. Res., 16,* 226 (1978).

210. R. C. Murthy, W. Barkley, L. Hollingsworth, and E. Bingham, *J. Am. Coll. Toxicol., 2,* 193 (1983).

211. J. A. Graham, F. J. Miller, M. J. Daniels, E. A. Payne, and D. E. Gardner, *Environ. Res., 16,* 77 (1978).

212. T. Spiegelberg, W. Kördel, and D. Hochrainer, *Ecotoxicol. Environ. Safety, 8,* 516 (1984).

213. R. J. Smialowicz, R. R. Rogers, M. M. Riddle, and G. A. Stott, *Environ. Res., 33,* 413 (1984).

214. R. J. Smialowicz, R. R. Rogers, M. M. Riddle, R. J. Gardner, D. G. Rowe, and R. W. Luebke, *Environ. Res., 36,* 56 (1985).

215. C. N. Leach, Jr., J. V. Linden, S. M. Hopfer, M. C. Crisostomo, and F. W. Sunderman, Jr., *Clin. Chem., 31,* 556 (1985).

216. G. Rubányi, L. Ligeti, and A. Koller, *J. Mol. Cell. Card., 13,* 1023 (1981).

217. J. V. Linden, S. M. Hopfer, H. R. Gossling, and F. W. Sunderman, Jr., *Ann. Clin. Lab. Sci., 15,* 459 (1985).

218. S. M. Hopfer, J. V. Linden, W. N. Rezuke, J. E. O'Brien, L. Smith, F. Watters, and F. W. Sunderman, Jr., *Res. Commun. Chem. Pathol. Pharmacol., 55,* 101 (1987).

219. K. Szabó, I. Balogh, and A. Gergely, *Injury, 16,* 613 (1985).

220. F. W. Sunderman, Jr., *Ann. Clin. Lab. Sci., 13,* 1 (1983).

221. C. N. Leach, Jr., and F. W. Sunderman, Jr., *N. Engl. J. Med., 313,* 1232 (1985).

222. E. Nieboer and A. A. Jusys, in "Chemical Toxicology and Clinical Chemistry of Metals" (S. S. Brown and J. Savory, eds.), Academic, London, 1983, pp. 3-16.

11

Analysis of Nickel in Biological Materials

Hans G. Seiler
Institute of Inorganic Chemistry
University of Basel
Spitalstrasse 51
CH-4056 Basel, Switzerland

1. GENERAL ASPECTS OF ANALYSES OF NICKEL IN BIOLOGICAL MATERIALS

In all trace and ultratrace analyses the reliability and from this the relevance of analytical results depend not only on the sensitivity, precision, and reproducibility of the chemical analytical work

but they are influenced by all interfering events the specimen, i.e.,
the sample to be analyzed, has undergone from specimen collection to
analytical measurement. The general concept is that the lower the
concentration of the element to be determined and the higher the
ubiquitous occurrence of this element, the better must be the precautions to prevent contaminations. Nowadays very sensitive and
precise methods for the determination of nickel are available and
losses can be prevented by adequate sample preparation methods. The
main source for erroneous results is contamination.

In spite of the rather low natural occurrence, i.e., 0.02% of
the upper earth's crust, there are serious sources of potential contaminations [1,2] by the increasing widespread use of nickel and its
alloys with different other metals, like stainless steel, monel,
hastalloys, etc. Nickel is also employed as a catalyst for hydro-
genation and dehydrogenation, and for color pigments, e.g., yellow
nickel titanate.

Nickel and its alloys are preferred construction materials
because of their good corrosion resistance against atmospheric influences and different chemicals. This corrosion resistance is not
based on a highly positive electrochemical potential but on the formation of a very thin protective oxide film on the surface. This
oxide film can be dissolved by different chemicals, above all by
substances forming stable complexes either with nickel or with other
constituents of the respective alloy, e.g., with chromium in stainless steel. Corrosion is enhanced by the formation of local elements
in solutions of low oxygen concentration in the presence of electrolytes and/or complexing agents. Most biological materials contain
electrolytes as well as complexing agents. Many metallic implements
used for specimen collection, analysis, and chemical work in general,
like scalpels, spatula, needles of syringes, parts of mixers, etc.,
but also installations like fume hoods, ventilation channels, shafts
of stirrers, supports, etc., are made from stainless steel or nickel-
containing alloys. Most chemicals needed for chemical treatment have
come into contact with nickel during production. All this may give
rise to contaminations inside the laboratory, in addition to the

ambient air and above all dusts, which may contain considerable quantities of nickel depending on the specific environment [1,3-6].

In general, the concentration of nickel in biological materials
being rather low, all possibly occurring contaminations must be taken
into consideration. Thus the ambient conditions may even become pro-
hibitive for the execution of analyses for nickel. In the past, many
published results of analyses in biological materials showed too high
contents of nickel mainly due to contaminations [7]. If the expected
nickel content is in the same order of magnitude as in the ambient
air (urban and industrial environment), all laboratory work should
be done in a clean-air room or within a laminar flow workstation
fitted with a high-efficiency particulate air filter. By using laminar flow workstations one must be aware that installations of apparatus and manipulations inside the workstation may disrupt the laminar
flow and thus generate turbulences by which dust from outside can
penetrate. The efficacy of the workstation should be tested period-
ically. All materials, apparatus, and chemicals needed for handling,
treatment, and measurement as well as the environment in the laboratory must be tested for not contaminating the substance to be analyzed.
This will be of major importance in such cases where specific substances are to be isolated from a great bulk of material by different
stages of preparations.

2. CHOICE OF IMPLEMENTS AND CHEMICALS

As mentioned, materials and chemicals coming into contact with the
biological material to be analyzed should be selected above all with
regard to possible contaminations, but also to losses and interfering
interactions. Their exact composition, mode of fabrication, and chemical behavior should be well known. But even the best information
does not eliminate the need to scrupulously test all materials and
chemicals for their suitability, not only with respect to nickel but
also with respect to possible alterations of the biological material.
Every new lot of materials and chemicals must be retested for suita-

bility before use. In sight of the low content of nickel in biological materials and the possibly occurred contamination during
fabrication, even reagents of so-called highest purity must be
further purified. Therefore chemicals should not only be selected
with respect to their chemical properties but also with regard to
the ease and quality of their purification. Hence chemicals which
are distillable under normal laboratory conditions or easy to sub-
limate are to be preferred. Techniques for preparing exceptionally
high-purity acids have been reported by several authors [8-10]. In
order to prevent contaminations by opening the bottles and inserting
pipettes, liquid chemicals and solutions of reagents are preferably
stored in containers fitted with dispensers.

Generally, polyethylene containers have proved to be adequate
for specimen collection and the storage of biological materials.
But polyethylene is a collective name for a great variety of materials with different characteristics. Therefore containers from
polymers must scrupulously be tested for suitability because there
is a very great probability of having come into contact with nickel
during fabrication by molds or catalysts. Special attention must be
given to colored plastic stoppers, the pigment may contain nickel.
Rubber stoppers should never be used. All containers must be carefully cleaned by soaking with diluted HNO_3 (10% v/v) during several
days, renewing the acid every day, followed by a thorough rinsing
with high-purity water [11]. If the biological material may deteri-
orate due to acids, the containers must be thoroughly soaked with
high-purity water because acids are able to penetrate polymers and
are only slowly released. Cleaning with concentrated oxidizing
acids is not advisable, as these may create active adsorption sites
in the polymers causing possible losses of nickel from the stored
specimen. Common detergents, mostly sulfonated fatty acids, contain
considerable quantities of nickel from fabrication and should not be
used for cleaning purposes.

Constructional materials for the apparatus, the vessels, and
the tools for specimen collection, sample preparation, and measurement must obviously not be contaminating or absorbing and must be

resistant to specimens and chemicals. Their surface must be smooth
and easy to clean. Pure silica shows the best properties. Glassware
of high silica content like Pyrex may also be used but must first be
leached with boiling concentrated HNO_3. It must be remembered that
nickel-containing materials are used in the fabrication of glassware.
Tools like knives, spatulas, etc., should be made from pure titanium,
silica, or thoroughly cleaned high-purity plastics. The equipment
should allow several stages of the analytical procedure—from specimen
collection to measurement—to be carried out in the same vessel or
apparatus.

3. SPECIMEN COLLECTION; SAMPLING

Specimen collection and sampling are the most important but also the
most critical steps in every analysis. No analytical result can be
more relevant than the sample from which it has been obtained. The
specimen and the sample derived from it must exactly reflect the
properties of the investigated matrix. Thus thorough knowledge of
the investigated biological material is indispensable. Therefore the
planning of all steps of an analysis, but above all of the specimen
collection, should only be undertaken in close collaboration between
a scientist in the field of speciality and the analyst. In the past
a great number of data of dubious value have been produced by analysts
without sufficient knowledge about the special biological system as
well as by biologists and physicians without analytical training. It
should be remembered that errors in the chemical analytical work can
be recognized, evaluated, and largely eliminated by controls using
certified biological reference materials and statistical quality controls of the methods employed. But there is no method of control and
correction for the so-called preanalytical phase consisting of specimen collection, mailing, storage, homogenization, and aliquotation.
Any alteration of the biological material during this phase may modify
the result in vitro; these are the so-called interference factors
which can only be minimized by scrupulous standardization of all

manipulations observing all possible sources of errors [12]. A
major requirement for nickel analysis in biological materials is
that the specimens be taken under noncontaminating conditions, also
avoiding all kinds of possible losses. Whenever possible the analyst
or a person well instructed by him should execute the specimen collection in order to be able to recognize and prevent possible contaminations and losses. In any case, specimen containers and collection
implements should be prepared and made available by the analyst.
Disposable specimen collection equipment should be controlled for
suitability. For ethical and professional reasons the collection
of specimens of human blood and tissue can only be carried out by
medical staff; these persons must be well informed about the analytical requirements. Sterility does not exclude the possibility of
contamination.

With respect to their composition, liquid specimens of biological materials—mostly body fluids or the juice of plants—may
be considered more or less homogeneous at the moment of collection.
But soon after collection biological fluids may undergo different
alterations by coagulation, precipitation, bacterial actions, etc.
Therefore preservatives must be added immediately, which must be
scrupulously controlled for their nickel content. As nickel has
widespread use, exogenous contaminations can only be prevented by
adequately cleaning the subject from which the specimen is collected
and by using specially prepared tools and containers. Dust may be
one of the most important contaminants. The physiologic facts of
the subject from which the specimen is collected should also be
taken into consideration [13,14].

When collecting blood, the skin around the spot of collection
should thoroughly be washed because nickel-containing particulate
matter may strongly adhere to the skin. The position of the person
during collection should be defined because by the flow of extracellular fluid into the blood vessels in recumbent position the protein in the blood plasma is diluted [15]. The usually employed disposable needles made from stainless steel must be replaced by plastic
intravenous cannulae or by needles of silver or titanium. The blood

is transferred from the syringe to disposable containers of polyethylene which have been washed with 1 M HNO_3 and thoroughly rinsed
with high-purity water. Rubber-stoppered tubes should be avoided
[16]. Sodium citrate has proved to be a suitable anticoagulant.
As the genuine specimen for nickel analysis is serum, the blood
should be centrifuged immediately after the arrival in the laboratory in hermetically sealed tubes to avoid external contaminations
from the aerosol of nickel-containing particles generated by the
centrifuge motor. Serum specimens are removed from blood clots with
acid-washed Pasteur pipets and voided into thoroughly cleaned plastic
tubes. For a short period they can be stored at $+4°C$; for longer
periods $-20°C$ is advised [17].

Urine is the preferred specimen for monitoring human exposures
to nickel compounds, since collection of urine is noninvasive and
nickel concentrations are higher by nearly one order of magnitude
in urine than serum. In the specimen collection of urine several
additional aspects have to be considered. The composition of urine,
a final product of the metabolism, varies over a wide range. It is
influenced by physiologic factors, as well as by nutrition, state of
health, and physical and psychic stress. The composition varies not
only from one individual to another but also within the different
excretions of the same individual. Therefore the desirable specimen
would be the collection of all excretions during 24 hr, but as this
is difficult to realize, generally spontaneous excretions are collected. In order to be able to compare different specimens they
should be well defined with respect to time, e.g., the second excretion in the morning or at the end of a working shift [12,18]. Calculations from one spontaneous excretion to the total excretion of a
day are usually made by determining the creatinine content or the
density.

The main risk for exogenous contamination in the specimen
collection of urine is the subject himself. Therefore a whole-body
washing prior to collection and noncontaminating clothes are indispensable. The collection should take place in a noncontaminated
area. The urine is usually voided into adequately cleaned plastic

containers with large openings. Immediately after collection a
suitable aliquot of the urine is transferred to a specimen tube and
acidified to a pH <2 with high-purity HCl or HNO_3. This acidifica-
tion prevents the coprecipitation of nickel with urine sediments as
well as bacterial decomposition. Acidified specimens of urine may
be stored for several days at +4°C in the dark; for longer periods
-20°C is advised.

In contrast to liquids, tissues cannot be considered homogeneous, neither to their nickel content nor to the composition of the
biological matrix. Therefore the accuracy of the analysis is closely
correlated with the kind and the size of the specimen. Preferred
materials are the organs in which nickel is accumulated. The content
in these organs being significantly elevated, the risks to obtain
erroneous results due to contamination is reduced [18]. As the distribution of nickel within an organ is seldom uniform, the specimen
should be as large as possible. Analytical results from small parts
of organs, e.g., material from biopsy, should be specified by histo-
logic investigations. The reliability of the analytical result may
also be affected by the inhomogeneity of the biological matrix, e.g.,
adherent parts of tissue with very different nickel content or variations of the water content [13].

The weight of specimens and samples to which the nickel content
is related is another serious problem. The so-called fresh weight is
difficult to define after specimen collection. This can be overcome
to a certain extent by using tared specimen containers with tight
stoppers. Thus loss of water is minimal and the true weight of the
specimen can be determined. Normally no preservatives are added to
tissue specimens. In order to prevent deterioration and losses of
water they are immediately cooled to -20°C for storage.

In the collection of animal or human tissues, normally carried
out in laboratories or hospitals, considerable attention must be
given not to contaminate the specimen with nickel. All precautions
concerning contaminating tools and chemicals but also the environment
must be scrupulously observed. The surface of the body from which
the specimen is excised must be thoroughly cleaned. A frequent source

of losses is the outflow of intracellular liquid during excision.
Material from an autopsy must be taken as soon as possible post
mortem because deterioration of the tissue starts immediately after
death, thus altering the contents of elements [19,20].

Specimens of plants must mostly be collected in the field,
where they were exposed to the environment. Particles of dust or
soil may adhere strongly to their surface, especially if leaves,
stems, or fruits are covered with tiny hairs or wax. These particles
create a high risk for contamination. Small plants should be taken
entirely into the laboratory and decontaminated by thorough rinsing
with distilled water before cutting, in order to prevent losses of
extravascular juice and contamination by penetrating wash water.
Plants should never be immersed in water for long periods because
water may penetrate the plant or elements inherent to the plant may
be extracted. The use of detergents is not advisable. Adherent
water is wiped with filter paper, without pressing or rubbing. The
time elapsed between collection in the field and the laboratory
workup should be as short as possible. Plants must be transported
in cooled tanks. After decontamination, the parts of interest are
cut directly into an adequate container using a very sharp tool of
noncontaminating material, e.g., knives or scissors made of titanium
or silica. Metal-free plastic gloves should be worn [6].

To obtain specimens from large plants which cannot be taken in
their entirety into the laboratory, rather large parts should be collected in the field and decontaminated in the laboratory under adequate conditions. The genuine specimen should consist of only those
parts that are at a sufficient distance from the original surface of
cut made in the field, so that interfering events from decontamination as cited above cannot occur [6].

Prior to storage of tissue it is important to know whether the
entire specimen or an aliquot sample representative of the mean value
of the specimen will be submitted for analysis. In the first case no
special precautions are needed. When a representative aliquot is to
be taken, a homogenization step prior to storage is essential. During
freezing and thawing the structure of the specimen will be altered by

the rupture of membranes and outflow of intracellular liquid. Homogenization is mostly carried out in mixers with rotating knives. Often
the composition of the different materials from which the mixer is
made is unknown. The most contaminating items are the knife and the
shaft because there will always be some abrasion. They should never
be made from stainless steel or other nickel-containing alloys, but
rather from titanium or tantalum. All homogenized tissue specimens
are preferably dried or lyophilized before storage to better maintain
homogeneity and diminish bacterial interactions. Contaminations by
drying are mainly due to the material of the drying apparatus. Specimens for analyses for nickel should never be dried in a stainless
steel oven [18,21].

Sampling is the last step of the preanalytical phase that can
possibly affect the accuracy of the analytical result via interference factors. Its precision depends above all on the homogeneity of
the specimen and the size of the sample. In general, correctly
homogenized and lyophilized tissue specimens do not undergo changes
in their composition with respect to homogeneity during storage.

However, liquid specimens may have become inhomogeneous during
storage in spite of added preservatives. Nickel often tends to be
contained by or adsorbed to precipitated particles. Therefore homogeneity must be restored as best as possible. This is of highest
importance for analytical methods which allow the direct measurement
of nickel in very small sample volumes (10-50 µl) without prior sample
preparation. But inhomogeneity may also be an interfering factor for
analytical procedures with prior sample preparation, normally using
larger samples. Furthermore, it should be kept in mind that all
pipetting devices are calibrated with water. Biological fluids differ
from water with respect to density, viscosity, and surface tension.
Therefore, the accuracy of the delivered volume should be tested by
weighing.

The genuine specimen of blood being serum, there are fewer
problems to expect with respect to homogeneity. After warming up
to room temperature, it is sufficient to gently shake the specimen
while avoiding foaming.

In contrast, urine specimens are often inhomogeneous after storage, especially when they were frozen. As nickel tends to bind to the sediment or to adsorb to the wall of the container, much attention must be given to restoring homogeneity. Many investigations showed that the only method is to warm up the specimen to room temperature and to shake it vigorously for at least 30 min by means of a mechanical shaker. Immediately prior to pipetting a sample, the specimen must be shaken once more by hand. This demonstrates that true homogeneity cannot be achieved, and consequently the volume of the sample should not be too small [12].

4. SAMPLE PREPARATION

This is the first step of the analytical phase in which interference factors of unknown effectiveness can be excluded. All random and systematic errors can now be recognized and overcome by means of the statistical quality control and the use of certified reference materials [22]. The often used method to control the effectiveness of preparation methods by adding a defined quantity of the element of interest in the form of an inorganic salt to the biological matrix and to determine its recovery is of dubious value. Considerably different behaviors are to be expected of the "spiked" and the inherent element during the whole procedure [13].

Sample preparation comprises every treatment the biological material must undergo between sampling and the definitive determination of the element. The purpose of sample preparation is to bring the sample and the element of interest in a chemical and physical form suitable to the method of determination, eliminate disturbing substances, and in some cases enrich the element of interest. Every preparation procedure has a certain potentiality for contaminations and losses. Contaminations arise mainly from the employed chemicals, the air in the laboratory, and the implements. Losses are mainly due to improper separation methods, the formation of volatile [e.g. $Ni(CO)_4$] or insoluble compounds, and vessel transfers. These aspects should dominate the selection of a preparation method. Furthermore, the procedure of choice depends on the nature of the biological matrix, the expected concentration range of nickel, and the method used for the determination. The smaller the consumption of reagents and the fewer operations needed, the better will be the method. Typical kinds of sample preparation methods for the determination of nickel in biological materials are solubilization and dilution with suitable reagents, complexation followed by extraction, or solid separation and mineralization. Often a combination of the different methods is used. There is no determination method for nickel which needs no sample preparation. Even electrothermal atomic absorption methods with direct injection of liquid samples without preliminary treatment include a charring step which may possibly give rise to losses by volatilization of nickel.

4.1. Mineralization

The organic constituents of the biological matrix can be eliminated either by dry ashing or by wet mineralization.

In the classic method of dry ashing the only reagent is oxygen. This method of mineralization is very simple, large numbers of samples can be treated at the same time, and the quantity of the sample seems not to be limited. Suitable materials for crucibles are platinum, pure silica, and with some restrictions glass with high silica content (e.g., Pyrex). Contaminations result mainly from the material of the crucible and the furnace, and from the air in the laboratory. Losses may be due to volatilization and the formation of refractory materials resistant to solubilization by acids (e.g., silicates). Volatilization of nickel by the formation of $Ni(CO)_4$ may occur in the early stage of incineration when there is an oxygen deficiency in those parts of the sample which are distant from the surface. Therefore thick layers of biological material should be avoided. Furnace temperatures up to 780°C have been applied without any observable loss of nickel [23]. But it should be remembered that the reaction is

exothermic and the temperature within the sample cannot be controlled.
At higher temperatures the formation of refractory materials is favored. Special attention must be given if crucibles from Pyrex or
silica are used for the incineration of biological materials rich in
alkali metal ions because by the formation of basic melts the surface
of the vessels can be attacked giving rise to the formation of insoluble silicates. The latter may also result from the incineration of
biological material with elevated silica content. In order to prevent
losses, such ashes must additionally be treated with HF. A simple,
inexpensive device for ashing small samples of metalloproteins (e.g.,
jack bean urease) in pure oxygen atmosphere consists of a Pyrex test
tube into which extends in a certain distance from the bottom a gas
delivery tube with small holes directed horizontally so that the flow
of oxygen does not disturb the ash [24]. By the use of pure oxygen,
the applied temperature can be lowered (460°C) and by this the possibly occurring interferences from glassware are diminished. In addition, the continuous flow of oxygen prevents contaminations by the
ambient air. This device may also be useful for ashing small samples
of other biological materials. Low-temperature ashing (T < 150°C)
can be achieved by using activated oxygen [25,26]. The resulting
ashes are normally dissolved in diluted acid suitable for the measurement or a necessary subsequent treatment.

In order to diminish the risk of loss of nickel by volatilization or by the formation of insoluble compounds, wet mineralization
is often preferred to dry ashing. By the action of different oxidizing agents, the volatile reaction products of the organic constituents
of the biological matrix as well as those of the reagents and their
excess can be eliminated, whereas the metal ions remain in solution.
Sometimes this method is used only to bring the matrix in homogeneous
solution without completely eliminating the organic constituents.
Contaminations are mainly due to the reagents and the material of
the vessels and apparatus. The best material for vessels is pure
silica, as most wet mineralizations are done in strongly acidic
medium. Glassware like Pyrex can also be used, but must scrupulously
be cleaned and tested in order not to be contaminating. The risks

for losses of nickel is fairly low as the formation of volatile or insoluble compounds as well as adsorption to the walls of the vessels is diminished.

The normally used oxidants are HNO_3, H_2O_2, $HClO_4$, whereas H_2SO_4 serves above all as solubilizer and catalyst. The order of consumption is oxidants > solubilizers > catalysts. The choice of the reagents should be governed by the following points: (1) The purity of the reagents, their easiness of purification, and their consumption, for these determine the possible contaminations. The reagent blank is decisive for the determination limit of the analytical method. (2) The possible formation of interfering substances with constituents of the matrix, e.g., formation of sparingly soluble calcium sulfate from matrices rich in calcium mineralized with sulfuric acid. Calcium sulfate can cause losses of nickel by adsorption. (3) The reagent's suitability for the determination method and the ease with which the excess reagent can be removed, e.g., in ETAAS residual H_2SO_4 and $HClO_4$ may produce strong fumes during atomization which disturb the measurement. (4) Possible hazardous reactions— under certain circumstances concentrated $HClO_4$ as well as H_2O_2 can give rise to violent explosions with organic substances.

Solid samples of biological materials should never be treated directly with concentrated acids. The surface becomes hydrophobic and the number of reactive sites is diminished, thus resulting in increased reagent consumption and lengthier mineralization. Whenever possible, mineralization should start with diluted acids to hydrolyse the biological material. After this, the water is evaporated until the concentration of the oxidant is sufficient to initiate the oxidation of the organic material. This procedure can only be performed in so-called open systems [6]. It cannot be applied to closed systems working with elevated pressure. These systems consisting of a stainless steel pressure container into which a sealed vessel of suitable material (Teflon, silica or glassy carbon) is inserted are not to advise for mineralizations destined for determinations of nickel because there is a potent risk for contamination by the pressure container [27]. Besides, this method cannot be applied to samples

which must be measured with voltammetric methods because the mineralization is incomplete.

Much simpler sample preparations for the determination of
nickel in serum and whole blood are the deproteination of the matrix
with nitric acid [17], and the dilution of plasma and serum (1:1)
with 10^{-3} M HNO_3 containing 0.1% Triton X-100 [28] before quantitation by means of Zeeman-corrected ETAAS.

4.2. Complexation, Separation, and Enrichment

In all cases where the concentration of nickel is too low to be
determined with sufficient accuracy or where the accompanying substances would interfere with the determination, a separation and/or
enrichment of nickel from the bulk is necessary. Separations are
usually performed by complexation of nickel with suitable ligands
forming very stable complexes which are only sparingly soluble in
aqueous media but fairly soluble in water-immiscible organic solvents.
Such ligands are dimethylglyoxime (DMG) [29,30], α-furildioxime [31],
cyclohexanedione-1,2-dioxime (Nioxime), 4-methyl-cyclohexanedione-
1,2-dioxime, 4-isopropylnioxime, ammonium diethyldithiocarbamate
(ADEDC) [32], ammonium dibenzildithiocarbamate, ammonium pyrrolidinedithiocarbamate (APDC) [33], and hexamethyleneammonium/hexamethylene-
dithiocarbamidate (HMA/HMDC) [34,35]. The complete formations of
Ni-dioximates and Ni-DEDC demand pH values between 8 and 10, whereas
Ni-PDC and Ni-HMDC are formed at pH 3-5. These ligands are not specific to nickel. They also form complexes with other metal ions.
Obviously these ligands may be a source of contamination, and must
therefore be scrupulously controlled.

Preferred organic solvents for the extraction of the different
nickel complexes are trichloro- and tetrachloromethane for the Ni-
dioximates. The Ni-dithiocarbamates are usually extracted into
isobutyl methyl ketone (IBMK), xylene and diisopropyl ketone, or
into mixtures of these. The ratio of the solubilities of the Ni
complexes in the organic solvent to that in the aqueous medium is

indicative for the attainable enrichment. The low solubilities of
Ni-PDC and HPDC in aqueous acid medium can be used to coprecipitate
nickel with an excess of reagent and to separate the solid by centrifugation [27].

5. DETERMINATION METHODS

The different methods of determination generally used in trace element analyses are colorimetry, flame atomic absorption spectrometry
(FAAS), electrothermal atomic absorption spectrometry (ETAAS), voltammetry (VA), inductively coupled atomic emission spectroscopy
(ICPAES), and neutron activation analysis (NAA). The most potent
methods for the quantitation of nickel in biological materials are
ETAAS and VA. FAAS and colorimetry can only be used for materials
with a rather high nickel content or after an efficient enrichment
procedure. The sensitivities of ICPAES and NAA being comparatively
low, they are seldom used. Gas chromatographic methods for the
quantitation of nickel have also been reported [23].

5.1. Colorimetry

In spite of its limited sensitivity, this oldest method for the
determination of small quantities of substances may still be useful
in such cases where the nickel content of the sample is in the μg
range or higher. Its main advantage is that the needed apparatus,
a spectrophotometer, is available in every laboratory. This method
demands for a complete destruction (mineralization) of the organic
matrix. Among the different reagents giving colored complexes with
nickel, the dioximes are the preferred ones because of their selective reactions with nickel. The most applied is DMG; other oximes
can also be used but show no advantages [36].

A very satisfactory method is based on the development of a
brownish color when DMG is added to a basic solution of a nickel
salt which has previously been treated with an oxidizing agent such

as bromine. This reaction has been discovered by Feigl [37] and first quantitatively used by Rollet [38]. Two colored complexes can be formed, containing nickel and DMG in the molar ratios 1:2 and 1:4. The 1:4 complex containing Ni^{3+} is a monovalent anion. This complex is stable for several hours. The uncharged 1:2 complex only formed in ammoniacal medium is unstable, decomposing or being converted into the 1:4 complex [38]. A convenient procedure consists in dissolving the mineralized residue of the organic material in hot HCl and to evaporate the excess acid. The residue is dissolved in 1.3 M NH_3 (pH $\approx$ 11.7) and Ni^{2+} oxidized with bromine water to Ni^{3+}. After the addition of a 1% solution of DMG in ethanol, the brownish color develops and remains stable for several hours. The absorbance at 465 nm is directly proportional to the concentration of nickel with an apparent molar absorption coefficient $\varepsilon = 1.3 \cdot 10^4$ M^{-1} cm^{-1}. This method has been successfully applied to the quantitation of nickel in jack bean urease [24]. In the presence of larger quantities of other metal ions which form colored complexes with DMG like Cu, Co, and Fe, interferences may occur. These can be prevented by a previous extraction of the NiDMG into chloroform and subsequent washings with diluted ammonia [38].

5.2. Atomic Absorption Spectrometry

Atomic absorption spectrometry (AAS) makes use of the classical principle that an atom can absorb light of the same wavelength as it would emit at the transition from its excited state to the ground state. The element to be determined in the analytical sample is thermically atomized in the light beam emitted from a light source of the same element. The measured absorbance is proportional to the respective atoms and follows the law of Lambert-Beer over a more or less wide concentration range depending mainly on the efficiency of the light source. Thermic atomization can be performed by flames (FAAS) or by electrical heating (ETAAS). AAS is the most widely used method for the determination of Ni in biological materials.

It does not require a complete elimination of the organic constituents of the matrix. In many cases it is sufficient to bring the
sample into homogeneous solution.

Nowadays even small solid samples can directly be analyzed
using special techniques. With respect to sensitivity there must
be a distinction between FAAS and ETAAS. Whereas FAAS without previous enrichment of Ni can be applied to samples with Ni contents
down to the lower ppm range, ETAAS permits the quantitation of Ni
below the ppb range. Among the different spectral resonance lines
of Ni only those of 232.003 nm and 341.5 nm are of analytical interest. Using the very sensitive 232-nm line, the slit width should be
0.2 nm or below; otherwise the two strong emission lines at 231.7
and 232.14 nm will interfere causing a considerable curvature of the
calibration curve as well as a loss of sensitivity. But even with
0.2-nm slit width this effect becomes pronounced at higher concentrations of Ni. In these cases the less sensitive 341.5-nm line
(about 20 times less sensitive than the 232-nm line) is to be preferred as it permits a slit width up to 0.7 nm and shows a better
signal/noise ratio [39].

For the quantitation of Ni in biological materials, FAAS is
nearly only applicable after sample preparation steps like mineralization, extraction, and above all enrichment (see Sec. 4). Oxidizing
fuel lean flames (acetylene/air or hydrogen/air) must be used. The
exact composition of the combustible mixture and the position of the
burner with respect to the beam of the hollow cathode lamp are of crucial importance to avoid interferences—enhancements or suppressions—
from other metals (e.g., Co, Cu, Cr, Fe, Mn, Zn) [40,41]. These parameters are inherent to the respective instrument and to the composition
of the sample solution. They must be optimized and tested for effectiveness for every new kind of sample composition. In order to obtain
accurate results, it is indispensable to adapt the composition of the
calibration standards as well as possible to the composition of the
sample or to apply the standard-addition technique. In this connection it should be remembered that organic solvents and ligands used
for extractions and enrichment are fuels and alter the composition

of the combustible mixture. Beside this, the viscosity and surface
tension of the solution influences the volume aspirated in the time
as well as the dispersion of the droplets in the spray nozzle.
Obviously these effects are of major importance for the analytical
results. Interferences from acids like HCl, HNO_3, H_2SO_4, and H_3PO_4
were not observed. In every case the absorbance of the solvent must
be corrected. The determination of the reagent blank of the whole
analytical procedure is indispensable. FAAS has been successfully
applied to different biological materials, e.g., blood, urine, urease
[42,24]. Its advantage is the great number of samples which can be
measured in a short time.

The most convenient atomic absorption method for the determination of Ni in biological materials is ETAAS. In this method the sample is introduced into a graphite tube, preferably pyrolytically
coated, which is electrically heated to the desired temperature (up
to 3000°C) for atomization. By its high sensitivity samples with
very low concentrations of Ni can be measured without preliminary
enrichment, thus avoiding additional possible contaminations. This
increased sensitivity is achieved by a more complete atomization, by
a higher local concentration and a longer duration of the atoms in
the light beam compared with FAAS. The normal cycle of an ETAAS
determination is: injection of the liquid sample (10-100 µl) or
introduction of a solid, drying, thermic decomposition, atomization,
and cleaning by heating to an elevated temperature to eliminate residues. The temperatures and durations of the different steps depend
on the apparatus and on the kind and size of the sample. Common
values for the determination of Ni are: drying 100-200°C during
20-60 sec; charring 900-1200°C during 10-20 sec; atomization 2500-
2700°C during 3-6 sec; cleaning 2700°C during 2-3 sec. In order to
prolong the lifetimes of the graphite tubes these steps occur in an
argon atmosphere. There is a certain risk for loss of Ni by the
formation of volatile $Ni(CO)_4$ in matrices which can develop CO during
drying and charring. To enhance the sensitivity by prolongation of
the duration of the atoms within the graphite tube the gas flow of
argon is reduced or stopped during atomization. Interferences from

the matrix as well as those from reagents are normally eliminated or reduced by the underground and baseline compensation with deuterium or hydrogen lamps. These systems are only able to correct weak interferences. Therefore sample preparation becomes often necessary not only to enrich the Ni concentration but to eliminate possible interferences. Disturbances caused by HNO_3 and $HClO_4$ can be overcome by the addition of ascorbic acid to the sample [27]. A new compensation system using the Zeeman effect enables to correct strong interferences as they are encountered in the direct determination of Ni in biological fluids without preliminary sample preparation. Using this system, the risks for contaminations by sample preparations besides dilution can be eliminated [35]. ETAAS has widely been used for the determination of Ni in different biological materials like body fluids [16,17, 27,34,35,43-46] and plant materials [27,47]. An ETAAS method has been accepted as a reference method by the International Agency for Research on Cancer [46].

5.3. Voltammetry

The most sensitive method for the quantitation of Ni is adsorption differential pulse voltammetry (ADPV). Voltammetric methods are based on the potential-dependent redox reaction of the species to be analyzed at the surface of a suitable electrode. The measuring equipment consists of the electrolytic cell and an apparatus permitting the application of a controlled potential to the working electrode and to measure the current caused by the redox reaction. The normally used electrolytic cell consists of three electrodes (working electrode WE, reference electrode RE, and auxiliary electrode AE) and the supporting electrolyte. The working electrode is an easily polarizable microelectrode of an inert material in the investigated potential range at which the redox reactions occur. For the determination of Ni electrodes in the form of a hanging mercury drop (HMDE) or a thin film of mercury on a suitable support like glassy carbon (MFE) are mostly used. The reference electrode is a non-

polarizable electrode with constant potential against which the
potential of the working electrode is measured (mainly an AgCl/Ag
electrode in 3 M KCl or a saturated calomel electrode). The auxiliary
electrode is made from an inert material (Pt or glassy carbon) and
serves as counterelectrode to the WE. In addition the cell is fitted
with a gas inlet tube and a mechanical or magnetic stirrer.

The direct reduction of Ni^{2+} to Ni as it occurs in the classic
polarography at the dropping mercury electrode is not sensitive enough
for determination of Ni in biological materials. The often used technique of anodic stripping voltammetry (ASV) in which the metal ion of
interest is enriched at the WE by the reduction to the element during
a certain time at a sufficient negative potential under formation of
an amalgam with the material of the electrode cannot be applied to
Ni because Ni does not form an amalgam.

However, it was found that Ni can be accumulated by adsorption
at the solution/electrode interface at a potential more positive than
the reduction potential of Ni^{2+} in the presence of a suitable complexing agent like dimethylglyoxime in the supporting electrolyte.
After a defined time of accumulation with mechanical stirring the
potential of the WE is continuously changed to more negative values.
When the reduction potential of Ni^{2+} is reached, the Ni ions in the
adsorbed complex are irreversibly reduced and the resulting current
is measured [48]. Other complexing agents such as heptoxime, nioxime,
and 4-methylnioxime have also been assayed, but the best results are
obtained with DMG [23]. In order to guarantee the completeness of
the complexation, the supporting electrolyte must be buffered to pH 9.
By this process the sensitivity of the voltammetric determination of
Ni could be considerably enhanced and makes it today the most sensitive determination method for Ni. But the adsorption process is com-
peted by all substances which can also be adsorbed or which influence
the electrode/solution interface. Therefore a complete mineralization
of the biological material is indispensable. In order to suppress the
influences of the capacitive residual current and to enhance the sensitivity, the measurement is normally made in the differential pulse
mode. In this measuring mode the potential of the WE is linearly

changed with time to more negative (positive) values. In defined intervals (trigger time) the pulse, a small potential (10-50 mV), is superimposed to the linearly changed potential during a short period (e.g., 50 msec). Just before and a short time after the application of the pulse the respective currents are measured during a very short period and the difference of the currents is registered in function of the linearly changed potential.

Voltammetric measurements are very well reproducible and can be repeated without any observable change. Therefore these methods are especially destined for the calibration by the standard addition procedure because the Ni in the standard encounters the exactly identical conditions as the Ni from the original sample.

A typical determination procedure for Ni in biological material by ADPV would consist of the following steps: The acidic solution of the mineralized sample or an aliquot is transferred to the electrolytic cell and buffered with ammonia or triethanolamine to pH 9. If a precipitation takes place, 200-300 mg of citric acid monohydrate is added and if necessary the pH readjusted to 9. Then 0.5 ml of a 1% solution of purified DMG in ethanol is added and the solution is deaerated by bubbling through a moderate stream of purified nitrogen during 10-15 min. The polarograph is set to the DPP mode. The WE is a HMDE with a surface of 2 mm^2 which has been formed immediately before starting the enrichment procedure. This is of great importance because adsorption also takes place without applied potential to the WE and thus erroneous results may be obtained when the time between the formation of the mercury drop and the start of the enrichment is out of control [49].

The conditions for the determination are:

1. Enrichment and starting potential: -0.4 V against AgCl/Ag
2. Pulse amplitude: 50 mV
3. Trigger time: 0.4 sec
4. Enrichment time: 3 min with stirring, 20 sec without stirring
5. Linear voltage sweep: 10 mV/sec
6. Sweep range: -0.8 V

Enrichment with stirring is started by applying the starting potential to the WE. After 3 min the stirrer is stopped and after a further 20 sec the voltage sweep is started. The current voltage curve
shows a very sharp current peak at approximately -1 V whose height
is directly proportional to the concentration of Ni in the cell.

ADPV allows the determination of Ni down to the ppt range.
This method has been applied to different biological materials like
body fluids [50-53] and food [23,48].

Voltammetric methods have the disadvantage of being time con-
suming and needing complete mineralization. They have the advantage
of being very reproducible, very sensitive, and the equipment is
cheap in comparison to that of AAS.

ABBREVIATIONS

AAS	Atomic absorption spectrometry
ADEC	Ammonium diethyldithiocarbamate
ADPV	Adsorptive differential pulse voltammetry
AE	Auxiliary electrode
APDC	Ammonium pyrrolidinedithiocarbamate
ASV	Anodic stripping voltammetry
DMG	Dimethylglyoxime
DPP	Differential pulse polarography
ETAAS	Electrothermal atomic absorption spectrometry
FAAS	Flame atomic absorption spectrometry
HMA/HMDC	Hexamethyleneammonium/hexamethylenedithiocarbamidate
HMDE	Hanging mercury drop electrode
HPDC	Pyrrolidinedithiocarbaminic acid
IBMK	Isobutyl methyl ketone
ICPAES	Inductively coupled atomic emission spectrometry
MFE	Mercury film electrode
NAA	Neutron activation analysis
Ni-DEDC	Nickel diethyldithiocarbamate
Ni-HMDC	Nickel hexamethylenedithiocarbamidate

Ni-PDC   Nickel pyrrolidinedithiocarbamate
RE   Reference electrode
VA   Voltammetry
WE   Working electrode

REFERENCES

1. F. W. Sunderman, Jr., in "IARC Monographs on Environmental
    Carcinogens: Selected Methods of Analysis", Int. Agency Res.
    Cancer, Lyon, 1986, Vol. 8, pp. 79-92.

2. F. W. Sunderman, Jr., in "Encyclopedia of Occupational Safety
    and Health" (L. Parmeggiani, ed.), International Labor Office,
    Geneva, 1983, pp. 1438-1440.

3. P. Grandjean, in "Nickel in the Human Environment" (F. W.
    Sunderman, Jr., ed.), Int. Agency Res. Cancer, Lyon, 1984,
    pp. 469-485.

4. B. G. Benett, in "Nickel in the Human Environment" (F. W.
    Sunderman, Jr., ed.), Int. Agency Res. Cancer, Lyon, 1984,
    pp. 487-495.

5. B. Sansoni and G. V. Iyengar, in "Elemental Analysis of Bio-
    logical Materials", International Atomic Energy Agency, Vienna,
    Tech. Rep. Ser. No. 197, 57, 1980.

6. H. G. Seiler, in "Metal Ions in Biological Systems" (H. Sigel,
    ed.), Marcel Dekker, New York, Vol. 20, 1986, p. 305.

7. J. Versieck, Crit. Rev. Clin. Lab. Sci., 22, 97 (1985).

8. J. R. Moody and E. S. Beary, Talanta, 29, 1003 (1982).

9. J. M. Mattinson, Anal. Chem., 44, 1715 (1972).

10. R. P. Maas and S. A. Dressing, Anal. Chem., 55, 808 (1983).

11. D. P. H. Laxen and R. M. Harrison, Anal. Chem., 53, 345 (1981).

12. J. Angerer, K. H. Schaller, and H. G. Seiler, Trends Anal. Chem.,
    2, 257 (1983).

13. D. Behne, J. Clin. Chem. Clin. Biochem., 19, 115 (1981).

14. K. Heydorn, E. Damsgard, N. A. Larsen, and B. Nielsen, in
    "Nuclear Activation Techniques in the Life Sciences", Inter-
    national Atomic Energy Agency, Vienna, 1978, pp. 129-142.

15. H. Jürgensen and D. Behne, J. Radioanal. Chem., 37, 375 (1977).

16. S. S. Brown, S. Nomoto, M. Stoeppler, and F. W. Sunderman, Jr.,
    Clin. Biochem., 14, 295 (1981).

17. F. W. Sunderman, Jr., C. Crisostomo, M. C. Reid, S. M. Hopfer, and S. Nomoto, *Ann. Clin. Lab. Sci.*, *14*, 232 (1984).

18. M. Stoeppler and H. W. Nürnberg, in "Metalle in der Umwelt" (E. Merian, ed.), Verlag Chemie, Weinheim, FRG, 1984, p. 45.

19. G. V. Iyengar, *J. Pathol.*, *34*, 173 (1981).

20. G. V. Iyengar, K. Kasperek, and L. E. Feinendegen, *Radioanal. Chem.*, *69*, 463 (1982).

21. J. R. Moody, *Trends Anal. Chem.*, *2*, 116 (1983).

22. K. H. Schaller, J. Angerer, G. Lehnert, M. Valentin, and D. Weltle, *Arbeitsmed. Sozialmed. Präventivmed.*, *19*, 79 (1984).

23. A. Meyer and R. Neeb, *Fresenius Z. Anal. Chem.*, *321*, 235 (1985).

24. N. E. Dixon, R. L. Blakeley, and B. Zerner, *Can. J. Biochem.*, *58*, 469 (1980).

25. J. E. Patterson, *Anal. Chem.*, *51*, 1087 (1979).

26. G. Kaiser, P. Tschöpel, and G. Tölg, *Fresenius Z. Anal. Chem.*, *253*, 177 (1971).

27. Jin Long-zhu and Ni Zhe-ming, *Fresenius Z. Anal. Chem.*, *321*, 72 (1985).

28. J. R. Andersen, B. Gammelgaard, and S. Reimert, *Analyst*, *111*, 721 (1986).

29. E. B. Sandell and R. W. Perlich, *Ind. Eng. Chem.*, *Anal. Ed.*, *11*, 309 (1939).

30. H. Christopherson and E. B. Sandell, *Anal. Chim. Acta*, *10*, 1 (1954).

31. D. Mikac-Devic, F. W. Sunderman, Jr., and S. Nomoto, *Clin. Chem.*, *23*, 948 (1977).

32. O. R. Alexander, E. M. Godar, and N. J. Linde, *Ind. Eng. Chem.*, *Anal. Ed.*, *18*, 206 (1946).

33. S. Nomoto and F. W. Sunderman, Jr., *Clin. Chem.*, *16*, 477 (1970).

34. A. Dornemann and H. Keist, *Fresenius Z. Anal. Chem.*, *300*, 197 (1980).

35. J. Angerer, K. H. Schaller, and M. Fleischer, in "Analyses of Hazardous Substances in Biological Materials", Vol. 1 (J. Angerer and K. H. Schaller, eds.), VCH Verlagsgesellschaft, Weinheim, FRG, 1985, p. 177.

36. R. C. Ferguson and C. V. Banks, *Anal. Chem.*, *23*, 1486 (1951).

37. F. Feigl, *Ber. deut. Chem. Ges.*, *57*, 758 (1924).

38. E. B. Sandell, "Colorimetric Determination of Traces of Metals", 3rd ed., Interscience, New York, 1959, pp. 665-681.

39. B. Welz, "Atom-Absorptions-Spektroskopie", 2nd ed. Verlag
     Chemie, Weinheim, FRG, 1975, pp. 180-181.

40. W. B. Barnett, *Anal. Chem.*, *44*, 695 (1972).

41. L. L. Sundberg, *Anal. Chem.*, *45*, 1460 (1973).

42. J. Angerer and K. H. Schaller, in "Analysen in Biologischem
     Material", Vol. 2 (D. Henschler, ed.), Verlag Chemie, Weinheim,
     FRG, 1981; Met. Nr. 1: Nickel in Blut.

43. M. Stoeppler, in "Nickel in the Environment" (J. O. Nriagu,
     ed.), John Wiley, New York, 1980, pp. 661-822.

44. F. W. Sunderman, Jr., A. Marzouk, M. C. Crisostomo, and D. P.
     Weatherby, *Ann. Clin. Lab. Sci.*, *15*, 299 (1985).

45. F. W. Sunderman, Jr., S. M. Hopfer, M. C. Crisostomo, and M.
     Stoeppler, *Ann. Clin. Lab. Sci.*, *16*, 219 (1986).

46. F. W. Sunderman, Jr., in "IARC Monographs on Environmental
     Carcinogens: Selected Methods of Analysis", Int. Agency Res.
     Cancer, Lyon, Vol. 8, 1986, pp. 319-334.

47. R. J. Green and C. J. Asher, *Analyst, 109,* 503 (1984).

48. B. Piklar, P. Valenta, and H. W. Nürnberg, *Fresenius Z. Anal.
     Chem.*, *307*, 337 (1981).

49. R. Kissner and H. G. Seiler, unpublished results.

50. C. J. Flora and E. Nieboer, *Anal. Chem.*, *52*, 1011 (1980).

51. M. Uto, Y. Uto, and M. Sugawara, *Fresenius Z. Anal. Chem.*, *321*,
     68 (1985).

52. S. B. Adeloju and A. M. Bond, *Anal. Chim. Acta, 164,* 181 (1984).

53. P. Ostapczuk, P. Valenta, M. Stoeppler, and H. W. Nürnberg, in
     "Chemical Toxicology and Clinical Chemistry of Metals" (S. S.
     Brown and J. Savory, eds.), Academic, London, 1983, pp. 61-64.

Author Index

Numbers in parentheses are reference numbers and indicate that an author's work is referred to although his name may not be cited in the text. Underlined numbers give the page on which the complete reference is listed.

A

Abbracchio, M. P., 102(58), 103 (59,64), 118, 119
Abdulwajid, A. W., 176(45), 265
Abeles, R. H., 227(253), 274
Achten, G., 367(51), 395
Adalis, D., 388(206), 402
Adams, M. W. W., 248(409), 251 (429), 282, 283; 286(4), 289 (37,38), 302(51), 311, 313
Adams, R. M., 370(76), 396
Adeloju, S. B., 425(52), 428
Adman, E. T., 177(50), 233(275), 234(289), 235(292), 266, 275, 276
Aggag, J. M., 247(393), 281; 287(10), 311
Aguirre, R., 246(375), 280; 288 (17), 290(17), 296(17), 298 (17), 301(17), 311
Ahmad, H. M., 66(128), 67(128), 88
Ahsanullah, M., 36(20), 38(20), 45
Aisen, P., 178(58), 236(58,319), 266, 277
Aitio, A., 95(28), 98(28), 109 (28), 110(28), 117; 362(22), 389(22), 390(22), 393
Aizawa, H., 372(93), 397
Akamatsu, N., 379(131), 398
Akers, W. A., 367(48), 395
Åkeson, Å., 233(266), 275
Akman, Y., 71(155), 89
Alagna, L., 242(348), 279

Albert, D. M., 387(200), 401
Alberts, G. S., 207(152), 270
Albracht, S. P. J., 52(25), 83; 105(87), 106(87), 112(87), 120; 246(386-388), 248(387,406,407), 252(406), 253(387,406,407), 254 (387,441), 255(453), 256(453), 281, 282, 283, 284; 288(15,21), 289(36), 294(45), 302(15,21), 305(15), 306(15,60), 311, 312, 313, 314
Albrecht-Ellmer, K. J., 246(388), 281
Alden, R. A., 235(294), 276
Alesenko, A. V., 332(4), 356
Alexander, O. R., 417(32), 427
Allen, A. M., 367(48), 395
Allpass, P. R., 376(112), 380 (112), 381(112), 387(197,200), 388(201), 398, 401
Al-Mahrouq, H., 94(11), 116
Al-Rabaee, R. H., 380(146), 399
Alston, K., 178(56), 266
Al-Tawil, N. G., 369(73,74), 396
Amacher, D. E., 379(130), 398
Ames, B. N., 107(99), 120; 386 (182), 400
Amsler, P. E., 327(49,50), 330
Anderegg, G., 211(155), 270
Andersen, A., 95(24), 117; 373 (105,106), 397
Andersen, I., 94(18), 116; 388 (205), 402
Andersen, J. R., 370(78), 396; 417(28), 427
Andersen, K. E., 368(67), 395

429

Bruschi, M., 234(289), 247(402), 276, 281
Bryant, V., 36(24), 38(24), 45
Buchet, J.-P., 386(185), 401
Buckingham, D. A., 188(94), 189 (94), 190(102), 191(108), 268
Buffle, J., 169(9), 264
Bunting, J. W., 145(84), 150(84), 162
Burce, G. L., 214(179), 271
Burch, G., 247(399), 248(399), 255(399), 281
Burch, M. K., 175(39), 265
Burns, R. G., 4(5), 5(5), 24(5), 26
Burns, V. M., 4(5), 5(5), 24(5), 26
Burridge, J. C., 61(91), 62(91), 86
Burton, M. A. S., 18(27), 19(27), 20(27), 27; 56(49), 57(49,58), 60(49), 84
Busch, D. H., 207(146), 213(171, 172), 255(456), 256(172,456, 457), 270, 271, 284
Butcher, R. J., 198(113), 268
Butt, T. R., 177(55), 266
Butzow, J. J., 384(172), 400
Byerly, L., 104(71), 119
Byers, B. R., 231(261), 275

C

Caerteling, G., 255(451,452), 284
Callan, W. M., 22(47), 28
Callot, H. J., 196(111), 197 (111), 268
Calnan, C. D., 367(59), 368(63), 395
Cameron, A. G. W., 4(4), 26
Cammack, R., 246(373-375,378), 247(373), 248(373), 250(419), 252(373,374), 280, 282; 288 (16,17), 289(32,33), 290(17, 32,33), 291(32,33), 292(33), 294(32,33), 296(17,32,33), 297 (16,49), 298(17), 300(16,33), 301(17,32), 303(16), 305(58), 311, 312, 313, 314
Camner, P., 388(209), 402
Campbell, H. D., 224(236), 274

Campbell, J. A., 66(129), 67 (129), 88
Campbell, P. C. G., 39(32), 45
Campbell, R., 36(24), 38(24), 45
Cannon, H. L., 21(29), 27; 67 (132), 88
Canovas-Diaz, M., 52(29), 83
Canters, G. W., 177(50), 266
Cantoni, O., 102(58), 118; 383 (156,157), 399
Capobianco, J. A., 41(40), 46
Caprioli, R. M., 381(140), 399
Carlbom, U., 233(265), 275
Carbone, P. P., 212(163), 260 (163), 261(163), 271
Carlin, R. L., 198(114), 268
Carper, W. R., 224(235), 274
Carter, C. W., 235(294), 276
Cartier, A., 372(86,89), 396
Cartwright, I. L., 385(179), 400
Carvalho, S. M. M., 94(10), 116
Cary, E. E., 50(4), 51(17), 82, 83
Casey, C. E., 19(37), 27; 110 (109), 121
Casto, B. C., 381(139,142), 399
Cathcart, R., 107(99), 120
Cavelier, C., 367(57), 395
Cavins, J., Jr., 41(38), 46
Cayley, G. R., 124(3), 125(3), 159; 200(119), 269
Cazzulo, J. J., 224(232), 273
Cecutti, A., 19(36), 22(36), 23 (36), 27; 98(45), 103(45), 110 (45), 112(45), 118; 363(23), 376(23), 380(23), 393
Chace Lottich, S., 185(78), 267
Chakravorty, A., 106(89), 120
Chamberlain, P. I., 155(93), 162
Chamberlin, M., 144(79), 147(79), 148(79), 150(79), 162; 204(140), 270
Chambers, R. R., 218(206), 272
Chan, S. I., 237(324,325), 278
Chance, B., 237(328), 278
Chanchalashvili, Z. I., 332(2,3, 6), 340(2), 356
Chandra, S. V., 23(65), 29
Chaney, R. L., 52(30), 62(30), 83
Chang, J. W., 146(88-90), 162
Chapman, E. T., 363(27), 394

Subject Index

Red beet, 64
Redox potentials, 106, 107, 155,
   213, 216, 251, 253, 256, 298,
   301, 302, 304-307
Reductases
   methyl-, see Methylreductases
   nitrate, 228, 236
Reference materials, biological
   standards, 413
Refineries, 100, 362
   electro-, 362, 375, 386
   nickel, 363, 373, 375, 386
Renal (see also Kidney)
   cancer, 378
   nickel excretion, 93, 94
   toxicity, 386
Respiratory failure, 97, 364
Respiratory tract, 92-95, 108,
   114, 388
   allergens, 372
Rhacomitrium lanuginosum, 20
Rhizobium japonicum, 246
Rhodium, 50
Rhodopseudomonas, 250
   capsulata, 246
Rhodospirillium rubrum, 235
Rhus
   vernicifera, 237
   wildii, 72
Rhyolite, 5, 6
Ribonuclease A, 184
Ribonucleic acid, see RNA
Ribosomes, 181
Rinorea bengalensis, 21
Rivers (see also Water)
   Meuse, 39
   nickel in, 13
   sediments, 16
RNA (see also Nucleic acids),
   105, 172, 243, 382
   nickel in, 23, 105
   polymerases, 23
   synthesis, 384
Rocks (see also Ores and individual names)
   cobalt in, 5
   igneous, 5, 6
   metamorphic, 6-8
   nickel in, 5, 6, 25
   sedimentary, 6-8
Rodents, 108, 111, 376, 380,
   387

Rotifers, 34
Rubredoxin, 234, 291
Rudites, 6, 7
Ryegrass, 55

S

Saccharides, 176, 177
Saccharomyces cerevisiae, 237,
   379
Salicylaldehyde (and anion),
   188, 189
Saliva, 98
Salmo
   gairdneri, see Rainbow trout
   salmar, 37
Salmonella typhimurium, 379
Samples (see also Specimen
   collection)
   preparation, 413-418
Sandstones, 6, 7
Saprolites, 6, 7
Sarcomas (see also Tumors), 380
   Kirsten, 355
   mice, 332
   rat, 332
Sarcosine (and residues), 127,
   128
Scatchard plots, 339
Scenedesmus, 37
   obliquus, 37, 40
Schiff bases, 189, 198
Schists, 7
Schuachardtite, 4
Scintillation, β counting, 173
Sea urchin, 221, 222, 228
Seawater (see also Waters)
   fish, 21
   nickel in, 13, 14, 39, 168
Sebertia acuminata, 21
Sediments, 38
   drainage, 14, 16
   nickel in, 39, 40
Selenide, nickel, 3, 4, 383
Selenium (different oxidation
   states), 9, 251, 289-291,
   294, 295, 299-301, 307-309
   ^{77}Se, 308
Selenocysteine (and residues),
   251
Semicarbazide, 261

Urea, 33, 51, 226, 261
  hydrolysis, 257
  metabolism, 50
Urease, 32, 50, 51, 105, 179,
  226, 256-263, 421
  amino acid sequence, 257-260
  ethanolysis, 218, 219
  jack bean, see Jack bean
  nickel sites, 241, 242
  rumen bacterial, 22
  substrates for, 260-262
  synthesis, 50
Uremia, 22, 111
Uric acid, 107
Uridine, 318, 321, 322
Uridine 5'-monophosphate, see
  5'-UMP
Uridine 5'-triphosphate, see
  5'-UTP
Urine (see also Body fluids and
  Excretion), 114, 421
  creatinine, 409
  nickel analysis, 409, 413
  nickel concentrations, 386
  nickel in, 19, 22, 94-96, 98,
  99, 175, 364
Uterus contraction, 104
5'-UTP, 321, 322, 324
UV absorption or spectra (see
  also Absorption bands and
  spectra), 105, 146, 201, 209,
  323, 324, 336, 337, 353

V

Vaccinium angustifolium, 65
Valine (and residues), 127, 128,
  134
Vanadates, 57
  nickel, 4
Vanadium (different oxidation
  states), 57, 169, 182, 236
Vanadium(II), 200
Vanadium(III), 197
Vegetables (see also individual
  names), 95, 362
Vermiculite, 4
Veronal, 101, 102
Vertebrates, 113, 170, 171, 177,
  225
  nickel deficiency, 19

Vibrio succinogenes, 246
Viruses (see also individual
  names), 385
  avian myeloblastosis, 221, 222
  oncogenic, 355
  Rauscher, 351
Viscosity of DNA, 335, 345
Visible spectra, see Absorption
  bands and spectra
Vitamins (see also individual
  names), 180, 353
Volcanoes, 17
Voltammetry, 417, 422-425
  adsorption differential pulse,
  422, 424, 425
  anodic stripping, 423
  cyclic, 107

W

Wad, 8, 12, 16
  nickeliferous, 3
Waste disposal, 11, 63
Waters (containing), 62
  cobalt, 15
  hardness, 33
  nickel, 13-17, 25, 31-44
Welding, stainless steel, 363
Wheat seeds, 50
Willemsite, 4
Wolinella succinogenes, 302
Workplace, 93, 94, 100

X

XANES, 242
Xanthine
  dehydrogenase, 236
  hypo-, see Hypoxanthine
  oxidase, 107, 108
Xanthobacter autotrophicus, 247
Xenopus laevis, 242
Xylosma, 73
X-ray
  crystal structures, 104, 135,
  142, 143, 203, 232-235, 237,
  239-241, 253
  extended absorption fine structure spectroscopy, see EXAFS